Fisheries and Blue Economy

Fisheries and Blue Economy

Guest Editor

Ben Drakeford

MDPI

Basel • Beijing • Wuhan • Barcelona • Belgrade • Novi Sad • Cluj • Manchester

Guest Editor
Ben Drakeford
Centre for Blue Governance
University of Portsmouth
Portsmouth
United Kingdom

Editorial Office
MDPI AG
Grosspeteranlage 5
4052 Basel, Switzerland

This is a reprint of the Special Issue, published open access by the journal *Fishes* (ISSN 2410-3888), freely accessible at: https://www.mdpi.com/journal/fishes/special_issues/8MO30MN301.

For citation purposes, cite each article independently as indicated on the article page online and as indicated below:

Lastname, A.A.; Lastname, B.B. Article Title. *Journal Name* **Year**, *Volume Number*, Page Range.

ISBN 978-3-7258-3175-3 (Hbk)
ISBN 978-3-7258-3176-0 (PDF)
https://doi.org/10.3390/books978-3-7258-3176-0

Contents

About the Editor . . . . **vii**

Preface . . . . **ix**

Michael Bennett, Antaya March and Pierre Failler
Blue Economy Financing Solutions for the Fisheries and Aquaculture Sectors of Caribbean Island States
Reprinted from: *Fishes* **2024**, *9*, 305, https://doi.org/10.3390/fishes9080305 . . . . **1**

Veronica Mpomwenda, Tumi Tómasson, Jón Geir Pétursson and Daði Mar Kristófersson
From Co-Operation to Coercion in Fisheries Management: The Effects of Military Intervention on the Nile Perch Fishery on Lake Victoria in Uganda
Reprinted from: *Fishes* **2023**, *8*, 563, https://doi.org/10.3390/fishes8110563 . . . . **40**

Chang Xu, Yang Liu and Zhaobin Pei
Research on Legal Risk Identification, Causes and Remedies for Prevention and Control in China's Aquaculture Industry
Reprinted from: *Fishes* **2023**, *8*, 537, https://doi.org/10.3390/fishes8110537 . . . . **58**

Wendong Zhu, Wenhui Sun, Dahai Li and Limin Han
Spatial–Temporal Characteristics and Influencing Factors of Marine Fishery Eco-Efficiency in China: Evidence from Coastal Regions
Reprinted from: *Fishes* **2023**, *8*, 438, https://doi.org/10.3390/fishes8090438 . . . . **82**

Mialy Andriamahefazafy
Governing Distant-Water Fishing within the Blue Economy in Madagascar: Policy Frameworks, Challenges and Pathways
Reprinted from: *Fishes* **2023**, *8*, 361, https://doi.org/10.3390/fishes8070361 . . . . **102**

Michela Martinelli, Lorenzo Zacchetti, Andrea Belardinelli, Filippo Domenichetti, Paolo Scarpini, Pierluigi Penna, et al.
Changes in Abundance and Distribution of the Sea Pen, *Funiculina quadrangularis*, in the Central Adriatic Sea (Mediterranean Basin) in Response to Variations in Trawling Intensity
Reprinted from: *Fishes* **2023**, *8*, 347, https://doi.org/10.3390/fishes8070347 . . . . **121**

Jorge Ramos, Pedro G. Lino, Jaime Aníbal and Eduardo Esteves
(Un)wanted Fish: Potential Consumers' Acceptability of Landings in the Portuguese Case
Reprinted from: *Fishes* **2023**, *8*, 324, https://doi.org/10.3390/fishes8060324 . . . . **137**

Peiwen Wang, Isabel Mendes and Ramona Franić
Investigating the Relationship between Aquaculture Investments, Training, and Environmental Factors in Guangdong: An Alternative Perspective
Reprinted from: *Fishes* **2023**, *8*, 237, https://doi.org/10.3390/fishes8050237 . . . . **149**

Temiloluwa Jesutofunmi Akinyemi, Denis Worlanyo Aheto and Wisdom Akpalu
Sampling Error and Its Implication for Capture Fisheries Management in Ghana
Reprinted from: *Fishes* **2022**, *7*, 333, https://doi.org/10.3390/fishes7060333 . . . . **164**

Manikandan Arumugam, Sudharshini Jayaraman, Arun Sridhar, Vignesh Venkatasamy, Paul B. Brown, Zulhisyam Abdul Kari, et al.
Recent Advances in Tilapia Production for Sustainable Developments in Indian Aquaculture and Its Economic Benefits
Reprinted from: *Fishes* **2023**, *8*, 176, https://doi.org/10.3390/fishes8040176 . . . . **176**

About the Editor

Ben Drakeford

Ben Drakeford is an Associate Professor in Marine Resource Economics at the Centre for Blue Governance, University of Portsmouth. He earned his PhD in Fisheries Economics from the University of Portsmouth. His research interests lie in the broad field of blue economy development, with specific interest in fisheries and aquaculture economics and management. He has led research in blue economy development in developed and developing countries.

Preface

The blue economy concept has developed strongly over the last decade or so. By 2030, the global blue economy is expected to be worth more than USD 3 trillion. Fisheries represent one of the oldest sectors of the blue economy, and involve crucial activity for income generation, employment, food security and poverty reduction in some of the world's poorest countries. Fish remains one of the world's most traded food commodities, with per capita consumption continuing to increase. However, global fisheries face significant challenges, including the impacts of marine litter, declining stocks, and illegal, unreported and unregulated (IUU) fishing activity, all coupled with increasing demand from the world's growing population. While fisheries are expected to remain a vital sector, particularly for developing countries, further sustainable expansion of the sector will require transformative blue policies.

Ben Drakeford
Guest Editor

fishes

MDPI

Article

Blue Economy Financing Solutions for the Fisheries and Aquaculture Sectors of Caribbean Island States

Michael Bennett *, Antaya March and Pierre Failler

Centre for Blue Governance, University of Portsmouth, Portsmouth PO1 2UP, UK; antaya.march@port.ac.uk (A.M.); pierre.failler@port.ac.uk (P.F.)
* Correspondence: mbenntt97@gmail.com

Abstract: This study reviews various financing solutions available for fisheries and aquaculture development in Caribbean small island developing states (SIDS) and Barbados, Grenada, and St. Vincent and the Grenadines. Previously identified financing needs within the fisheries and aquaculture sectors have been matched with the most suitable financing mechanisms. However, the use of blue levies is recommended and applicable in almost every scenario, as they allow these sectors to drive their own development in financing research and conservation projects to their own benefit. The use of "blue tokens" with sufficiently low repayment coupons allows development projects to gather public support for fisheries, thereby increasing the likelihood of the project being successful through community buy-in. The possibility of natural capital being traded as public equities as "Natural Asset Companies" provides the opportunity for development projects to fund themselves. The review concludes that natural capital can be leveraged as the base through which public-private partnerships (PPPs) can facilitate optimal delivery of ecosystem services, benefit multiple stakeholders, and provide numerous development opportunities. An enabling environment for debt and lending with low-interest loan repayments is also applicable to almost every scenario, as it facilitates access to capital finance for infrastructure development and the acquisition of increasingly sustainable fishing equipment. Steps towards generating an enabling environment for financing fisheries and aquaculture in the Caribbean region are also discussed. The establishment of dedicated financing institutions, PPPs, and sufficient data reporting infrastructure for the fisheries and aquaculture industry are essential for driving development in these sectors. Likely, the largest limiting factor in financing Caribbean fisheries and aquaculture industries is a lack of awareness of the range of finance and financing mechanisms available to stakeholders, as well as an enabling environment for financing blue Economy sectors. This review is thus intended to aid financing institutions, Blue Economy developers, and specifically Caribbean fisheries and aquaculture stakeholders and Caribbean governments by raising awareness of the financing mechanisms available, encourage the incorporation of their use in the fisheries and aquaculture industries in the Caribbean, and encourage policymakers to create an enabling environment for financing development in these crucial sectors.

Keywords: Caribbean Blue Economy; fisheries and aquaculture; natural assets; SIDS financing; sustainable finance

Key Contribution: Review and matching of the most suitable finance and financing solutions to previously identified development needs for the fisheries and aquaculture sectors of Caribbean Island States. Their associated implementation and necessary structures for incentivizing their use through an enabling environment are also discussed.

Citation: Bennett, M.; March, A.; Failler, P. Blue Economy Financing Solutions for the Fisheries and Aquaculture Sectors of Caribbean Island States. *Fishes* **2024**, *9*, 305. https://doi.org/10.3390/fishes9080305

Academic Editor: Dimitrios Moutopoulos

Received: 22 May 2024
Revised: 18 July 2024
Accepted: 23 July 2024
Published: 3 August 2024

1. Introduction

In recovering from the restrictions placed on countries due to the COVID-19 pandemic, the development of the Blue Economy, and by extension, fisheries and aquaculture, has been heralded as an option for the greatest sustainable development [1]. The rise of the

Blue Economy is seen as an avenue through which countries can coordinate and develop their aquatic resources for sustainable economic, environmental, and social development, especially small island developing states (SIDS) such as those in the Caribbean [1] and in particular Barbados, St. Vincent and the Grenadines, and Grenada.

Currently, fisheries in Barbados, Grenada, and St. Vincent and the Grenadines land approximately 6000 tonnes of fish per annum and are not able to satisfy local consumption needs [2]. Fish imports have increased over the last decade due to high demand, with fish consumption at a high international level in Barbados (around 40 kg/capita/year) and mid-level in Grenada and St. Vincent and the Grenadines (27 and 20 kg/capita/year respectively; Ref. [2]). Population forecasts show a slight increase in the population of these 3 countries by 2030 and then a decrease to reach their 2015 levels [2]. This indicates that the pressure on fish resources as a result of demand is not likely to increase. Moreover, the economic and social importance of aquaculture and inland fisheries is currently low in these countries [3]. Marine capture fisheries are of greater importance as they employ a significant fraction of each country's labour force (23,500 people in total), which generates significant national revenue [4]. However, the failure to integrate fishery data into economic value (especially from artisanal, subsistence and small-boat fishing) [3] risks a potential underestimation of the contribution these fisheries sub-sectors have in contributing to the national Blue Economy.

Coastal and marine habitats provide multiple important ecosystem services which contribute to the Blue Economy sectors in the three countries [5]. In terms of fisheries and aquaculture in the Caribbean region, marine and coastal habitats (namely mangroves, seagrass beds and coral reefs) provide noteworthy functions to capture fisheries, specifically as breeding grounds, nurseries and feeding grounds that contribute to productivity. Activities or phenomena that degrade or reduce these habitats (i.e., a reduction in either area or health) will likely diminish the provisioning of these services, as a healthy environment underpins the economic activities dependent on it [5]. Moreover, a single ecosystem concurrently contributes to several ecosystem services [5], emphasising the potential significance of these systems to a country's well-being and national Blue Economy development. The economic price of replacing these ecosystem functions and services when a degraded ecosystem can no longer deliver them will be extremely high [4]. A rough estimate of the total contribution of mangroves, coral reefs and seagrass beds is USD 800 million annually for all three of the countries [6]. Moreover, the natural capital (living natural resources such as plants, animals, and ecosystems) in Grenada, Barbados, and St. Vincent and the Grenadines is being depleted, owing to anthropogenic drivers, specifically coastal development, overfishing, introduction of invasive species, pollution, and the impacts of climate change [3,7–9].

Currently, there is insufficient infrastructure in place for the economic-value development of the fishery sector, nor for the expansion of the aquaculture sector, particularly where sector-specific policies are not aligned to the Blue Economy concept nor integrated into national or regional development planning [4]. The Blue Economy in the three case-study countries is still in the early stages of development [4]. Despite the significant natural resources of Barbados, Grenada, and St. Vincent and the Grenadines, as well as the opportunity to implement integrated Blue Economy approaches, the rate of adoption has been relatively slow. Currently, insufficient infrastructure is in place for the development of the fisheries and aquaculture sectors of Grenada, Barbados, and St. Vincent and the Grenadines [4]. However, the potential of these sectors is significant, but only if the appropriate development investments are made [4]. In addition, the current regulatory and policy environment is also inadequate for attracting investment and funding for the Blue Economy (and thus fisheries and aquaculture) [4].

Significant development funding, including official development funding (ODA) and other official flows (OOF), has been directed to the Caribbean region (Table 1; Ref. [10]). The Caribbean region receives an estimated 8.8% share of the total global ODA funding (Table 5 p. 20 in [11]). However, Grenada, St. Vincent and the Grenadines have received relatively small amounts of development assistance compared to other countries in the

Caribbean (Table 1). There is no recorded data available for development finance allocated to Barbados after 2010 [10]. Since receiving development funding support, neither Barbados, Grenada, nor SVG have developed stable, productive national economies nor sustainable local societies. This may be due to the comparatively limited funding received by these countries, which highlights an opportunity for significant national development through increased funding support in these countries. There is no pattern in this data (Table 1) to suggest that the amount of ODA funding is likely to increase significantly going forward, further highlighting the need for innovative financing mechanisms to be developed in these countries. In implementing such mechanisms, these countries could generate development finance themselves and reduce their reliance on ODA funding support for national development goals. The development of the Blue Economy of Grenada, Barbados, and St. Vincent and the Grenadines could generate the synergies needed to facilitate long-term finance [4].

Table 1. Total official flows (sum of ODA and other official flows) to Caribbean states from 2016 to 2022 (represented in millions of USD, from [10]).

Country	2016	2017	2018	2019	2020	2021	2022
Grenada	6.41	1.09	0.65	1.65	3.42	3.30	2.26
St. Vincent and the Grenadines (SVG)	2.99	4.30	0.63	10.92	3.75	5.91	9.12
Montserrat	44.13	41.76	39.04	36.36	47.19	37.37	42.74
Haiti	744.43	712.668	635.033	463.29	483.40	434.89	450.21
Caribbean (total)	367.91	1254.48	1221.13	748.764	984.413	1292.16	1104.86

The aquaculture and fisheries sectors of Barbados, St. Vincent and the Grenadines (SVG), and Grenada are currently underdeveloped yet have the potential to contribute to each country's Blue Economy once commercially viable [3,4]. The financing needs for Caribbean fisheries and aquaculture sectors have recently been identified by March et al. [4], contextualising the challenges and opportunities in the sector. This paper continues to build on the work of March et al. [4] by discussing solutions and options for public and private financing for aquaculture and fisheries development in Barbados, Grenada, and St. Vincent and the Grenadines (as part of the 'SDG Joint Fund Programme: Harnessing Blue Economy Finance for SIDS Recovery and Sustainable Development' consultancy) as this has not yet been conducted in the available literature, such that the "right investments can be made". The paper also discusses steps towards generating an enabling environment for blue financing in the Caribbean. While the paper focuses on Barbados, St. Vincent and the Grenadines (SVG), and Grenada, its content is not limited to these countries and may be applicable to SIDS in the greater Caribbean region as a whole. This review is intended to aid financing institutions, Blue Economy developers, and specifically Caribbean fisheries and aquaculture stakeholders, governments, and decision-makers.

2. Approach

This review supplements and directly builds upon this previous research [4], which contextualised the fisheries and aquaculture sectors of Grenada, Barbados, and St. Vincent and the Grenadines. Together, the current work and March et al. [4] constitute the findings of the 'SDG Joint Fund Programme: Harnessing Blue Economy Finance for SIDS Recovery and Sustainable Development' consultancy. This manuscript's content is based on a literature review of relevant documents relating to the financing of fisheries and aquaculture sectors, as well as stakeholder engagement and consultation through workshop events. The consultation process involved correspondence and meetings with parties with vested interests at a national level, comprising state and various sector representatives at the industrial, semi-industrial, and small-scale levels, who are able to implement blue financing strategy and identify gaps and needs in the current system (under the 'SDG Joint Fund Programme: Harnessing Blue Economy Finance for SIDS Recovery and Sustainable Development'). The consultation was necessary for the identification of projects that would be suitable for blue financing solutions and was conducted from October 2022 to December 2022. A series of regional workshops were also hosted during this period and included the countries of Barbados, St. Vincent and the Grenadines, and Grenada, as well as regional development organisations, specifically the United Nations Environment Programme (UNEP), United Nations Development Program (UNDP), and the Fisheries and Aquaculture Organisation (FAO). Participants varied over the course of the workshop series. However, workshops consisted of a minimum of 20 different stakeholders per workshop. In-person workshops were run over the course of a workday and involved presentations and Q&A sessions concerning prepared questionnaires that were developed for different stakeholder groups (as above).

March et al. [4] identified financing needs for the fisheries and aquaculture sectors of these countries, highlighting the development needs of the national aquaculture and fisheries sectors, the lack of adequate policy and regulatory frameworks for the sectors as well as the structuring of the respective national blue economies, and highlights the specific opportunities for the development of the fisheries and aquaculture sectors of these countries. The authors state that dedicated blue financing funds and mechanisms are needed for the development of the currently immature aquaculture sector [4], which can be applied to the fisheries sector as well. The authors identify the need for an enabling environment for such development and the necessity of contextual analysis to develop tailored financing options and solutions for the fisheries and aquaculture development within each respective country's context [4]. The content of this manuscript is in direct response to the needs and opportunities identified by March et al. [4] by proposing the "how" of mobilising finance for facilitating the development of the fisheries and aquaculture industries in Grenada, Barbados, and St. Vincent and the Grenadines.

Section 3 reviews proposed financing solutions (specifically) available for the common development needs and priorities for fisheries and aquaculture development (as identified in [4]). The proposed range of financing solutions (non-exhaustive) was synthesised from the author's own experience in developing national Blue Economy strategies and from the advisory documents for other island nations (such as Madagascar in [12] and Seychelles Blue Economy Action Plan, in [13], Africa Blue Economy Strategy, in [14], among others). Non-return-seeking finance solutions (like public finance) have been omitted from this review as they are unlikely to incentivise investment by not providing any benefits for potential investors who engage with them. Mutual benefits (i.e., benefits for both/multiple parties involved) form the basis of the mechanisms discussed herein. Furthermore, given that investor returns are fiscally based, fisheries and aquaculture development are generally assumed to be reflected by changes in industry/sector productivity (which is easily quantifiable as catch/production data) as a result of improved ecosystem health. However, industry development may also include other indicators, such as increases in job creation or in the nutritional quality of fish products. The financing solutions are summarised in Table A1: Summary of various financing mechanisms for use in fisheries and aquaculture sectors of Caribbean states. (Appendix A), with their optimal use case being contextualised with the previously identified common financing needs among the case study countries in Table A2: Linking financing instruments with financing needs in the fisheries and aquaculture sectors in Barbados, Grenada, and St. Vincent and the Grenadines (Appendix A).

Section 4 discusses the development of an enabling financing environment for fisheries and aquaculture, the features of which are recommended to be developed in each of the case study countries. Section 5 concludes the paper, emphasising key take-aways in financing fisheries and aquaculture development in Grenada, Barbados, and St. Vincent and the Grenadines.

3. Finance vs. Financing Solutions

A framework that supports sustainable long-term financial support for fisheries and aquaculture development requires the distinction between finance and financing and the integration of the two in a clear, targeted manner.

Finance is the mobilisation of fiscal resources, but ("throwing money/currency at the problem") should not be viewed as a once-off solution to solve the challenges that face the fisheries and aquaculture industry. For finance to be invested in sustainable fishing practices, the goal/purpose of new and sustainable fishing techniques needs to be communicated to current and prospective fisheries stakeholders, as well as the legitimacy of sustainable management practices [15] to increase investor confidence and buy-in. The legitimacy of sustainable management practices and fishing methods can be communicated by providing evidence of widespread consensus (regional/global or among the scientific community) as to the practices used, a historical track record of mutual benefit (between investors and the industry), and the presentation of reliable data that represent the impact of the sustainable methods in question by reliable indicators in the relevant industry. Examples can include the amount of national public finance that is allocated to the development of the national fisheries and aquaculture sectors, ODA funding, and development project funding for fisheries and aquaculture projects [15]. Public authorities are encouraged to support the transition of current fishing mechanisms to more sustainable practices, as well as promote an environment which upholds and maintains the regulations and development of fisheries and aquaculture. Much of the funding that is available to support sustainable (fisheries and aquaculture) development is from grants for time-limited projects, which can inhibit the momentum of innovation [15]. A further potential disadvantage of short-term funding is that the experience and expertise built up in short-term project teams are not retained, and thus, the stakeholder relationships that build trust between government bodies and public sector officials at the termination of the project [15,16]. For these reasons, it is unlikely that finance alone will be able to incentivise consistent development in aquaculture and fisheries and identify the need for long-term financing structures and frameworks [15,16].

Financing includes the fiscal mechanisms or arrangements that facilitate returns on investment (whereas finance itself does not inherently seek any returns [15,16]. These include debt and equity instruments, insurance products, securities, derivatives, and others. These mechanisms encourage the maintenance and development of whatever initiatives they are invested in, which will consistently produce a return on investment, making them more suited for long-term sustainable development projects.

The consistent supply of seafood and the concurrent conservation of marine ecosystems will require financing from both public and private fisheries and aquaculture stakeholders. The conservation of natural ecosystems and exploitation of marine capital can be both revenue sources and costs, and they are in need of financing [15,16]. Capture fisheries and aquaculture (as well as other terrestrial-based industries) exert pressure/strain on marine ecosystems. The aim is thus to create financing mechanisms and frameworks that facilitate industry initiatives/operations to limit negative ecosystem impacts and promote environmental sustainability. The protection of ecosystems provides critical services (such as supplying safe nursery environments for juvenile fish) for the production of seafood for a country's people, even if conservation and maintenance costs are likely to be greater than the revenue produced. This is particularly important for Caribbean countries as the region's economy is largely centred around tourism and ecotourism, which is reliant on inherent natural capital. Prioritising the funding and financing of natural assets of Barbados, Grenada, and St. Vincent and the Grenadines not only benefits the fisheries and aquaculture sectors but also synergises with the tourism sector and the development thereof.

3.1. Fiscal Policy for Fisheries and Aquaculture Development

Fiscal policy covers state spending and the state's generation of revenues. State governments also have multiple sources of revenue: taxation, fees (from government authorisation documents like licences), state asset ownership, state-owned enterprises

(SOEs), and relations of debt and credit [17]. Budgetary governance is also classified as fiscal policy and deals with the administrative and institutional systems that control the fiscal flows of the state [18].

Table 2 summarises different financing structures that may be applicable to the development of Caribbean fisheries and aquaculture. These can be applied to the relevant environments and ecosystems that are aligned with the interests of aquaculture and fisheries development. State-owned enterprises with environmental objectives, state-mandated tourist fees and revenues generated, and leases and licences are all (directly or indirectly) part of environmental fiscal policies. However, different parts of public fiscal policy can be revenue generators and liabilities at different times and at different levels of the organisation. For example, state-owned enterprises with positive profits generate revenue for governments, but they can also be a liability due to their indebtedness. The specific application of fiscal policy thus requires an in-depth understanding of the relevant industry (fisheries and aquaculture) for optimal use as a development mechanism, including government support of lending activities that promote sustainable development or minimise environmental harm.

Table 2. Government financing structures that are applicable to the development of Caribbean fisheries and aquaculture.

Type of Arrangement	Description
Environmental taxes	An environmental tax is meant to target activities that cause environmental degradation or pollution, but this does not imply that the tax has a proven effect in terms of minimising environmentally degrading or polluting activities. Rather, the minimum definition of an environmental tax is that it increases the costs of pollution activities [19]. Nonetheless, the actual application of an environmental tax is likely to have the effect of incentivising some activities whilst disincentivizing others.
Leases, licences and fees	Whereas a tax is compulsory, a lease, licence or fee can be seen as a form of exchange between a public authority and a (legal) person, which grants the latter a certain right or privilege. Another difference between this category of public fiscal revenues and taxes is that these leases, licences and fees remain stable regardless of the incomes that the payer may generate as a result.
State-owned enterprises (SOEs)	SOEs are rarely thought of as part of fiscal policies. Nonetheless, they are and many of the types of companies that have historically been state-owned around the world have some clear environmental impacts. Public utilities companies such as water companies and energy companies quite explicitly have an environmental aspect since they usually have environmental quality or resource efficiency as key objectives. Other types of SOEs, like transportation companies, may increasingly incorporate environmental objectives. SOEs can be run with a focus on generating profits, but they can also seek to strategically increase public goods and services through their operations. SOEs can be a source of revenue when they generate profits while their inventory and equipment can be seen as capital goods. Meanwhile, SOEs can also hold liabilities in the form of debt.

Table 2. *Cont.*

Type of Arrangement	Description
Debt and lending	Debt can be an income for public entities as well as a cost and liability. Lending from development banks under national control, lending from sovereign wealth funds and debt in the form of deferred taxes (with potential interest added) are examples of the state as creditor and debt as a source of public incomes. By contrast, sovereign debt is a continuous source of public expenditures. Lending practices, both when the state is the creditor and debtor, can be explicitly targeted towards the objectives of sustainable aquaculture and fisheries development. (see Section 3.3)
Public finance support (government subsidies)	Public finance such as grant funding or philanthropic contributions (while not directly a financing mechanism as previously defined) can be used as a financing mechanism to garner private investment. This can be performed in the form of government subsidies, where public finance is directed towards development mechanisms or initiatives that also require further private funding to be implemented. The total cost of the fiscal contribution from the private sector is lowered, thus incentivising private investment in an industry (fisheries and aquaculture). This mechanism can thus facilitate affordability of industry (aquaculture and fisheries) development for the private sector.

Blue Levies and Stakeholder Taxation

Several countries impose levies that are targeted at environmental protection, usually in the form of environment or tourism enhancement levies. Similarly, this kind of mechanism can be used to enhance the fisheries and aquaculture industries in various ways, such as environmental protection, MPA management, and infrastructure development. These levies are usually applied in different ways, at different rates and at different levels of organisation, like on importation, consumption, accommodation, service, or travel. For example, the British Virgin Islands (a known SIDS country) imposed an Environmental and Tourism Levy of US$10 to be paid on arrival at all ports of entry [15]. The tourism, hospitality, cruise and charter, extraction, and ports and ship-building industries could be potential targets from which the proceeds of blue levies could be directed to support fisheries and aquaculture in Barbados, Grenada, St. Vincent and the Grenadines.

Environmental taxes (or levies) could be applied to organisations that rely on natural resources on which fisheries and aquaculture industries are reliant as well and are known to negatively affect those environments. However, the degree to which this occurred may not necessarily have been quantified [20]. The fisheries industry is known to pollute the marine environment, such as due to the use of old and outdated equipment that breaks and is lost at sea [21–23] and thus should not be exempt from this kind of tax (they could also be taxed for unacceptable bycatch of non-targeted species). The proceeds from these taxes could then be used to reverse the degradation and further improve the environment or intensify the ecosystem service these areas provide to fisheries [20]. Similarly, an organisation can be incentivised to proactively develop these environments of importance and, in exchange, receive specific tax exemptions, thereby fostering an enabling environment of coastal systems protection, which would ultimately benefit the fisheries and aquaculture sector. Potential industries that can be targeted are tourism, terrestrial agriculture, fisheries and aquaculture, and maritime transport. A regional environmental tax could be established to support fishing grounds shared by multiple states or multiple sectors, ensuring that no one country or industry is more negatively affected by another (similar to the idea of the "tragedy of the commons" with many actors depleting unregulated finite resources [24]). This would require transparency and control over the number of vessels from each country allowed to access such a particular fishing ground. Various mechanisms for managing and regulating shared resources have been discussed elsewhere [25].

Environmental taxes and fees were, on average, 1.19% of the Latin American and Caribbean (LAC region) GDP in 2019 [19]. This level was slightly lower for Barbados, which only generated taxes equivalent to 0.81% of its GDP ([19]; numbers are not included for Grenada and St. Vincent and the Grenadines). Caribbean environmental taxes, as a share of GDP, were roughly equal to the entire LAC average and were greater than that of South America and below Central American and Mexican tax levels [19]. There is an opportunity for increased environmental taxes and fees in countries like Barbados for development revenue generation, where existing taxes and fees fall below regional averages or when the current taxes as a proportion of national GDP are generally limited (for example, less than 3% of national GDP).

3.2. Compensation Mechanisms for Natural Capital

Compensation can be classified as a type of fee payable for the renting of a resource or for degrading a natural asset or ecosystem (but this needs to be quantified). The design of compensation mechanisms can vary immensely. These mechanisms can be designed to directly benefit the environments that underpin extractive industries, such as capture fisheries and unsustainable aquaculture, particularly payment for ecosystem services (PES schemes) and biodiversity offsetting. By contrast, where compensation is linked to environmental outcomes, levies, fees, or licences are not necessarily linked to environmental outcomes, but their proceeds may still be directed to benefit fisheries and aquaculture (non-environmental benefits).

Payment for ecosystem services (PES) schemes are known by five characteristics [26]: "(1) a voluntary transaction where (2) a well-defined ecosystem service (ES, or a land-use likely to secure that service) (3) is being 'bought' by a (minimum one) ES buyer (4) from a (minimum one) ES provider (5) if and only if the ES provider secures ES provision (conditionality)." PES schemes thus entail the voluntary buying and selling of ecosystem services based on the seller protecting an ecosystem.

Although PES implementation mechanisms for marine sustainability have been limited in the Caribbean region, they can be designed to benefit the ecosystems that provide the underlying support for fisheries in the Caribbean region. Most historic PES schemes are drawn from the forestry industry and indicate that PES can work as a subsidy [27], encouraging resource users to implement environmentally sound practices [28]. This can take the form of fisheries stakeholders using fishing equipment that is less damaging to the environment (such as moving to trawling as opposed to dredging). Despite most PES programmes being thought of as a market-based approach, most programmes are based on compliance with government regulations [29]. However, these subsidy-PES programmes are only as good as the (government) authorities that implement them since PES requires enforcement of regulations. Subsidy-based PES schemes are potentially more difficult for cash-strapped governments to maintain.

Carbon sequestration, biodiversity protection, watershed protection and ecosystem beauty are central types of ecosystem services [26], and these benefit several industries beyond just aquaculture and fisheries (replenishing fish stocks and providing safe nursery grounds). Biodiversity offsetting, carbon and other nutrient trading credits, as well as listing natural environments or protected areas on the stock exchange, are other forms of PES that are able to finance the natural environments that support and underpin fisheries and aquaculture. Table 3 summarises a few examples of PES for marine environments that support fisheries and aquaculture (but more detailed information is presented in Trends [30]).

As an example, the CARIPES project (started in 2011) aims to facilitate the active participation of coastal fishermen in the conservation and sustainable use of available marine resources in Caribbean marine protected areas (MPAs). The project further aims to leverage the use of local fishermen's knowledge of the coastal and marine biodiversity within pre-established MPAs, facilitate resilience development among marine ecosystems towards global climate impacts, and develop appropriate PES schemes in the Caribbean through the use of such local knowledge. The project has been active in the islands of Grenada, St Eustatius, and Martinique and further endeavours to generate avenues for the development of compensation mechanisms (payments for ecosystem services) while supporting fishers reaping the associated economic benefits of such protected areas. Such a project could be replicated on other islands of concern, such as Barbados and St. Vincent and the Grenadines.

Table 3. Examples of payment for ecosystem services for marine environments (based on Trends [30]).

PES Type	Elaboration	Examples
Regulated markets	Cap and trade markets require resource users to hold a purchasable right to the resource they use. It sets a limit to resource use and allocates a tradable share of the resource to an asset owner. A licence is not necessarily tradable and the issuer of a licence does not necessarily set a limit to damages to an ecosystem, but it still constitutes a transaction based on compliance.	Fishing can in different ways be governed through mechanisms that can reasonably be seen as PES. On the most simple level, recreational fishers are in some jurisdictions required to pay for fees or hold annually paid licences. Another example are individual transferable quotas (ITQs) that are used to allocate annual fishing rights within an exclusive economic zone (like quota's per fishing ground). ITQs can then be traded and leased amongst fishers. By using the regulated ITQ marketplace, a sustainable fishery and ecosystem services market can be achieved [31], through the trading of rights to exploit the fish provisioning service of an ecosystem within identified sustainable harvesting limits. The rights to a proportion of the total fish population (fishing quota) has to be identified beforehand and the total harvestable quota must be within the sustainable harvesting limits of the ecosystem, for this mechanism to be effective. Government mandated fees related to ecosystem services can likewise be seen as a form of PES based on compliance. This can for example be tourist fees associated with entering (marine) protected areas, that in this case are of importance to fisheries and aquaculture.
Voluntary transactions	Voluntary PES transactions can involve private and public sector actors alike. Governments, private individuals, NGOs and development organisations can all pay private actors to change practices or avoid harm to ecosystems.	The Marine Legacy Fund of Tanzania is an example of voluntary PES. It is a revolving fund whose original sources of revenue as well as its spending can be seen as forms of PES. It gains revenue from tourist fees, fossil fuel taxation and fishing licences. However, the fund uses this revenue to finance the protection of coastal habitats and important marine sectors. Whereas its sources of revenue are based on compliance, its spending can be seen as voluntary PES. This kind of structure can be applied to countries individually, or regionally, due to shared fish stocks (or other shared resource use).

Stakeholders (public or private) that make use of the same natural environments on which fisheries and aquaculture rely (eco-tourism, for example) can be mandated to contribute to its optimal natural functioning, maximising the ecosystem services they provide to the aquaculture and fisheries industries, which would also benefit other sectors such as the tourism industry.

3.2.1. Biodiversity Offsetting

Biodiversity offsetting is a form of compensation that is based on counterbalancing any lost biodiversity from a development project by investing in equivalent biodiversity somewhere else, thus aiming to maintain biodiversity despite economic development (or environmentally harmful extraction processes). It is relevant to consider biodiversity offsetting here as it has previously been used to preserve marine biodiversity and wetlands [32]. The ideal application of biodiversity offsetting is the application of the "mitigation hierarchy" when considering the predicted negative biodiversity impacts of development initiatives [33,34]. If developers follow the mitigation hierarchy (avoidance—minimisation—restoration—offsets), they should try to avoid negative impacts in the first place, secondly consider the minimising of impacts, thereafter restoring any negative impacts stemming from the development, and finally using biodiversity offsets to compensate for the unavoidable biodiversity loss stemming from development projects [33]. Thus, biodiversity offsetting is only meant to counter biodiversity loss, which cannot be prevented [34].

In the case of other extractive industries, a portion of revenue earmarked for biodiversity offsetting could be invested into supporting the fishing grounds that a state or region relies on through examples such as investing in sustainable fishing programmes, the acquisition of sustainable fishing equipment, financing the prevention of illegal, unreported, and unregulated (IUU) fishing. The associated biodiversity loss that is found with dredging, trawling, or new aquaculture infrastructure developments can be offset by investing a portion of the proceeds into less damaging subsectors, such as domestic handline fisheries, sustenance fisheries or developing restocking programs that use the available nursery habitats already present on Barbados Grenada, or St. Vincent and the Grenadines (such as the development of mangroves and seagrasses). Furthermore, different levels of compensation can be mandated due to the perceived value of the affected (extracted or displaced) species or environment in question.

However, determining an equivalent unit of biodiversity is extremely difficult, relies on simplifying conventions, and includes several moments of uncertainty. One issue, for example, is whether to consider biological diversity at a species level or at a system level. In other words, is an equivalent sum of biodiversity required or is an equivalent ecosystem required? To even approximate the latter is certainly difficult, but even if offsetting is limited to a species level, another question that emerges is whether compensation should be of the exact same species or if another species of equal importance can be considered for compensation (for an in-depth analysis of difficulties pertaining to biodiversity finance, see [35]). For example, for every shark that is caught and succumbs to bycatch, should 50 individuals of its primary prey species be required as compensation for (restocked) or the protection of another shark? It is likely that biodiversity offsetting is further complicated by the extraction of mid-trophic level species, as the effects on trophic levels above and below it could destabilise a functioning ecosystem (in a multi-directional "trophic cascade", despite being defined as top-down interactions [36]), and potentially risk the collapse of an entire fisheries sub-sector. For this reason, it is advisable that ecosystems as functioning units be considered [37,38]: where two ecosystems are available as fishing grounds, one should be protected and demarcated as an MPA, whereas the other may be afforded less regulation. This would be considered a form of biodiversity offsetting if the productivity of each system were initially similar (this further emphasises the need for accurate and detailed ecosystem evaluation).

Accurate offsetting requires establishing baseline levels of biodiversity at the sites that provide offsets. A significant risk is that baselines may be set too low and thus give developers an unwarranted amount of credits. If an area is invested in due to offsetting (an MPA, for example), the response of the investment needs to be equitable to the initial biodiversity loss. This requires sufficient long-term data on that area to assess historical performance or attributes, and where this is not available, data would first have to be collected. This limits the potential number of sites for immediate biodiversity offsetting, as only sites with historical data would be able to provide accurate indications of developmental productivity (historical fisheries catch data may be particularly useful). However, cost-efficiency for environmental changes in biodiversity also changes [39]. Thus, these become economic issues and require funding and financing themselves. Given that the extent of habitats that provide key ecosystem services (such as mangroves) is low and decreasing in Barbados, Grenada, and SVG [6,40–42], extensive monitoring is necessary to establish what the historical coverage of such a valuable habitat was. This enables an appropriate response elsewhere, like conserving and fostering the development of the coral reefs in areas where they are relatively underdeveloped (as in Barbados and Grenada).

The difficulties of appropriately managing a biodiversity offsetting programme imply that taxation and fees are better as a form of environmental compensation and compensation for a developer's renting of the ocean as a public resource. The fee could be applied as part of licensing programmes and could secure an appropriate minimum public revenue. One of the potential benefits of biodiversity offsetting is that funding from development projects becomes earmarked specifically for biodiversity purposes. The challenge for regular fees and taxation is to ensure the political will and administrative capacity to ensure an appropriate share of the public revenue is directed towards the improvement and protection of ecosystems that support fisheries and aquaculture.

3.2.2. Carbon and Other Nutrient Trading Credits

Coastal environments are likely to offer nutrient capture and sequestration services (such as blue carbon) in addition to the benefits they provide to fisheries and aquaculture [5]. These services can be capitalised upon and leveraged as pollutant offsets (on international and domestic markets) from which revenue can be generated to further support the protection, optimisation of services, and maintenance of the environments themselves or be directed towards development in other areas of fisheries and aquaculture.

Blue Carbon entails the use of coastal and marine ecosystems as vehicles for carbon capture and sequestration. Mangrove forests, salt marshes, algae, wetlands and even whales contribute to carbon storage in coastal and marine environments [43,44]. Blue carbon, thus, refers to processes where biological organisms permanently store carbon as long as the organism (tissues) remains intact and alive. Beyond this broad conceptualisation, blue carbon is usually used to refer to the active promotion of blue carbon processes in order to generate carbon credits through carbon sequestration for use in international climate commitments or carbon markets. Since the majority of carbon sequestration happens in coastal and marine areas, their applicability to SIDS states and their support for the sustainable development of SIDS states are considerable. Many of the ecosystems that promote blue carbon (such as mangroves) also provide other co-benefits like nursery/feeding grounds for fish and wave attenuation, which contribute to the reliance and longevity of infrastructure along SIDS' coastlines. While historically, there have not been many examples of development projects tapping into the nutrient credit markets in the Caribbean region, the conservation and development of the mangrove habitats on the islands of Barbados, Grenada, and St. Vincent and the Grenadines present an ideal opportunity to access nutrient trading markets (such as the carbon credit market) which can further finance the development of these key ecosystems, thereby stimulating the performance of the fisheries and aquaculture sectors.

Applicable to other nutrient offset projects, a central concern with any type of carbon mitigation programme is that it is inherently reliant on carbon accounting methodologies and frameworks [45], irrespective of whether these are used for national development contributions (NDCs), voluntary markets or compliance markets. Many blue carbon projects face data limitations (due to determining the appropriate and market-accepted measurement techniques of these ecosystems) and may limit the amount of attention and funding these projects are able to garner to either fund their own development or the industries the ecosystems underpin (fisheries and aquaculture). The implication is that priority is not inherently given to the carbon sequestration projects that lead to the optimal carbon outcomes, but rather that prospective investments may be preferentially directed to the projects where the extent of carbon capture and sequestration impact can most easily be traced and tracked, at the lowest costs.

The market uptake for carbon credits and carbon offsets has been somewhat limited, one reason being the high demand in compliance markets. Market conditions for carbon credit prices have continuously been changing. The price of carbon in the EU emissions trading system (ETS) has been increasing since the end of November 2020, at around 22 Euros per tonne of carbon dioxide (CO_2). Prices peaked at around 96 Euros in February 2022 but fell later in the same year [46]. Whereas the carbon market fluctuates somewhat regularly, other nutrient credit markets may be more stable. Other elements, such as nitrogen, may be charged and capitalised on in a similar manner as carbon, thus potentially increasing the value of any one ecosystem. By harnessing the multiple ecosystem services it provides on trading markets, especially considering the effective bioremediation services that ecosystems like mangroves and seaweeds provide in reducing the negative effects associated with complex pollutants or fertiliser runoff (i.e., addressing multiple pollutants concurrently). However, given that nutrient offsetting (such as carbon offsetting) is still a nascent industry, a stable market price for specific nutrient offset credits would allow for the necessary confidence and risk assessment in employing this type of funding scheme for revenue generation, as well as increasing investor confidence.

Any type of blue carbon project is likely to involve either NDCs or the voluntary carbon market. Towards the end of 2021, carbon prices were rising on voluntary carbon offsetting markets, which several stakeholders expected to continue in 2022 [47]. Major companies are making net zero pledges and increasing their voluntary commitments to mitigating climate change privately. This opens up the possibility for the ecosystems of Caribbean countries to capitalise on the voluntary blue carbon market. An increase in the demand for blue carbon projects is thus expected. However, a concern has been the lack of quality, verified projects and carbon offsetting. The classification of an environment that provides such carbon removal and sequestration services (such as declaring a marine environment as an MPA) may increase the perceived reliability with which the offsetting can be expected to be available year on year (if that environment were involved in a carbon sequestration and offsetting programme). Such protected status further facilitates buyer confidence that the service will be improved upon in future and that nutrient credits may become cheaper in the future (similar to developing more sophisticated computer product offerings year on year). Declaring where proceeds are re-invested in a transparent manner may further encourage buy-in from offset buyers through contributions to development in other areas, such as NDCs or fisheries and aquaculture development. In addition, this may incentivise continued participation of stakeholders from such industries in blue carbon projects, as it would benefit their own sector as well (i.e., added benefits to carbon offset investing or investing in blue carbon development projects if the habitat in question also provides benefits such as fisheries stimulation). However, voluntary offsetting is an additional expense relative to the operations of the companies buying the offsetting, which limits their incentive to invest—if faced with an economic slump (or the risk of one), there is a chance that companies will discontinue their voluntary offsetting arrangement or switch to cheaper, lower quality credits.

When considering nutrient offsets (like blue carbon), it is essential that existing local practices and livelihoods (social development) are taken into account. Research on carbon projects shows that local community involvement facilitates project success, as compared to if it were not present [48]. This includes free, prior and informed consent (FPIC), but successful project implementation cannot be limited to a formal exercise of securing FPIC. It needs broad support from a community that can see themselves in the project, as well as the potential for livelihood development. For example, blue carbon projects that seek to preserve seagrasses may collaborate with fishers that usually pass through marine areas with seagrasses. Collaboration with local fishermen and women could be more cost-effective and provide additional income streams if integrated into a seagrass monitoring scheme.

3.2.3. Natural Capital as Publicly Traded Equities

The Intrinsic Exchange Group (IEG) has collaborated with the New York Stock Exchange (NYSE) to create a new asset class: Natural Asset Companies (NACs). The purpose of such NACs is to maximise the performance or delivery of the natural asset they are associated with, either through ecosystem services provisioning, the use of the asset for restorative or regenerative agriculture (including aquaculture), or hybrid cases. NACs have the explicit mandate to actively manage, maintain, restore and grow the value of the natural capital they are associated with. A NAC may also use payment-for-ecosystem services mechanisms such as producing carbon credits, other nutrient credits, and biodiversity credits (i.e., verified ecosystem services delivery) in which the NAC, as a company, trades. The NAC itself is listed on the stock markets and, based on its performance, may attract investment support.

Prospective NACs are evaluated by the IEG and then listed for trading on world platforms, enabling the conversion of natural assets (such as publicly owned land) into revenue. This process has the potential to facilitate environmental, social, and industrial benefits at scale, contributing towards a shift to a more sustainable and circular economy [49]. One such example is in Costa Rica, where IEG is collaborating with the local government to explore the creation of a NAC to value and finance conservation and social priorities and meet national and global commitments (e.g., High Ambition Coalition 30 × 30 goal). The coral reefs that support the livelihoods of many locals in Barbados, Grenada, and particularly St. Vincent and the Grenadines (having the largest expanse of coral reefs among the three) can be registered under public-private NACs (between governments and the national population) to generate revenue for the development and conservation of these valuable ecosystems from the stock markets.

However, the evaluation criteria and indicators of performance used to assess the natural assets in question need to be standardised and recognised (i.e., agreed upon). This constitutes a major hurdle as discrete natural assets (for example, in different countries) face different stressors (whether ecological, climate, social, or political), which determine their functioning and performance. Thus, establishing an optimal functioning baseline of performance becomes inherently difficult and is subjective to what historical data is available for each region, as well as to whoever is performing the evaluation.

Nonetheless, the sustainable management of the environmental areas that underpin fisheries and aquaculture industries (such as fishing grounds and MPAs or other natural assets) such that the maximum amount/number of benefits are realised is thus incentivised by potentially global markets through the use of NACs. This mechanism presents a catalysing mechanism for fisheries and aquaculture development while concurrently incentivising sustainable and social development as well.

3.3. *Debt for Nature Swaps and Debt Buy-Backs*

A debt-for-nature swap (also known as a "debt buy-back") can be defined as a scenario where a creditor forgives debt owed to them in exchange for a commitment by the debtor to use the outstanding service payments for a particular investment [50]. This can be explained differently as a creditor agreeing to sell a portion of the debtor's debt for an agreed purpose under agreed conditions. The redemption of debt can thus be conducted at a discount. The service repayments can be invested into whatever project or initiative is agreeable to both parties in the transaction, such as for fisheries and aquaculture industries or environmental and social-based projects. Furthermore, third parties can facilitate such transactions by providing a loan or guarantee to the debtor. These can be used to finance marine-related development, including fisheries and aquaculture, as well as reduce debt repayments. Debt swaps can present an enticing opportunity for the governments of (SIDS) countries hoping to simultaneously facilitate development and reduce the country's historical debt.

Historically, debt swaps have not been considered for the protection of marine environments (MPAs). However, this changed with the Seychelles Debt Swap of 2015, which was the most successful debt-for-nature swap for a SIDS [51]. This deal, facilitated by The Nature Conservancy (TNC), converted US $21.6 million of sovereign Seychelles debt to the Paris Club of Creditors. The issuance was made possible due to World Bank expertise and guarantee, a Global Environment Fund (GEF) non-instrument grant to reduce the coupon rate, and technical expertise from the then Prince of Wales International Sustainability Unit. The bond was purchased by three impact investors, and the blue bond proceeds led to the establishment of an environmental trust fund—Seychelles Conservation and Climate Adaptation Trust (SeyCCAT) and committed the Republic of Seychelles to protect 30% of its EEZ as marine-protected areas (MPAs). The many ecosystem services and industry opportunities that MPAs can provide to a country's people and economy indicate that deals such as these have the potential to galvanise a country's future development in a sustainable way. Given the similar context that many SIDS countries share with one another, such debt swap structures can be applied to Barbados, Grenada and SVG, addressing both the historical national debt and nature conservation challenges they face simultaneously.

Many Caribbean countries (Haiti, Jamaica, and others) have been involved with debt swaps since the 1990s, but these have rarely contributed to a significant reduction of debt in the region [52]. For example, in 2012, Antigua and Barbuda negotiated a 'debt for climate adaptation with coastal zone management swap' with Brazil for USD 18 million, but it did not come to fruition due to delays with the Brazilian Parliament [52].

Challenges that arise with the mobilisation of debt swaps include the composition of creditors (where the heterogeneous composition of sovereign credit transactions can make it difficult to make enough actors agree on the terms and conditions of debt swaps). As is often symptomatic of debt-swap deals [53], another challenge is the size of the deal. As a country develops and creditors become more confident of a country's ability to repay their loans, the creditors become less willing to sell the loans at a discount, debt being swapped at lower valuations and thus resulting in smaller deals. However, there are exceptions: a recent Belize debt-for-nature swap raised US $364 million to buy back US $553 million of debt by the Belize government (at a discounted rate). The debt conversion was made possible by a loan and guarantees from TNC, Credit Suisse, and the Inter-American Development Bank (IDB). The scale of the transaction benefited from high discounts on Belizean debt and new structuring practices [54].

Political controversies may also bring the legitimacy of debt swap deals into question and thus cast doubts on the legitimacy of any resulting development. Political controversies ascribed to debt swaps can be seen as a question of whether the high indebtedness of SIDS is legitimate in the first place or if a debt swap is a solution to immediate economic and environmental problems. If, on the one hand, existing debt levels are taken for granted, a debt swap may be seen as a pragmatic tool to achieve different policy objectives. By contrast, if the original debt is essentially considered politically illegitimate, actors may see a debt swap as equally illegitimate. The *de facto* loss of sovereignty that is associated with government debt swap may be another point of contention. When the debt swap is used to create an environmental trust fund, a way to ensure that the government is at arm's length of the trust is to not have a government majority on the board [55]. However, this makes it more likely that a deal is going to be perceived as illegitimate [55]. A debt swap can best be seen as a means of establishing and incentivising policy consistency on the part of the government. Whether or not this should be thought of as a *de facto* loss of sovereignty is a political question. On the one hand, it is a commitment that the government voluntarily enters into. On the other hand, it makes it difficult for a government to change course, which can be beneficial as different government offices come and go during elections/regime changes.

The efficacy of debt swaps as a solution for reducing national historical debt is controversial and debated, as debt conversions do not necessarily lead to long-term debt sustainability (see [56]). Perry et al. [57] provide further context for the source of historical debt among Caribbean island states, as well as comment on the socio-cultural and political implications of debt swaps (and other financing mechanisms herein discussed). However, the author does not provide pragmatic alternatives to overcoming historical debt beyond demanding recompense for the injustice of a colonised past. Long-term sustainable debt reduction through debt swaps would more likely arise from investing in long-term projects which have the potential to facilitate a country's future economic development in a sustainable way, as well as enhancing the country's future resilience to the phenomena that create the debt in the first place (such as resilience to destructive weather). One such example is the stimulation of the fisheries and aquaculture industries through the development of nature-based solutions, such as MPAs, which protect ecosystems with multiple benefits (such as wave damping, blue carbon, and providing fish nursery and feeding grounds). Such solutions that involve local communities are also more likely to last as local buy-in in the initiative/project would more likely maintain any development (or identify alternative funding and financing) even if debt repayments were late or discontinued. However, this would require informed communities to have access to possible financing mechanisms, potentially through a local financing unit. Despite the challenges associated with debt swaps, they have the potential to resolve some of the historical debt of some Caribbean SIDS as well as other national development needs but are likely not the only solution to this challenge. The use of debt swaps, together with other financial mechanisms and arrangements, likely offers the greatest possibility of success in resolving national historical debt.

3.4. Blue Bonds and Other Sovereign Bonds

Blue bonds have been promoted as a means of financing Blue Economy developments amidst the fiscal constraints that SIDS are facing. Similarly, blue bonds can be used to finance the development of fisheries and aquaculture. Blue bonds, like other bonds, are tradable fixed securities issued by an authority to raise funds on global markets and increase the issuer's debt. The World Bank defines blue bonds as "a debt instrument issued by governments, development banks or others to raise capital from impact investors to finance marine and ocean-based projects that have positive environmental, economic and climate benefits" [58]. Issuances of blue bonds are conducted based on a per-case basis and case-specific environmental and economic returns, but generally the criteria for bonds to be considered "blue" require that the investment be used for oceanic or marine resource development.

Barbados is among some of the first countries to partner with The Nature Conservancy (TNC) on a Blue Bonds project after Seychelles. A novel co-guarantee structure with a $50 million guarantee from TNC, alongside a $100 million guarantee from the Inter-American Development Bank (IDB), was used to facilitate a $150 million debt conversion that will facilitate the expansion of Barbados' marine protected areas from virtually zero to approximately 30% and improve management for all marine waters within its jurisdiction [59]. This project is expected to free up approximately $50 million to support environmental and sustainable development actions in Barbados over the next 15 years, making both the country and its people more resilient in the context of climate change. Barbados worked with Credit Suisse, who acted as Global Lead Arranger, to raise approximately $150 million through a dual currency term loan facility (with CIBC FirstCaribbean as Domestic Lead Arranger). This Blue Loan funded the buyback of a portion of Barbados' existing debt and was partially funded through the implementation of Blue Bonds in capital markets. The new financing featured a lower interest rate than the old debt, with both TNC and IDB each providing repayment guarantees on the country's behalf, and 100% of the resulting cost savings will be directly allocated for marine conservation [59].

Using blue bonds as a financing mechanism supports an environment that stimulates the development of a country's entire Blue Economy by presenting financial capital to private actors wishing to make a sustainable change at low risk to investors [60]. When considering a bond issuance, it is critical that the bond structure is fit for purpose (in this case, the development of fisheries and aquaculture industries), ensures the highest degree of environmental and social impact, and that the issuer receives the lowest possible interest rates on repayments. March et al. [61] discuss the challenges involved when designing blue bond financing for Caribbean SIDS, using the Bahamas as a case study example. Three different bond structures may be applicable for the development of Caribbean fisheries and aquaculture sectors: catastrophe bonds, environmental impact bonds, and use-of-proceeds bonds (see also [62,63]).

The mechanism of catastrophe bonds is that a trigger level (like the wind speed of a hurricane) for a specific area is determined before the phenomenon occurs. If the trigger level is surpassed, the insurer pays out to the insured party. However, this bond differs from insurance in that it pays out before the phenomenon has struck, whereas insurance pays out afterwards [61]. One challenge of catastrophe bonds is that they only imply that very specific events are being insured and that these events have to fulfil specific conditions. Furthermore, no payout is required if the trigger level is not surpassed, but damage is still widespread because of the phenomenon. Environmental bonds raise capital, but the return on investment is based on the success of an environmental programme or project, as defined by pre-determined key performance indicators (KPIs). Before using such a financing mechanism, it is advised that the developmental sector has clearly defined key performance indicators (KPIs) to provide added confidence and clarity for potential investors [61]. In the case of fisheries and aquaculture development, this could be achieved by achieving a maximum sustainable yield of a fish stock within two years, for example, or the protection of 20 hectares of mangroves as fish nursery grounds. A use-of-proceeds bond entails the upfront promise that proceeds will be used towards blue development (not necessarily environmental development). These bonds are at risk of "environmental non-performance", where returns as benefits of the environment do not materialise, but the economic returns do [61]. This type of bond may be particularly suitable for the development of fisheries and aquaculture infrastructure (equipment, value-addition practices, workspaces, etc.), given that no explicit environmental benefit is mandated.

3.5. Blue Tokens and FinTech

"Fintech" or financial technology refers to the use of new technology to improve management and access to financial operations and processes. It involves the use of specialised software, algorithms (machine learning), and artificial intelligence to achieve improved management of finances [15]. This has expanded into the insurance and investing industries. Fintech has enabled improved compliance and faster transactions and has further allowed financing to be raised on open markets with far less friction or difficulties.

Blue tokens are a proposition where fintech and blockchain technology are used to raise money for blue (fisheries and aquaculture) development projects [15]. An issuer could set an amount they would like to raise, for example, US$10 million, with an initial fixed repayment coupon. The initial price of each token could be set at US $10 (predetermined), with one million tokens being issued on a secure blue token market or platform. Any investor who has been approved through rigorous identity checks, like Know Your Client (KYC) and anti-money laundering (AML) checks, can then buy tokens and either hold them to maturity or trade them among other investors on the blue token platform. A Blue Economy credit rating agency (alluded to before) could also rate the issuance (initially and later annually) for development outcomes and financial viability, thereby giving investors maximum information to assist with their investment decisions [15].

The use of blue tokens could democratise investments, making the opportunity to invest in the development of the Blue Economy, or in this case, fisheries and aquaculture industries, accessible and giving all stakeholders a real stake in the Blue Economy. With the establishment and development of blue bonds, blue tokens can thus give the citizens of Barbados, Grenada, and St. Vincent and the Grenadines the ability to invest in the future of their nations as big ocean developed states (BODS) rather than SIDS [15]. They are also able to invest in blue tokens from other countries' blue bonds, fostering development support where such structures are in place, even if not in place in their own nation yet. This may be applicable for blue bonds issued towards the development of coral reefs in SVG: the largest coral reefs among the three counties are likely to facilitate the greatest economic returns (among the reefs of the three countries), garnering more attention than others in terms of interest for development.

3.6. Insurance

Instead of being a source of new finance, insurance can be viewed as a tool with which to support the financing of fisheries and aquaculture projects. Insurance can create confidence for a potential project developer or investor in that it limits the risks that the project may face, potentially reducing the costs of capital investment. The role of risk management, risk pooling and risk transfer has become important for any potential development in the Caribbean as the intensity of natural disasters like hurricanes increases [15]. Marine insurance can be explored and tailored to specific industries within the Blue Economy, like fisheries and aquaculture.

Parametric insurance is one type of ex-ante disaster financing and makes payments based on the intensity of a disaster event and the amount of loss calculated using a model previously agreed to by both parties [15]. This type of insurance is different from indemnity settlements in that there is no on-site assessment of individual losses but rather depends on a triggering mechanism (based on variables out of the control of both the policyholder and insurance issuer). This may be of interest to stakeholders in the fisheries and aquaculture sector, where infrastructure is swept away or is unrecoverable for assessment of damages, or instances where a natural environment that underpins the industry gets damaged, and objective ecosystem valuation becomes skewed by lack of historical data (however in such a scenario, ownership or shared ownership may have to be allocated and proven, like through an NAC). Another example is the Caribbean Oceans and Aquaculture Sustainability Facility (COAST), a parametric insurance facility developed jointly by the Caribbean Catastrophe Risk Insurance Facility—Segregated Portfolio Company (CCRIF-SPC), the United States Department of State, the World Bank, TNC and the Food and Agriculture Organisation. COAST targets the fisheries sector specifically and is geared towards addressing the impacts of natural hazards on the food security and livelihoods of those working in the fisheries sector of the Caribbean. This policy was first issued and piloted in July 2019 for Grenada and Saint Lucia. The Caribbean would benefit from drawing lessons learnt from these examples and scaling up insurance for other Blue Economy areas [15].

Public social protection programs, as well as private and community savings arrangements, can function as insurance, too. Such savings clubs can be used to finance different needs, including insurance, but savings can itself function as an economic buffer for times of crisis or a short-term economic downturn [54]. Savings and insurance are, of course, not new sources of financing, but they can create support during periods of hardship. These savings programmes can be managed and used to recover local fishing and aquaculture projects that provide direct benefits to the local communities before contributing to the industry at large (such as local job creation and food provisioning). Types of development projects that may be of high priority after a catastrophe include fisheries (small and large-scale), shipbuilding and repair, and natural ecosystem recovery (particularly those that underpin essential fisheries). Such projects lead to revenue generation that could expedite the recovery from natural disasters sooner.

Many different insurance products exist (catastrophe and resilience bonds), each with its own advantages and potential pitfalls. The demand for high premiums on insurance products is likely to present a barrier to entry to their use in Caribbean SIDS, where finance and financing are already in short supply. Insurance companies that assign their risks on a biannual or annual basis also assume greater risk as the insurance products would have to cover longer expanses of time, increasing the likelihood of a higher number of severe weather events and thus payouts [64].

4. Generating an Enabling Environment for Financing Fisheries and Aquaculture in the Caribbean

Generating an enabling environment to encourage investment and financing, thereby reducing barriers to spontaneous entrepreneurial development in the private sector, will have knock-on benefits across the entire Blue Economy in the Caribbean as a whole [65], including the fisheries and aquaculture sectors. As the Caribbean is a developing region, there is limited private sector finance support available from governments in the region, as funds are directed to more critical or prioritised development needs elsewhere. It is thus necessary to put into place the structures that can facilitate the long-term financing of sustainable (environmentally, socially, and economically) fisheries and aquaculture development projects that arise from the industry itself, such that new development is incentivised through potentially greater fiscal returns for the investment made. In addition, the development of an enabling environment is proposed as the solution for overcoming the capacity constraints and regulatory barriers to new development in the fishery and aquaculture sectors of Barbados, Grenada, St. Vincent and the Grenadines [4]. Together, these (following) structures contribute to an enabling environment that facilitates industry development for the fisheries and aquaculture industries and the Blue Economy and have been curated from the national development strategies of other successful island nations and advisory documents (for example, such as from Africa Blue Economy Strategy [14]; The Seychelles Blue Economy Action Plan [13]; Madagascar Blue Economy Strategy for Fisheries and Aquaculture [12]). However, their effectiveness may vary and is specific to different national development contexts, including different fisheries and aquaculture sectoral contexts.

4.1. Sustainable Blue Economy Finance Principles

Banking mechanisms for blue development, such as for fisheries and aquaculture, are lacking in Grenada, Barbados, and SVG, as are the associated governmental policies. A transparent policy framework that follows blue financing principles has the potential to increase investor confidence, thus increasing the availability of working capital to industries. Underdeveloped financial markets and large historic national debt only further emphasise the need for an enabling and sustainable investment policy framework [66,67].

Emphasising sustainable Blue Economy finance principles would encourage the use of sustainable finance mechanisms for the fisheries and aquaculture sectors. These principles were launched in 2018 and present the world's first guiding framework for banks, insurers and investors to finance sustainable development [68]. The framework promotes Sustainable Development Goal (SDG) 14 (Life Below Water) and establishes specific standards related to sustainable development in financial ocean sectors. The European Commission, WWF, the World Resources Institute (WRI) and the European Investment Bank (EIB) collaborated to develop these principles, which are also hosted by UNEP FI as part of the Sustainable Blue Economy Finance Initiative [68]. These principles can be integrated into the policy and regulatory frameworks of Barbados, Grenada, and St. Vincent and the Grenadines. However, regional or national efforts can be directed towards the development of financing principles that align with NDCs.

4.2. Integrating Aquaculture and Fisheries into the Greater Blue Economy

At present, current policies operate disjunctively, with insufficient overarching frameworks to support financial investment or mitigate the risks of climate change and habitat health reduction to the fisheries and aquaculture industries. The maximum potential of fisheries and aquaculture and their capacity to add to the development of the broader Blue Economy is not reflected in the current unintegrated (and not sufficiently holistic) policy structure. Existing frameworks can be characterised as a disjunct assemblage of fisheries agreements and programmes, and they are often dated and thus inappropriately formulated national laws at various levels for current development challenges. New legislation has been drafted (e.g., in Barbados), but these have not yet been approved by the cabinet, with fisheries management being excluded from coastal and marine resource management, other environmental agendas, and tourism management [4]. Furthermore, the significance of the ecosystem services that natural environments provide to the fisheries and aquaculture sectors (such as habitat provisioning, nursery grounds provisioning, and fisheries stimulation) remains ignored, resulting in gaps in implementation and duplication of development efforts across institutions and agencies, at national and regional scales. The lack of sustainable blue financing and an overarching integrated governance framework to coordinate various pressures on coastal and marine resources continues to hamper development efforts in valorising the fishery and aquaculture sectors [69–71]. Further barriers to policy integration are due to a lack of consistent and integrated data collection with which to inform policy-making decisions (at both national and the subsequent regional scale), lack of coordination among governance structures leading to duplication of efforts, and differing prioritisation of development goals due to different national development needs.

The fisheries and aquaculture industries should remain a priority when considering Blue Economy development because fisheries will continue to provide the majority of animal protein to the people of Barbados, Grenada and St. Vincent and the Grenadines, as well as supplying a majority of jobs (ever more so with the increasing spread of aquaponics and aquaculture). These sectors have the potential for considerable wealth creation opportunities should value addition be facilitated within each country prior to export [72]. Such wealth creation is possible within the Blue Economy concept, as it promotes the development of biodiversity within coastal habitats by facilitating the development of solutions that are beneficial for both biodiversity conservation and climate change mitigation and adaptation. Furthermore, fisheries have the opportunity to play a central role in the conservation and rehabilitation of key ocean habitats as the main observers of changes in the open sea. These are key indicators for the status of ecological and ecosystem health, as they observe associated changes in fisheries' productivity (a less productive fishery relative to the past, indicating that the ecosystem may be degraded).

The development of the Blue Economy in Barbados, Grenada, and SVG is nascent [4]. Despite the significant natural resources of Barbados, Grenada, and St. Vincent and the Grenadines, as well as the opportunity to implement holistic Blue Economy approaches, the rate of adoption has been slow. Grenada is the first Organisation of Eastern Caribbean States (OECS) member country to have developed a vision for its blue growth economy. The county's blue growth vision is to become a world leader and international example of blue sustainability by optimising its coastal, marine, and ocean resources. Comprehensive, holistic preparation, design, and capacity development are needed to consolidate and coordinate sectors and industries to create synergies and development considerations between sectors. There is significant potential for generated synergies to facilitate finance for the development of the Blue Economy in each country, further emphasising the importance of mediating collaboration.

The Ocean Governance Committee (OCG) is in place in most OECS Member States. They have, in most instances, identified and mandated the creation of National Coordination Agencies that work closely with the OECS Ocean Governance and Fisheries Unit (which leads and coordinates activities at the regional level). OECS Member States are at various stages of establishing national OGCs to serve as standing committees of public sector departments, statutory bodies and non-governmental organisations. Their goal is to facilitate inter-sectoral coordination on ocean governance issues. They may thus have an important role to play regarding the management of shared fisheries stocks and the conservation of the natural environments that underpin those important fisheries. Although the OCGs offer regional consistency and the ability to provide some support to nations seeking to implement Blue Economy approaches, the mechanism still requires enhancement. Barbados, despite being in the eastern Caribbean, is not a part of the OECS [73], thus limiting regional consistency by hindering the establishment of a Barbadian OCG and the associated benefits therefrom. The Regional Ocean Governance Team (OGT) (funded by the World Bank) provides technical support to the national technical committees through the OECS committee on matters related to ocean governance and the Blue Economy.

4.3. Finance and Financing Coordinating Facility

A regional Caribbean Blue Economy financing facility (or unit) may help address some of the challenges which are anticipated if financing for the fisheries and aquaculture industries is to be scaled up, as well as investments into the Caribbean Blue Economy as a whole [15]. One such example is managed by the International Union for Conservation of Nature (IUCN) and has a global reach: the Blue Natural Capital Financing Facility. A national or regional financing facility would be responsible for the coordination and development of financing solutions/instruments for development projects that align with the objectives of the Blue Economy and leverage existing resources more efficiently [15]. This would include activities such as market research and valuation studies to inform investment into the Blue Economy and, by extension, the fisheries and aquaculture industries as well. The facility would have the potential to serve as a connection hub for development partners and private funders and act on behalf of national or regional governments to issue financing mechanisms aligned with the development of Blue Economy projects. The agency would ideally also seek to contribute towards capacity building to aid in project implementation, as well as monitor the results and effectiveness of those projects for which it has put financing structures in place [15]. The facility could further be responsible for testing innovative finance and financing instruments, facilitated through structures like innovation networks, accelerators, and incubators, which could source local and global financing knowledge and information, leveraging it for the benefit of the Caribbean communities. For instance, the Caribbean Science Foundation uses various platforms to engage and educate aspiring engineers and scientists across the region. Similarly, "Ten Habitat", a start-up ecosystem, has mushroomed across the region, where potential entrepreneurs are supported and funded using a range of practical tools, including networking and mentorship [15]. Such a unit may be most useful when part of pre-existing government structures (i.e., not independent), thereby serving as an example of a successful public-private partnership (PPP) benefiting both parties involved.

A regional coordinating facility might best be situated in locations that are easily accessible to the rest of the Caribbean region. However, given the complexity of the Caribbean's governance structures, a regional coordinating facility for finance and financing may best be located where previous governance structures have been situated or within such structures themselves. Examples include the seat of the Caribbean Community and Common Market (CARICOM) secretariat in Georgetown (Guyana) or possibly the seat of the CRFM secretariat in Belize City (Belize). The benefit of such a placement, beyond mere ease of access, is that it allows for increased awareness of regional development programmes, initiatives and opportunities among multiple governance structures. This facilitates synergies and integration across multiple departments with an increased likelihood of actual impact throughout the Caribbean region. Such a regional facility will be able to better undertake the explicit actions and research specifically directed towards finance and financing Blue Economy development, which may fall out of the purview of CARICOM (which is concerned with more than just financing the Blue Economy) and the CRFM (which although concerned with financing development in the fisheries and aquaculture sectors, also concerns itself with other aspects of these industries such as coordinating catch data for example).

4.4. Development Tools and Reporting Infrastructure

4.4.1. Screening Tool

The Sustainable Seas Draft Blue Economy Investment (BEI) project screening tool/criteria (by Jonathan McCue, Service Agreement Number: BBRSO145916) has been created for assessing and prioritising different potential development projects to finance, specifically for the development needs of Barbados, Grenada, and St. Vincent and the Grenadines. The criteria used to assess each project are based on the cornerstones of Blue Economy development: environmental sustainability, social responsibility, and economic impact. The tool includes criteria for development projects across all Blue Economy sectors as well as fisheries and aquaculture industries. The use of this tool has the potential to facilitate the most significant available returns of investment for investors in determining appropriate development projects to finance while also increasing the likelihood of multi-sector development. The use of this tool should ensure the financing of projects with minimal environmental, social and economic impact, thus avoiding further accruement of debt.

4.4.2. Data Reporting and Infrastructure

The importance of data reporting for fisheries management (and thus ecosystem management) should be emphasised, as the effectiveness of any management decision is reliant on the data that represent the real-time scenarios affecting the industry. The integrity and honesty of the data being reported should thus be maintained and encouraged as much as possible.

The development of the fisheries and aquaculture industries necessitates national centralised data reporting structures. The Caribbean Regional Fisheries Mechanism (CRFM) is a regional-scale example. However, national-level platforms can more easily facilitate the reporting of fisheries catch data, which would allow for better management of local fisheries sub-sectors, address national-level issues and concerns, and more engaging management of national marine environments that underpin local fisheries. Providing a tool or platform for the hassle-free reporting of catch and aquaculture production data, as well as the methods and locations associated with those data, can facilitate the creation of tailored financing solutions for those practices or projects, allowing for more efficient industry development, management, and effectiveness (profitability).

Currently, monitoring approaches such as accounting for activities associated with the Blue Economy have not been conducted in a coherent manner. Blue accounting data need to be recorded and catalogued from isolated sources to produce a comprehensive view of the Blue Economy industries and their impact on the livelihoods of nationals of Barbados, Grenada, and St. Vincent and the Grenadines. There are also gaps where no data are recorded for certain sectors (like vessel maintenance). Such gaps lead to a skewed interpretation of the industry and facilitate unoptimised management decisions. Critically, ecological components of the Blue Economy need to be accounted for, particularly of environments of key development sectors such as fisheries and aquaculture, as the ecosystem services provided by these habitats directly facilitate the developmental prospects and functioning of these industries. The implementation of nationally determined contributions towards sustainable development (and climate change) will necessitate the need for blue and green accounting frameworks for assessing developmental change that relates to environmental sustainability (of those that underpin key industries).

Partnering with the Global Ocean Accounts partnership (GOAP) may offer benefits to national governments in the Caribbean in facilitating the establishment of ocean accounts and monitoring protocols for national Blue economy accounting. GOAP aims to build a global community of practice for ocean accounts (also known as "blue accounting") with an international network including national governments, the World Bank, the United Nations Economic and Social Commission for Asia and the Pacific (ESCAP) and specific fisheries departments like Fisheries and Oceans Canada. GOAP encourages countries and stakeholders to include the use of other indicators beyond only gross domestic product (GDP) in their measurement and recording of progress towards sustainable development, as the reliance on only GDP as a yardstick for development has led to previous misinformed policy decision-making (see [74] for a detailed analysis on this topic).

4.4.3. Spatio-Temporal Planning Tools and Prospective Approaches

The use of integrated and prospective approaches to the management of marine ecosystems is lacking in the three study countries. This is likely due to growing blue economies in their nascent stages in Barbados, Grenada, and St. Vincent and the Grenadines. Furthermore, the large marine ecosystem approach (specifically the Caribbean LME) has not been integrated or institutionalised into national environmental management frameworks. The adoption of such a paradigm would facilitate a better understanding of the dynamics of coastal and marine ecosystems. The use of approaches such as blue accounting and data reporting of key ecological indicators (biological productivity/fish biomass, pollution, ecosystem health) would facilitate improved resource management.

For countries like Barbados, Grenada, and St. Vincent and the Grenadines, where resources are limited, sustainable and optimal use of them is critical to economic prosperity.

As such, marine spatial planning (MSP) is particularly important and has been lacking in these three countries. "Marine Spatial Planning (MSP) is a public process of analysing and allocating the spatial and temporal distribution of human activities in marine areas to achieve ecological, economic and social objectives that have been specified through a political process" [75]. The use of MSP usually forms part of a country's national Blue Economy development strategy and implementation plan, particularly where shared resources are shared by multiple stakeholders (such as fisheries stocks). Despite potential challenges, the use of MSP has the potential to facilitate an enabling environment for national development and has been shown to be effective in other SIDS countries like the Seychelles.

4.5. Development of Sustainable Micro, Small, and Medium-Sized Enterprises (MSMEs)

Considering that many fisheries and aquaculture operations are more likely to operate at smaller scales than large commercial-scale fishing fleets, it is thus necessary to focus on the sustainable development of micro, small and medium-sized (MSME) enterprises in the fisheries and aquaculture sectors [76]. MSMEs that already operate on a successful (environmentally) sustainable business model are likely to perpetuate that model when operating at larger scales, as it has already proven to be profitable while benefiting the environment. Moreover, an enabling environment for the development of MSMEs encourages participation in the fisheries and aquaculture industries as support services are available for future business growth and expansion. The growth of small (subsistence) scale fishers is thus facilitated by an enabling environment for MSME, presenting a route for transitioning from a subsistence-based livelihood to entrepreneurship, which in turn may afford improvements in the livelihoods of Caribbean locals.

MSMEs can be considered as the "missing middle" and lack access to capital [76]. These include organisations that are raising USD 500 K–10 M (with revenues exceeding USD $250 K) annually. MSMEs are challenging for financiers because banks are not interested (as the company is likely unable to offer sufficient collateral for loans and presents a security risk for the investment), and MSMEs are too small for bonds or private placements. Microfinancing is too expensive, and MSME fragmentation presents an elevated risk, disincentivising investment [76]. Yet MSMEs represent the majority of employment in developing countries (such as in the Caribbean region) and represent the majority of industry stakeholders [76]. Moreover, investors require MSMEs as the mechanism with which to actualise returns and impacts of investments, such as through blue bonds, development banks, industry ventures, and impact funds (and many others [76]). An evaluation of the finance gap, such as that of Asian Blue SMEs (at the value of USD 2 trillion [76]), is a recommended starting point for MSME development in the Caribbean.

National or regional programmes that facilitate scaling offer readiness training and access to finance and financing, which present avenues with which to develop MSMEs to benefit the fisheries and aquaculture industry. A similar programme to SME BlueImpact Asia could be developed for the Caribbean region. The BlueImpact programme is an initiative to raise capital for sustainable SMEs in the Asia and Pacific Blue Economy by acting as a platform that connects qualified investors with Blue Economy SMEs. The programme thus supports catalytic funding and mobilises appropriate matching private sector enterprises [77]. The programme also aims to replicate successful business models across the Asian and Pacific region through this programme, thus facilitating Blue Economy development [77].

The integration of MSMEs (and small-scale fishers) into the broader Blue Economy can be accomplished by establishing a programme similar to 'Abalobi' (in South Africa). The non-profit organisation strives to connect individual small-scale fishers and fishing communities to buyers through its technology platforms. Their community-supported fishery model promotes fair market access, transparent supply chains, national food security, and "fish with a story" (presenting the people and communities behind the produce). The organisation has global international partners, such as the Seychelles Hook and Line

Fishers Organisation and the World Wildlife Forum, already replicating the model in other countries [78].

4.6. Public-Private Partnerships (PPP)

While no single definition satisfies all demands, (at a generic level) public-private partnerships (PPP) is an encompassing term that covers many forms of collaboration among public, private, and civil sectors [79]. The "Canadian Council for Public-Private Partnerships" defines a PPP as "a cooperative venture between the public and private sectors, built on the expertise of each partner that best meets clearly defined public needs for services or infrastructure through the transfer between partners of resources, risks and rewards". Two important elements of this definition are that the arrangement is to provide public services and that partners share any associated risk. The aim of PPPs is to structure the relationship between the public and private sectors, allocate the risks to the parties that are best able to manage them and add value to public services by using private sector skills and competencies [79]. However, it is important to make the distinction that PPPs are not incentives or subsidies given by the public sector to attract private investments.

PPPs are important as mechanisms for developing technologies where normal private sector market incentives fail. Furthermore, PPPs improve the quality of the delivered service by facilitating optimal participation of both sectors: the government acts as the regulator and is tasked with monitoring performance and the planning of services, whereas the private sector's attention is on managing and optimising the daily delivery of the service [79]. PPPs thus improve the cost-effectiveness of services (and risk management), the savings of which can finance other services or development in industries of importance.

Despite PPPs having the potential to bring about significant economic benefits, they are challenged by internal and external risks to both parties in the partnership. Of the many risks that exist, reputation damage is likely the most limiting in engendering future investor confidence and being offered future development project opportunities [79]. External risks can arise from events beyond the scope of the project as well as from changes in government, legislation or the political climate. Such risks may be addressed in PPP contracts but are fundamentally outside the project itself and may be beyond the control of the private sector [79]. Internal risks are particular to the project or the way it is constructed and operated and are generally under the control of the contracting parties. The contracting arrangement should explicitly allocate risk among the signatories to ensure no disagreements arise with the onset of internal risks. One of the critical internal risks is that demand will be insufficient to allow the project company to repay its financial obligations from project revenues. Given that demand risk is difficult to estimate, even more so in developing or recovering economies (common among SIDS states), the public sector commonly assumes that the private sector should shoulder demand risk. However, when it does so, the private sector is likely to ask for more support from the government in the form of subsidies, grants or guarantees to mitigate this risk [79].

There is a wealth of resources to provide assistance in the establishment of business-related PPPs. Development aid is available for the development of PPPs from organisations like the United Nations [79]. Such arrangements may be particularly suited to the development needs of fisheries and aquaculture in Barbados, Grenada, and St. Vincent and the Grenadines, as they have been shown to assist the governments of countries that have difficulty in fostering public sector investments to improve local infrastructure and product chains that meet international rules [79].

Most countries also have PPP knowledge centres or units from which assistance can be gained [79]. Such a unit is beneficial for leading the implementation of PPP programmes and training government authorities responsible for local implementation. It should give advice regarding the rules and regulations for PPPs in the country. Recommended as places to start searching for further information on general principles and approaches are the World Bank's Private-Public Partnership in Infrastructure program, Institute for Public-Private Partnerships, Inc., and Public-Private Infrastructure Advisory Facility [79].

Allam and Jones [80] propose a development framework specifically for SIDS to develop climate change and economic resilience, largely based on urban and cultural heritage PPP. However, given that many cities in Barbados, Grenada, and St. Vincent and the Grenadines are small, the framework may only be applicable for Bridgetown (Barbados), as it is the largest city of the three countries (with an estimated population of 98,500). Nonetheless, it may facilitate the development of fisheries through PPP as the city is likely a hub of trade (of fish, an important resource likely tied to the cultures of island inhabitants).

4.7. Enabling Policy Environment for Financing

The regulatory environment can have a significant impact in facilitating sustainable development of the fisheries and aquaculture sectors and the Blue Economy. An analysis of the existing regulations and policy environments for Grenada, Barbados, and St. Vincent and the Grenadines has been conducted [4]. In addition, aspects of the regulation and policy framework have been discussed, namely fiscal policy mechanisms for fisheries and aquaculture development (Section 3.1) and the implication of sustainable Blue Economy finance principles (Section 4.1). However, the development of a regulation and policy environment that enables and facilitates finance for the fisheries and aquaculture sector (and the Blue Economy at large) was not discussed but identified as a development priority [4].

An enabling financing environment alongside integrated national (and regional) Blue Economy frameworks can offer increased security for investments, incentivise investment and thus facilitate development for the fisheries and aquaculture industry [12–14]. We suggest that an ideal enabling regulatory environment may optimally facilitate the financing of sustainable industry development, such as fisheries and aquaculture, when constructed with consideration of the following:

- Facilitates and encourages the use of fiscal policy for sustainable development (Section 3.1)
- Supports the implementation of sustainable Blue Economy principles (Section 4.1)
- Facilitates local and international entrepreneurship
- Supports a mixed economy (with private—and state influence) and encourages free-market environmentalism.

The principles of the Blue Economy overlap with those of free-market environmentalism along the axes of environmental sustainability and economic sustainability, and free-market environmentalism does not specifically exclude social sustainability. Free-market environmentalism integrates the goals of preserving the environment with the concepts of a free-market economy. It recognises the potent incentives for conservation and environmental care that markets can offer, emphasising the use of private property rights and contracts as effective tools for environmental protection [81,82]. Consequently, we also suggest that policy and regulation surrounding property ownership rights be elucidated for the optimal leveraging of financing mechanisms (for example, payment-for-ecosystem services). We suggest that existing financing regulatory frameworks be reviewed to integrate the above considerations, thus developing a more enabling financing environment for the fisheries and aquaculture industries and the Blue Economy.

5. Conclusions

This review discusses various finance and financing mechanisms for supporting development projects in Caribbean fisheries and aquaculture industries. Developing financing mechanisms to support and drive the maximal sustainable exploitation of the environments that underpin fisheries and aquaculture will facilitate an increased contribution to national and regional blue economies, as well as avoiding exhaustion of the resources on which Caribbean livelihoods are dependent for years to come. Many of the fisheries and aquaculture industries of Caribbean countries face complex problems, but the improved use and management of natural environments and the development of PPPs as management tools for sectoral development have been suggested elsewhere [4]. The leveraging of natural capital forms the base around which the benefits of PPPs can be utilised for optimal delivery of ecosystem services and provide numerous opportunities for multi-sector development synergies that can ultimately benefit multiple sector stakeholders. However, policy regarding the ownership rights to natural environments would need further development and elucidation to determine who can leverage the multitudes of financing mechanisms suited to a specific ecosystem. The boundaries of ecosystems need to be well-defined and clear to all its users to limit confusion regarding ownership and property rights. Doing so is further necessary for leveraging natural capital for funds to finance Blue Economy development projects, as ownership of the natural capital (or services therefrom) may need to be used as collateral security. Further policy development may be directed towards determining whether the number of financing mechanisms employed per ecosystem should be capped or not. Regardless, the development of accurate ecosystem accounting programmes and mechanisms (to determine the value of ecosystem services) is essential in realising the benefits of sustainable management and ensuring returns for potential investors.

The various common financing needs of the fisheries and aquaculture sectors of Barbados, Grenada, and St. Vincent and the Grenadines [4] have been matched with the most suitable financing mechanisms (Table A2, Appendix A). Blue levies are generally recommended and applicable in almost every scenario, as they allow the fisheries and aquaculture industry to drive its own development in financing research and conservation projects (to its own benefit). The use of blue tokens with sufficiently low repayment coupons allows development projects to gather public support for fisheries, thereby increasing the likelihood of the project being successful through community buy-in. The possibility of natural capital being traded as public equities as "Natural Asset Companies" provides the opportunity for development projects to fund themselves. An enabling environment for debt and lending with low-interest loan repayments is also applicable to almost every scenario, as it facilitates access to capital finance for infrastructure development and the acquisition of increasingly sustainable fishing equipment. The development of an enabling environment through the development of dedicated financing institutions and PPPs and the establishment of sufficient data reporting infrastructure for the fisheries and aquaculture industry is essential for driving development in these sectors.

There are complex challenges for the development of the fisheries and aquaculture sectors to be addressed, such as overfishing, pollution, and the introduction of invasive species (and others). An injection of finance does not necessarily translate into meaningful development or solutions to these problems. However, finance is a tool to help mobilise action to facilitate the transition to a sustainable Blue Economy. Without finance or the mechanisms which generate finance, limited action takes place, particularly in terms of policy development, government coordination, and implementation and enforcement of development policies. The use of financing mechanisms has the potential to facilitate finance sustainability within sectors of the Blue Economy, enabling each industry to fuel its own development. In addition, the conditions that enable an increased degree of finance to support development through the Blue Economy need to be developed and maintained. These conditions include low corruption, efficient processing of finance transfers (through dedicated secure financing institutions), good governance, strong societal and environmental standards to guide industry development, and transparency of any non-private development. Developing methodologies, regulations, and policies to facilitate these conditions ensures that any finance that is injected into a sector and the Blue Economy results in tangible impact and sustainable development.

Financing mechanisms can facilitate the optimally sustainable exploitation of the fisheries and aquaculture resources in the Caribbean, thereby ensuring the ability of these industries to support developing Caribbean Blue Economies for future generations. One of the largest limiting factors in financing the fisheries and aquaculture industries in the Caribbean is likely to be awareness of the range of finance and financing mechanisms available to industry stakeholders, as well as an enabling environment for financing the Blue Economy sectors. This review is thus intended to aid financing institutions, Blue Economy developers, and specifically Caribbean fisheries and aquaculture stakeholders and Caribbean governments by raising awareness of the financing mechanisms available, encourage the incorporation of their use in the fisheries and aquaculture industries in the Caribbean, and encourage policymakers to create an enabling environment for financing development in these crucial sectors. The authors advocate for the establishment of financing mechanisms that generate finance for an industry as opposed to once-off finance contributions and support, as the former is more likely to have a consistent impact over the long term.

The methods used here, as well as March et al. [4], can be applied to the holistic Blue Economy development of other nations, including other SIDS, as follows:

1. The identification of capacity constraints and development opportunities (in alignment with national development priorities).
2. Identification of potential finance and financing mechanisms.
3. Analysis of the enabling environment at the national level.
4. Matching of suitable finance and financing mechanisms with previously identified financing needs in light of the current enabling environment.
5. Review of identified development solutions in light of other sectors' development prospects.

Discussion of the various elements of an ideal enabling financing environment usually occurs at a more general overarching level. However, the implementation of these various elements differs in their suitability and effectiveness in supporting development at a national or regional level. Given the need for the development of national Blue Economy strategies and frameworks to address the unique development needs of any one country (as these needs may differ between neighbouring countries), the appropriate resolution for implementing elements of an ideal financing environment may be best suited to the national level. In this manner, the enabling environment can be developed by supporting national Blue Economy strategies through synergies facilitating sustainable development. While the work in this paper completes only a part of what is required for meaningful development, future work can involve analysis of the local enabling environment at the national level of each of the countries discussed, such that the most suitable financing mechanism can be implemented effectively in addressing the identified financing needs for national fisheries and aquaculture sectors. Thereafter, the identified solutions should be reviewed in relation to the development needs of other Blue Economy sectors so that non-limiting solutions (to the development of other sectors) are implemented. In addition, future work should include a detailed analysis of the feasibility and potential risks (financial, environmental, and social) associated with the implementation of the mechanisms (at the national level) herein discussed.

This work focussed on the development of fisheries and aquaculture sectors as part of the Blue Economy due to these sectors having a disproportionately larger influence on the national economy and local livelihoods. However, the Blue Economy advocates for a holistic approach rather than a siloed or sectoral approach. We suggest that the development of priority sectors (such as fisheries and aquaculture for SIDS) be prioritised with the explicit consideration of the future development needs of other sectors in mind. While non-priority sectors may be underdeveloped, the development of priority sectors should not occur in such a way that it limits the development prospects of other sectors of the Blue Economy [65]. The most optimal method of concurrent holistic development of Blue Economy sectors with limited resources still remains a challenge, but the use of innovative sustainable financing mechanisms (as they develop) has the potential to contribute to sustainable development in Blue Economy sectors.

Author Contributions: M.B.: Investigation, Writing—Original Draft Preparation, Writing—Review and Editing. A.M.: Conceptualization, Methodology, Formal Analysis, Writing—Original Draft Preparation, Writing—Review and Editing, Project Administration. P.F.: Conceptualization, Methodology, Investigation, Writing—Original Draft Preparation, Supervision. All authors have read and agreed to the published version of the manuscript.

Funding: This research received no external funding.

Informed Consent Statement: Not applicable.

Data Availability Statement: All data generated from the study are available in the manuscript.

Conflicts of Interest: The authors declare no conflicts of interest.

Appendix A

Table A1. Summary of various financing mechanisms for use in fisheries and aquaculture sectors of Caribbean states.

Financing Instrument	Description	Requirements
Fiscal Policy	Fiscal policy covers state spending and the state's generation of revenues. Budgetary governance is also classified as fiscal policy, and deals with the administrative and institutional systems that control the fiscal flows of the state. Different parts of public fiscal policy can be revenue generators and liabilities at different times, at different levels of organisation. The specific application of fiscal policy thus requires an in-depth understanding of the relevant industry (fisheries and aquaculture) for optimal use as a development mechanism.	
Environmental Taxes	The minimum definition of an environmental tax is that it increases the costs of pollution activities [19]. An environmental tax is meant to target activities that cause environmental degradation or pollution, but this does not imply that the tax has a proven effect in terms of minimising environmentally degrading or polluting activities. The actual application of an environmental tax is likely to have the effect of incentivising some activities whilst disincentivizing others.	Public awareness Monitoring of pollution among stakeholders (Pollution accounts that track equipment lost at sea for example) Identification of the locus of implementation (individual, organisational, national, regional) Incorporation into legal structure (policy-making)
Leases, Licences, and fees	Whereas a tax is compulsory, a lease, licence or fee can be seen as a form of exchange between a public authority and a (legal) person, which grants the latter a certain right or privilege. These leases, licences and fees remain stable regardless of the incomes that the payer may generate as a result.	Public awareness. Easy-to-use infrastructure for applications, and issues. Controlled public register of licences and rights issued. Public enforcement mechanism (such as harbour officials checking permits for access to fishing grounds)
State-owned enterprises (SOEs) and public-private partnerships (PPP)	SOEs are a form of fiscal policy, where SOEs have some clear environmental impacts beyond only service delivery. Public utilities (water and energy companies) explicitly have an environmental aspect since they usually have environmental quality or resource efficiency as key objectives. Other types of SOEs, like transportation companies, may increasingly incorporate environmental objectives. SOEs can be run with a focus on generating profits, but they can also seek to strategically increase public goods and services through their operations. SOEs can be a source of revenue when they generate profits while their inventory and equipment can be seen as capital goods. Meanwhile, SOEs can also hold liabilities in the form of debt. PPPs are a cooperative venture between the public and private sectors, built on the expertise of each partner that best meets clearly defined public needs for services or infrastructure through the transfer between partners of resources, risks and rewards. The aim of PPPs is to structure the relationship between the public and private sectors to allocate the risks to the parties best able to manage them, and to add value to public services by using private sector skills and competencies. However, it is important to make the distinction that PPPs are not incentives or subsidies given by the public sector to attract private investments. Examples applicable to fisheries and aquaculture include organisations that facilitate data recording of fish catches, vessel regulatory bodies specifying monitoring equipment used, fisheries management organisations.	Identification of public needs (at national scale) and natural resources. Transparency in decision-making, operations, and impact to build public trust. Environmental monitoring and accounting. Identification of public needs and subsequent private sector expertise. Public buy-in. Transparency among public and private partners.

Table A1. *Cont.*

Financing Instrument	Description	Requirements
Debt and lending	An enabling environment for debt and lending presents an avenue for fisheries and aquaculture development. As fisheries and aquaculture equipment are often expensive (and become increasingly technical with the incorporation of sustainability objectives), many stakeholders globally make use of loans to develop and expand. An enabling environment such as low interest rates on loans and alternate avenues for debt repayments, may incentivise development in other sectors as well as aquaculture and fisheries. The state of national and global economies (such as recessions) needs to be considered and how this influences the borrowing of money (where in a recession, the borrowing of money is far less likely than in an economic boom/upturn). There is room for public influence (as well as regional organisations like development banks) in this regard to reduce loan and debt repayments through industry financing (subsidies).	Awareness among the industry as to the borrowing options available (through the establishment of coordinated information and development centres) Monitoring the development of the industry.
Blue levies and stakeholder taxation	Blue levies and stakeholder taxation are similar to environmental taxation, but are not limited to disincentivising pollution. These mechanisms are rather aimed at reinvesting revenue for industry development. These levies are usually applied in different ways, at different rates and at different levels of organisation, like on importation, consumption, accommodation, service, or travel. Sectors that rely on the environments that sustain fisheries and aquaculture or on the sector itself (such as tourism, hospitality, cruise and charter, extraction, and ports and ship-building industries) could be potential targets from which the proceeds of blue levies could be directed to support fisheries and aquaculture.	Transparency Monitoring of stakeholder relationships with the environment and/or fisheries and aquaculture industry. Mechanism or institution to implement, coordinate, and reinvest proceeds according to development needs.
Payment for Ecosystem Services (PES) Schemes	PES schemes are known by five characteristics [26]: "(1) a voluntary transaction where (2) a well-defined ecosystem service (ES, or a land-use likely to secure that service) (3) is being 'bought' by a (minimum one) ES buyer (4) from a (minimum one) ES provider (5) if and only if the ES provider secures ES provision (conditionality)". PES schemes thus entail the voluntary buying and selling of ecosystem services, based on the seller protecting an ecosystem. Where PES schemes are directly linked to environmental outcomes, levies, fees or licences are not necessarily linked to environmental outcomes, but their proceeds may still be directed to benefiting fisheries and aquaculture (non-environmental benefits). Despite most PES programmes being thought of as a market-based approach, most programmes are based on compliance with government regulations [29]. However, these subsidy-PES programmes are only as good as the (government) authorities that implement them since PES requires enforcement of regulations. Subsidy based PES schemes are potentially more difficult to maintain for cash-strapped governments. A requirement for all PES schemes is accurate environmental monitoring and accounting.	
Regulated markets	Cap and trade markets require resource users to hold a purchasable right to the resource they use. It sets a limit to resource use and allocates a tradable share of the resource to an asset owner. The rights to the ecosystem or ecosystem service can then be traded (if legal). Ecosystem service can then be exploited on regulated markets (for example, in the form of biodiversity or resilience credits as the measure of ecosystem service delivery). A licence is not necessarily tradable and the issuer of a licence does not necessarily set a limit to damages to an ecosystem, but it still constitutes a transaction based on compliance.	Controlled public register of licences and rights issued. National natural resources can be divided into concession areas which can then be exploited by the private sector on a rotational basis according to who owns the legal right to do so (such as the kelp-harvesting concession areas established in South Africa). Mechanism of access to acquire rights and licences and having this available to anyone. An example is an auctioning mechanism for available national rights to resource exploitation. Enforcement and monitoring of compliance and limitations of approved permits relating to shares of the natural resource (including parties involved with subletting). Requires easy-to-use mechanism. Sophisticated and sufficient legal policy framework regarding (temporary) ownership or rights to national natural resources.
Voluntary transactions	Voluntary PES transactions can involve private and public sector actors alike. Governments, private individuals, NGOs and development organisations can all pay private actors to change practices or avoid harm to ecosystems.	Platform to identify, mediate and increase awareness of voluntary opportunities available to stakeholders. Requires monitoring of fisheries and aquaculture sectors

Table A1. *Cont.*

Financing Instrument	Description	Requirements
Biodiversity offsetting	Biodiversity offsetting is based on counterbalancing any lost biodiversity from a development project by investing in equivalent biodiversity somewhere else, thus aiming to maintain biodiversity despite economic development (or environmentally harmful extraction processes). If developers follow the mitigation hierarchy (avoidance—minimisation—restoration—offsetts), they should try to avoid negative impacts in the first place, developers should secondly consider the minimising of impacts, thereafter restoring any negative impacts stemming from the development, and finally they can use biodiversity offsets to compensate for the unavoidable biodiversity loss stemming from development projects [33]. Thus, biodiversity offsetting is only meant to counter biodiversity loss that cannot be prevented [34]. Different levels of compensation can be mandated due to the perceived value of the affected (extracted or displaced) species or environment in question. The associated biodiversity loss that is found with unsustainable fishing practices (dredging, trawling), or new aquaculture infrastructure developments, can be offset by investing a portion of the proceeds into less damaging subsectors (domestic handline fisheries, sustenance fisheries or developing restocking programs).	Baseline levels of biodiversity at the sites that provide offsets (sufficient long term data on that area, to assess historical performance or attributes, and where this is not available data would first have to be collected). Mechanism or institution to implement, monitor and coordinate biodiversity offsetting with development needs, with the private sector.
Carbon and nutrient trading credits	Coastal environments are likely to offer nutrient capture and sequestration services (such as blue carbon) in addition to the benefits they provide to fisheries and aquaculture [5]. Many of the ecosystems that facilitate the generation of carbon and other nutrient trading credits, also provide other co-benefits like nursery/feeding grounds for fish and wave attenuation which contribute to the reliance and longevity of infrastructure along coastlines. These services can be capitalised upon, and leveraged as pollutant offsets (on international and domestic markets) from which revenue can be generated to further support the protection, optimisation, and maintenance of the environments themselves, or be directed towards other areas of fisheries and aquaculture. The conservation and development of the coastal (mangrove) habitats may present an ideal opportunity to access nutrient trading markets (such as the carbon credit market) which can further finance the development of these key ecosystems.	Dependent on environmental monitoring and blue accounting methodologies. Nutrient credit market access (voluntary offset market vs international compliance markets) Finding buyers for credits (can be facilitated by third party organisations). Certification of methods that generate credits by reputable and internationally recognised organisations (increases buyer confidence). Local community involvement and buy-in (as pertains to conservation, and development impact). Legal policy for clear ownership rights of ecosystem service and/or associated credits
Natural capital as publicly traded equities	The Intrinsic Exchange group (IEG) in collaboration with the New York Stock Exchange (NYSE) are pioneering the creation of a new asset class: Natural Asset Companies (NACs). The purpose of such companies is to maximise the performance of the natural asset they are associated with, whether this be through ecosystem services provisioning, restorative/regenerative agricultural use, or hybrid cases integrating both. These companies are evaluated by the IEG, and then listed for trading on world platforms, enabling the conversion of natural assets into financial capital. The sustainable management of the environmental areas that underpin fisheries and aquaculture industries (such as fishing grounds and MPAs) such that the maximum amount/number of benefits are realised is thus incentivised by potentially global markets.	Standardisation of assessment of natural assets. Establishing a baseline of optimal ecosystem performance (need historical data). Awareness raising. Legal policy for clear ownership rights of ecosystem or ecosystem service.
Debt swaps	A debt swap can be defined as a scenario where a creditor forgives debt owed to them in exchange for a commitment by the debtor to use the outstanding service payments for a particular investment [50]. The redemption of debt can thus be conducted at a discount. The service repayments can be invested into whatever project or initiative is agreeable to both parties in the transaction, such as for fisheries and aquaculture industries or environmental and social based projects. Debt swaps can present an enticing opportunity for the governments of (SIDS) countries hoping to simultaneously facilitate development and reduce the country's historical debt. However, debt swaps are not explicitly limited to fiscal policy as any private stakeholders can engage with this mechanism assuming mutual agreement to the (legal) terms of the arrangement.	Efficient mechanism or platform for arranging debt swaps. As they are time sensitive (unresolved debt accumulating with time), such a mechanism needs to be quick and easy (this has been a problem with historical national debt swaps and needing to be ratified in parliament first, leading to reduced effectiveness). Composition of sovereign creditors need to all agree on the terms and conditions of the deal. Transparency among parties and in terms of the progress of impact (development) achieved. Involvement and buy-in of local communities for environmental development, are likely to maintain impacts achieved regardless of late repayments or other complications. Consistency in the deal from the government side, regardless of changes in political composition/structure.

Table A1. *Cont.*

Financing Instrument	Description	Requirements
Blue (and other sovereign) Bonds	Blue bonds are "a debt instrument issued by governments, development banks or others to raise capital from impact investors to finance marine and ocean-based projects that have positive environmental, economic and climate benefits" [58]. Using blue bonds as a financing mechanism supports an enabling environment for stimulating the development of a country's entire Blue Economy, by presenting financial capital to private actors wishing to make sustainable change, at low-risk to investors [60]. When considering a bond issuance, it is critical that the bond structure is fit for purpose (in this case the development of fisheries and aquaculture industries) ensures the highest degree of environmental and social impact, and that the issuer receives the lowest possible interest rates on repayments. Issuances of blue bonds are conducted based on a per-case basis and case specific environmental and economic returns, but generally the criteria for bonds to be considered "blue" require that the investment be used for oceanic or marine resource development.	
Catastrophe bonds	A trigger level (like the wind speed of a hurricane) for a specific area is determined before the phenomenon occurs. If the trigger level is surpassed, the insurer pays out to the insured party. This bond differs from insurance in that it pays out before the phenomenon has struck, whereas insurance pays out afterwards [4]. Furthermore, no payout is required if the trigger level is not surpassed, but damage may still be widespread because of the phenomenon.	Transparent environmental monitoring. Widely informed and well-established environmental trigger levels.
Environmental bonds	Environmental bonds raise capital but the return on investment (ROI) is based on the success of an environmental programme or project, as defined by pre-determined key performance indicators (KPIs). In the case of fisheries and aquaculture development, this could be achieving a maximum sustainable yield of a fish stock within a two years for example, or the protection of 20 hectares of mangroves as fish nursery grounds.	Well recorded and documented development progress (as per KPIs). Evidence of longevity/persistence of development (such that any progress does not easily get reversed upon ROI). Timelines of assessments need to be sufficiently long to allow the environment to reflect the development changes (e.g., there may be a lag period involved in a fish stock producing at sustainable max yield, if fishing of the stock has recently been suspended). Examples of specific development options within fisheries and aquaculture may be beneficial for investors to choose from.
Use-of-proceeds bonds	A use-of-proceeds bond entails the upfront promise that proceeds will be used towards blue development (not necessarily environmental development). These bonds are at risk of "environmental non-performance" where returns as benefits of the environment do not materialise, but the economic returns do [61]. This type of bond may be particularly suitable for the development of fisheries and aquaculture infrastructure (equipment, value addition practices, workspaces, etc.), given that there is no explicit environmental benefit mandated.	Given that no specific environmental benefit is mandated, it may be beneficial to stipulate any development or acquisitions from a use-of proceeds bond that involves some measure of environmental afterthought; e.g., when developing fisheries and aquaculture infrastructure, it could be designed in a (environmentally) sustainable fashion. Transparency in use of proceeds.
Blue tokens and Fintech	"Fintech" or financial technology, refers to the use of new technology to improve management and access of financial operations and processes. It involves the use of specialised software, algorithms (machine learning), and artificial intelligence to achieve the improved management of finances [15]. Blue tokens is a proposition where fintech and block chain technology are used to raise money for blue (fisheries and aquaculture) development projects [15]. An issuer could set an amount they would like to raise, for example US $10 million with an initial fixed repayment coupon. The initial price of each token could be set at US $10 (predetermined), with one million tokens being issued on a secure blue token market or platform. Any investor who has been approved through rigorous identity checks, like know your client (KYC) and anti money laundering (AML) checks, can then buy tokens and either hold them to maturity or trade them among other investors on the blue token platform. A Blue Economy credit rating agency (alluded to before) could also rate the issuance (initially and later annually) for development outcomes and financial viability, thereby giving investors maximum information to assist with their investment decision [15].	Technological infrastructure (sufficient computing equipment, stable electric supply, stable internet connection, etc). Stakeholder awareness as to what is available. Digital security. Secure block chain technology and infrastructure. Blue token platform or market. Thorough identity verification (approval). Endorsement/certification would be beneficial (such as through a Blue Economy credit rating agency).
Insurance	Insurance can create confidence for a potential project developer or investor in that it limits the risks that the project may face, potentially reducing the costs of capital investment. The role of risk management, risk pooling and risk transfer has become important for any potential development in the Caribbean, as the intensity of natural disasters like hurricanes increase [15]. Many different insurance products exist each with their own advantages and potential pitfalls. The demand of high premiums on insurance products are likely to present a barrier of entry to their use in Caribbean SIDS where finance and financing is already in short supply.	Low premiums on insurance incentivise its use in lowering risk for investors. This may be achieved with various forms of fiscal policy (i.e., government support) or regional support (such as from development banks). Awareness among project developers and industry stakeholders. Risk management to be factored into development projects.

Table A2. Linking financing instruments with financing needs in the fisheries and aquaculture sectors in Barbados, Grenada, and St. Vincent and the Grenadines (formulated from March et al., 2023 [4]). Financing instruments are ranked by their potential effectiveness.

Financing Need	Description	Potential Solutions	Financing Instrument
Intra-regional trade and domestic market development	A minimal portion of exports from the three countries is traded within the region, while the bulk of exports surpass national imports. It is crucial for these countries to maintain trade with the global market but to reduce vulnerability to external disruptions by boosting intra-Caribbean trade and limiting exports. This approach will help ensure that each country's nutritional requirements and domestic market demands are fulfilled.	There is a need to simplify the transportation of fish products across borders to promote intra-regional trade. This could be accomplished by consolidating import/export declarations and phytosanitary inspections under a single government or regional agency. Setting a zero percent preferential tariff for intra-regional trade would ensure that fish imports/exports face no restrictions. Implementing a uniform system of conformity assessment procedures for testing, inspecting, and certifying fish products for import/export across all countries would minimize confusion about trade standards and ensure that all products on the market meet legislative and food safety requirements.	1. Blue levies 2. Fintech (blockchain) 3. PPP's and SOE's 4. Debt swaps 5. Blue bonds 6. Use-of-proceeds bonds
Value addition of fish products	The essential factor for maximising profit or gains from fish products is value addition, which also generates employment and foreign currency earnings. Countries need the technology and resources to meet processing, packaging, and marketing demands for target markets. However, this technology is often costly and needs to be imported, increasing the startup costs for developing value addition in Caribbean countries. Government planning should prioritise value addition in seafood value chains by (i) encouraging private investment in seafood value addition through zero-rating imported machinery (ii) recognising the importance of training initiatives for seafood producers to equip them with necessary skills and knowledge for various stages of the value chain (iii) studying, reassessing, and potentially redefining the seafood chain to address bottlenecks and operational challenges (iv) continuing market development and diversification, and (v) establishing information centers to provide operators at various chain nodes with the necessary information for planning and investment decision-making.	The three countries can consider producing ready-to-cook and ready-cooked meals, which are increasingly popular in developed countries. This approach would yield higher prices for their fish products compared to exporting unprocessed fish to American markets or other countries. An example is what Barbados is doing with its tuna sector: making its fishing fleet more efficient and sustainable by providing it with target-selective fishing equipment (also opens itself up to an increasing amount of fast growing markets); and processing tuna before export (e.g., boxed tuna loins). Enhancing value in the tuna sector would better support local hotels and restaurants, retaining more value on the island and reducing reliance on food imports.	1. Debt and lending 2. Blue tokens and fintech 3. Use-of-proceeds bonds 4. PPPs 5. Blue levies 6. Debt swaps 7. Voluntary PES transactions 8. Insurance
Improving ecosystem services delivery	Without healthy ecosystems, fishery resources provisioning declines. The coastal and marine ecosystems around Barbados, Grenada, and St. Vincent and the Grenadines are significantly degraded, severely impacting fisheries and fish catches. Mismanaged and unsustainable fishing practices, such as non-targeted bycatch, inefficient fishing gear, and overfishing, further harm ecosystem healthrazmakThere is an immense gap in the knowledge pertaining to the state of ecosystems. An effective understanding of ecosystems is needed to measure baselines and progress in this area. Evaluating ecosystem services and anthropogenic degradation of the environment, will assist in creating market-based mechanisms to pay for such services or compensate for any damages caused.	To sustain the capacity of ecosystems related to fisheries and aquaculture in providing services, it is essential to improve the valuation of these services and enhance restoration efforts. This includes well-managed fisheries, closed seasons for the recovery of natural stocks and the environment, sustainable resource extraction with appropriate fishing gear, and an ecosystem-based approach to sustainable resource management. Continuous evaluation of ecosystem services would provide a highly integrated, multi-sector management tool, combining knowledge from ecology, biology, economics, and social sciences. This valuation would be expressed in monetary terms, making it universally understandable. An example of such a tool is the Natural Capital Project's InVEST (Integrated Valuation of Ecosystem Services and Trade-offs), which offers a suite of open-access software tools for valuing natural capital. However, substantial data is required for its effective application, and a concerted effort in data collection is recommended alongside its use.	1. PPP's and SOE's. 2. Environmental taxes. 3. Blue levies. 4. Natural Asset Companies (NACs) 5. Voluntary PES transactions 6. Blue bonds 7. Environmental bonds 8. Catastrophe bonds 9. Use-of-proceeds bonds 10. Carbon and nutrient trading credits (requires environmental accounting to already be in place)

Table A2. *Cont.*

Financing Need	Description	Potential Solutions	Financing Instrument
Increasing the role of MPAs as fisheries management tool	Marine protected areas (MPAs) and harvest control are incentive-based fisheries management tools that have proven effective over time. Their benefits extend beyond MPA boundaries, including increased biomass and abundance, habitat preservation, reduced mortality, and enhanced growth and reproduction. The effectiveness of MPAs is context-dependent, and it is crucial to implement rules that control and limit fishing around their periphery to realise their full potential. MPAs also present significant opportunities for the eco-tourism sector to benefit from the services they provide. The development of MPAs allows for collaboration between fisheries, tourism, government, the private sector, and donor agencies to acquire resources, expertise, and technical knowledge related to the Blue Economy, helping to address sector challenges. Adequate support structures and education about eco-tourism opportunities may be necessary to implement more MPAs. The culture and traditional fishing methods of fisherfolk can be leveraged to facilitate this transition to eco-tourism by sharing and monetising these experiences for tourists.	A thorough understanding of species distribution and their habitat relationships is crucial for the success of marine protected areas (MPAs), yet this is often insufficient in existing protected areas worldwide, undermining their effectiveness. Each MPA must have its own tailored management plan developed and implemented. Key factors for effective MPAs as management tools include sustainable fisheries management, economic prosperity, the MPA's location, size, and habitats, its connectivity to other MPAs, and the quality of local stakeholders' participation in its management. MPAs should serve not only conservation goals but also the improved management of existing MPAs and the creation of a network of MPAs as part of a targeted fisheries management strategy. MPAs should be designed and managed in areas that significantly contribute to fishery resources. Stakeholders in the industry can be financially incentivised to adhere to and enforce MPA regulations, but a transparent platform or mechanism should be established before this can be effectively implemented.	1. PPP's and SOE's 2. PES schemes 3. Natural Asset Companies 4. Blue levies 5. Leases, Licences, and fees 6. Debt swaps 7. Blue tokens 8. Environmental bonds 9. Catastrophe bonds
Scaling and development of aquaculture and mariculture	The aquaculture and mariculture sectors of the three countries are underdeveloped, but present a means with significant potential to create revenue. With the collective fish stocks predominantly at or above their sustainable harvesting limits, fisheries offer decreasing capacity. Using investment directed at the development of the Blue Economy, upscaling aquaculture and mariculture offers the potential to optimise the benefits received from the development of the marine environment, create sustainable, quality employment, and offer high-value commodities for both export and the domestic market. This area also offers the potential to develop further economic opportunities up- and downstream of the mariculture and aquaculture ventures themselves, creating further livelihoods. The benefits from aquaculture development, including job creation, entrepreneurship, skills development, increased food security, and potentially circular resource use, can be realised more expediently than the transition to sustainable fisheries.	Investment should be directed towards the creation of an enabling environment in which aquaculture operations can flourish, with the required support structures in place (such as veterinary services and distribution chains) to encourage further development. Among some relevant initiatives, Earth Ocean Farms in Mexico use open sea aquaculture technology to cultivate *totoaba* (endangered species) and red snapper (*Lutjanus purpureus*) in their natural environment through cages that are introduced into deep waters with ideal conditions for their growth. Essential for further development of mariculture is the harnessing of external capabilities. Connecting with practitioners experienced in aquaculture activities would provide a valuable resource for the development of mariculture in Caribbean regions. These expert practitioners could deliver key pilot projects that would provide training and capacity building, while developing strategies that prioritise national development in aquaculture and mariculture. Cultivating these connections could provide a significant investment opportunity through international development agencies, while providing a passageway to engaging in purposeful development of mariculture in the Caribbean. Aquaculture financing could happen both at the level of individual projects and at a national level. On a project level, countries that have some aquaculture can mobilise financing for securing production and environmental quality of aquaculture. This can also be promoted through financing of research-based institutions that support aquaculture developers. On a national level, financing is needed for not only creating but, just as importantly, for ensuring that public authorities have the capacity to execute national policies. In Martinique, for example, local support for aquaculture development exists and is aided by local and international (EU) funding mechanisms. Aquaculture research institutions such as 'Delegation Ifremer des Antilles Francaise' (IFREMER) can facilitate collaboration projects to further the development of aquaculture on the island and in the region on a bigger scale.	1. Debt and lending 2. Blue bonds 3. Blue tokens and fintech 4. PPP's 5. Debt swaps 6. Voluntary PES transactions 7. Insurance

Table A2. *Cont.*

Financing Need	Description	Potential Solutions	Financing Instrument
Development of aquaponics and integration into BE development plans	As with aquaculture and mariculture, integrating aquaponics into Blue Economy development strategies alongside fisheries and aquaculture provides a pathway to realise the sustainable intensification of fisheries, aquaculture, food and agriculture. Aquaponics reintroduces biological complexity into agricultural systems, closely guided by knowledge co-creation and sharing processes that aim to maximise synergies. Tackling the region's high food import bill and improving food security is increasingly being explored through aquaponics. An aquaponics industry can facilitate the rearing of fish for high-value protein concurrently with a range of vegetables and other produce, which as an import substitution measure can help reduce dependence on these foreign imports. Depending on market trends, crop production can be rapidly accelerated according to the local, tourism, and export demands. However, it is important to have access to markets that are willing to pay higher prices for superior quality produce. Costs for both construction and operating of aquaponics is fairly high, and production expenses may thus not be recovered without access to high-value markets. The aquaponics initiative may not be profitable without access to, and leverage, in the markets. A great deal of aquaponic businesses have failed—typically due to poor business planning and marketing strategies, rather than production-related issues.	Community driven (potentially PPP) projects could drive development in aquaponics throughout the Caribbean. This may contribute to ensuring the nutritional needs of locals are met despite limited available resources, as well as the potential of generating revenue for community well-being (by selling to premium markets, with sufficient scale). While financing the acquisition of aquaponics infrastructure is straightforward, facilitating access to markets is arguably one of the greatest limiting factors in the development of aquaponics. Market access and transport throughout the Caribbean region are areas towards which investment can be directed as they impact the success (via scalability and long-term sustainability) of any aquaponics venture. In this regard, synergies with the tourism and hospitality (restaurants, hotels, etc.) industry can be developed ensuring a reliable supply of premium produce for local and international tourists. Investment into aquaponics research and training institutions is also advised, as these will yield important contributions such as the profitability of (the appropriate combination of) cultured species. Such centres could also facilitate training in business planning and marketing strategies, in addition to aquaponics production. An enabling environment for aquaponics development is more likely to attract foreign investment and external expertise. A unified system of conformity assessment procedures for testing, inspecting, and certifying aquaponics products for import and export would minimise confusion over trade standards and ensure that all products meet legislative and food safety requirements.	1. PPP's 2. Debt and lending 3. Blue bonds 4. Blue tokens and fintech 5. Debt swaps 6. Voluntary PES transactions 7. Insurance
Develop and integrate aquaculture with alternative and emerging industries	To achieve scalable growth in aquaculture, it is essential to invest in strengthening existing relationships with other sectors and fostering new ones. This approach opens up numerous opportunities through alternative and emerging industries related to aquaculture. There is potential to create synergies between aquaculture and the biomedical industry, where products or byproducts from aquaculture could offer additional resources for alternative biomedical applications. This can be thought of as increasing the value addition potential/capacity of the industry as a whole, whereby development in fisheries and aquaculture thus holds implications for other sectors of the Blue Economy.	For example, tilapia skin from aquaculture farms is used to treat first and second-degree burns in northern Brazil. Once considered a waste product, this skin is rich in collagen, helps protect against infection, speeds up healing, and reduces the need for pain medication. This use of tilapia skin offers a cost-effective way to integrate the aquaculture industry with the medical field, requiring no additional development beyond existing aquaculture practices to enhance food security, livelihoods, and economic growth. Species of Rhodophyta, such as Irish Sea Moss found in the Caribbean, contain unique compounds with various health benefits, making them valuable for biotechnological applications. Seaweeds also play a significant role in agriculture by increasing crop yields as fertilisers and reducing the chemical load on soils and crops. Expanding mariculture to include seaweed extracts in food processing, nutraceuticals, pharmaceuticals, and industrial applications presents significant investment opportunities within the Blue Economy, enhancing sector connectivity and income generation. This approach also addresses the issue of Sargassum seaweed influxes in the region. It is essential to establish appropriate regulations for using these algae in consumption and as fertilisers throughout the Caribbean, especially regarding processing requirements and removing potentially harmful substances from the biomass.The integration of products from aquaculture and aquaponics with emerging industries is research and time intensive. Partnerships with other sector stakeholders may thus be beneficial for developing future mutual benefit.	1. PPPs' and SOE's 2. Blue levies 3. PES schemes 4. Blue bonds 5. Use-of-proceeds bonds 6. Blue tokens and Fintech

Table A2. *Cont.*

Financing Need	Description	Potential Solutions	Financing Instrument
Integrating fisheries and aquaculture into the wider BE	In the context of Blue Economy development at the national level, fisheries and aquaculture should be prioritised. They will remain the primary sources of animal protein and employment for the population, especially with growing efforts to advance aquaculture and aquaponics. Fisheries, as key observers of marine changes, must play a central role in preserving important habitats and rehabilitating degraded ones. This aligns with the Blue Economy approach, which integrates marine biodiversity with coastal habitats to develop solutions beneficial for both biodiversity and climate change mitigation and adaptation. Currently, there are limited coordination mechanisms and no overarching entity to drive Blue Economy development, resulting in a dominance of sectoral approaches. This hinders the ability of countries to effectively design and implement Blue Growth policies and to protect the environment and enhance ecosystem health through the Blue Economy concept. Additionally, there is increasing confusion about the role of the state in each country due to a lack of clear commitment signals. Therefore, structuring the Blue Economy should be the top priority, preceding all other interventions. There is also a lack of an integrated and forward-looking approach to marine ecosystems and spatiotemporal management tools. Neither country has fully adopted the large marine ecosystem approach (Caribbean LME), which would help monitor the evolution of coastal and marine ecosystems using ecological indicators such as biological productivity (particularly fish biomass), pollution (including plastics and chemicals), and ecosystem health. The absence of this approach results in less effective resource and ecosystem management.	The establishment of supra ministerial Blue Economy coordination units in each of the countries respectively will facilitate the integration of the fisheries and aquaculture sectors into the wider BE, by being able to manage its impact relative to other sectors. Such a unit will also be better suited to develop coordination mechanisms between BE sectors with fisheries and aquaculture industries resulting in cross-sectoral synergies and increased development capacity. Improved understanding and comprehension of fisheries and aquaculture value chains and challenges facilitates integration into the wider BE, by being able to identify synergies and opportunities for cooperation with other sectors, and the fisheries and aquaculture sectors and markets of other countries. The development of integrated and spatiotemporal management tools can be coordinated and integrated into fisheries and aquaculture, by a BE unit. A Blue Economy unit with a holistic approach will also facilitate and inform policy decision-making more accurately than individual sector representatives. Adoption and awareness raising of the large marine ecosystem approach among fisheries and aquaculture stakeholders can facilitate a more sustainable and efficient use of available natural resources. Awareness-raising can be facilitated by regional organisations, BE units, NGOs, or the state.	1. Blue levies and stakeholder taxation 2. SOE's 3. Debt swaps
Data limitation among fisheries and aquaculture and lack of policy development	Blue Economy activities and components are currently not accounted for in a unified manner. At present, data must be gathered from various sources to obtain a comprehensive view of the Blue Economy's contributions to added value and job creation. For some sectors, such as ship maintenance, data is not even recorded. Implementing a national accounting system would facilitate the tracking of annual changes in economic sectors. Similarly, ecological aspects of the Blue Economy are not accounted for, despite the valuable ecosystem services provided by coastal areas and their role in mitigating hurricane impacts. However, with the implementation of nationally determined contributions, green and blue accounting should emerge as foundational tools for evaluating specific actions related to climate change. The lack of data results in uninformed decision making and inadequate fisheries and aquaculture policy making, negatively affecting the industry.	A national organisation for Blue Economy accounting (and environmental accounting) could regulate the accounting of BE activities. This thus ensures that the data from a wide variety of sources is verified and coordinated, meaning that it can be used to inform decision-making relating to climate-change, and industry development. Policy-making needs to be informed by available data, and where it is missing, similar examples (of other countries' fisheries and aquaculture sectors) can be used as proxies (if appropriately similar). Where such an organisation is missing, an easy to use platform for recording data in fisheries and aquaculture is necessary. The design of the platform should be such that it encourages unambiguous data input and should use data already generated/collected by industry stakeholders. Incentivising stakeholders to encourage honest data reporting of fishing activities (such as through tax exemptions or non-government subsidies) may contribute to filling the data limitation.	1. PPP's and SOE's 2. Blue levies and stakeholder taxation (and exemption) 3. Fintech 4. Use-of-proceeds bonds
An enabling environment for fisheries and aquaculture investment	Banking mechanisms are inadequate, and financial markets remain underdeveloped. In this context, a transparent policy framework adhering to emerging blue finance principles (such as those from UNEP-FI) could help build investor confidence. Creating a supportive environment for sustainable financing of fisheries within the broader Blue Economy will require focused attention, particularly in relation to addressing national debt.	The recent Caribbean Blue Economic Financing Project (Caribbean BlueFin) offers a chance to strengthen the capacity of selected countries and establish an environment conducive to private sector involvement and investment in the Blue Economy.	1. Debt and lending 2. Insurance 3. Fiscal Policy 4. Debt swaps

References

1. Bethel, B.J.; Buravleva, Y.; Tang, D. Blue Economy and Blue Activities: Opportunities, Challenges, and Recommendations for The Bahamas. *Water* **2021**, *13*, 1399. [CrossRef]
2. FAO. *FishStatJ—Global Production by Production Source 1950–2019*; FAO: Rome, Italy, 2022. Available online: https://www.fao.org/fishery/en/statistics/software/fishstatj (accessed on 4 April 2023).
3. FAO. *Fishery Country Profile*; Grenada; FAO: Rome, Italy, 2019; Updated December 2019. Available online: https://www.fao.org/fishery/en/facp/grd (accessed on 23 April 2023).
4. March, A.; Failler, P.; Bennett, M. Caribbean Fishery and Aquaculture Financing Needs in the Blue Economy: Identifying Opportunities and Constraints in Barbados, Grenada, and St. Vincent and the Grenadines. *J. Sustain. Res.* **2023**, *5*, e230004. [CrossRef]
5. UN. United Nations—Blue Economy: Oceans as the Next Great Economic Frontier. 2023. Available online: https://unric.org/en/blue-economy-oceans-as-the-next-great-economic-frontier/ (accessed on 3 March 2023).
6. Failler, P.; Pètre, É.; Binet, T.; Maréchal, J.-P. Valuation of marine and coastal ecosystem services as a tool for conservation: The case of Martinique in the Caribbean. *Ecosyst. Serv.* **2015**, *11*, 67–75. [CrossRef]
7. Jackson, J.; Donovan, M.; Cramer, K.; Lam, V. Status and Trends of Caribbean Coral Reefs: 1970–2012. IUCN. 2014. Available online: https://portals.iucn.org/library/efiles/documents/2014-019.pdf (accessed on 20 March 2023).
8. Patil, P.G.; Virdin, J.; Diez, S.M.; Roberts, J.; Singh, A. Toward A Blue Economy: A Promise for Sustainable Growth in the Caribbean; An Overview. World Bank Group. 2016. Available online: https://www.cbd.int/financial/doc/wb-blueeconomy.pdf (accessed on 20 March 2023).
9. Wilson, R. Impacts of Climate Change on Mangrove Ecosystems in the Coastal and Marine Environments of Caribbean Small Island Developing States (SIDS). 2017. Available online: https://www.seatoneconsulting.com/wp-content/uploads/2017/04/Impacts-of-Climate-Change-on-Mangrove-Ecosystems-in-the-Coastal-and-Marine-Environments-of-Caribbean-Small-Island-Developing-States.pdf (accessed on 21 March 2023).
10. OECD Data Explorer. Official Flows by Country and Region. From OECD.Stat. 2024. Available online: https://data-explorer.oecd.org/vis?df[ds]=DisseminateFinalDMZ&df[id]=DSD_DAC2@DF_DAC2A&df[ag]=OECD.DCD.FSD&df[vs]=1.1&dq=.DPGC.206.USD.Q&lom=LASTNPERIODS&lo=5&to[TIME_PERIOD]=false (accessed on 6 February 2024).
11. Lewis, F.; Saliman, A.; Peterson, E. Funding Trends 2023: Tracking the State of Global Ocean Funding. Our Shared Seas. 2023. Available online: https://oursharedseas.com/funding/ (accessed on 15 May 2023).
12. Failler, P.; Andriamahefazafy, M. *Madagascar Blue Economy Strategy for Fisheries and Aquaculture*; Government of Madagascar: Antananarivo, Madagascar, 2022.
13. Failler, P. *Seychelles Blue Economy Action Plan*; Department of the Blue Economy, Republic of Seychelles and United Nations Economic Commission for Africa: Addis Ababa, Ethiopia, 2020; 20p.
14. AU-IBAR. *Africa Blue Economy Strategy—Blue Governance Framework*; African Union: Addis Ababa, Ethiopia, 2020.
15. Ram, J.; Kaidou-Jeffrey, D. Financing the Blue Economy in the Wider Caribbean. In *The Caribbean Blue Economy*; Routledge: Abingdon, UK, 2020; pp. 210–225.
16. Clegg, P.; Mahon, R.; McConney, P.; Oxenford, H.A. (Eds.) *The Caribbean Blue Economy*; Routledge: Abingdon, UK, 2020.
17. Whiteside, H. Beyond death and taxes: Fiscal studies and the fiscal state. *Environ. Plan. A Econ. Space* **2021**, *55*, 1744–1761. [CrossRef]
18. Vidal-Hernández, L.; de Yta-Castillo, D.; Castellanos-Basto, B.; Suárez-Castro, M.; Rivera-Arriaga, E. Fiscal Economic Instruments for the Sustainable Management of Natural Resources in Coastal Marine Areas of the Yucatan Peninsula. *Sustainability* **2021**, *13*, 11103. [CrossRef]
19. Cepal, N. Revenue Statistics in Latin America and the Caribbean 1990-2019/Estadísticas Tributarias en América Latina y el Caribe 1990–2019. 2021. Available online: https://www.cepal.org/es/publicaciones/46811-revenue-statistics-latin-america-and-caribbean-1990-2019-estadisticas (accessed on 16 February 2024).
20. Attzs, M.; Maharaj, M.; Boodhan, G. *Survey and Assessment of Environmental Taxes in the Caribbean*; Inter-American Development Bank: Washington, DC, USA, 2014.
21. Ballesteros, L.V.; Matthews, J.L.; Hoeksema, B.W. Pollution and coral damage caused by derelict fishing gear on coral reefs around Koh Tao, Gulf of Thailand. *Mar. Pollut. Bull.* **2018**, *135*, 1107–1116. [CrossRef] [PubMed]
22. Unsbo, H.; Boltenstern, M.; Granberg, M.; Olshammar, M.; Karlsson, M. *Quantification and Environmental Pollution Aspects of Lost Fishing Gear in the Nordic Countries*; Nordic Council of Ministers: Copenhagen, Denmark, 2022.
23. Nelms, S.E.; Duncan, E.M.; Patel, S.; Badola, R.; Bhola, S.; Chakma, S.; Chowdhury, G.W.; Godley, B.J.; Haque, A.B.; Johnson, J.A.; et al. Riverine plastic pollution from fisheries: Insights from the Ganges River system. *Sci. Total. Environ.* **2020**, *756*, 143305. [CrossRef] [PubMed]
24. Thompson, N. *Thompson, William (1775–1833), Socialist and Economist*; Oxford Dictionary of National Biography (Online ed.); Oxford University Press: Oxford, UK, 2004.
25. Ostrom, E. *Governing the Commons—The Evolution of Institutions for Collective Action*; Cambridge University Press: Cambridge, UK, 2015; ISBN 9781107569782. Available online: https://www.cambridge.org/us/universitypress/subjects/politics-international-relations/political-theory/governing-commons-evolution-institutions-collective-action-1?format=PB&isbn=9781107569782 (accessed on 15 May 2023).

26. Wunder, S. *Payments for Environmental Services: Some Nuts and Bolts*; CIFOR Occasional Paper No. 42; CIFOR: Bogor, Indonesia, 2005. Available online: https://montagneinrete.it/wp-content/uploads/2024/03/op-42-wunder-on-payments_1495540914-1.pdf (accessed on 23 April 2023).
27. Engel, S.; Pagiola, S.; Wunder, S. Designing payments for environmental services in theory and practice: An overview of the issues. *Ecol. Econ.* **2008**, *65*, 663–674. [CrossRef]
28. Van Hecken, G.; Bastiaensen, J.; Huybrechs, F. What's in a name? Epistemic perspectives and Payments for Ecosystem Services policies in Nicaragua. *Geoforum* **2015**, *63*, 55–66. [CrossRef]
29. Salzman, J.; Bennett, G.; Carroll, N.; Goldstein, A.; Jenkins, M. The global status and trends of Payments for Ecosystem Services. *Nat. Sustain.* **2018**, *1*, 136–144. [CrossRef]
30. Trends, F. Payments for Ecosystem Services: Getting Started in Marine and Coastal Ecosystems. A Primer. 2010, pp. 4–5, Electronic Source. Available online: https://www.forest-trends.org/wp-content/uploads/imported/marine-coastal-pes-getting-started_2010-pdf.pdf (accessed on 15 May 2023).
31. Newell, R.; Sanchirico, J.; Kerr, S. An Empirical Analysis of New Zealand's ITQ Markets. In Proceedings of the IIFET 2002 Conference, Wellington, New Zealand, 19–22 August 2003.
32. Robertson, M.M. The neoliberalization of ecosystem services: Wetland mitigation banking and problems in environmental governance. *Geoforum* **2004**, *35*, 361–373. [CrossRef]
33. BBOP. The Mitigation Hierarchy—Business and Biodiversity Offsets Programme. 2023. Available online: https://www.forest-trends.org/bbop/bbop-key-concepts/mitigation-hierarchy/ (accessed on 11 April 2023).
34. Ten Kate, K.; Bishop, J.; Bayon, R. *Biodiversity Offsets: Views, Experience, and the Business Case*; IUCN & Insight Investment: Gland, Switzelrand, 2004; Online Report. Available online: https://www.iucn.org/sites/default/files/import/downloads/bdoffsets.pdf (accessed on 20 May 2023).
35. Dempsey, J. *Enterprising Nature: Economics, Markets, and Finance in Global Biodiversity Politics*; John Wiley & Sons: Hoboken, NJ, USA, 2016.
36. Ripple, W.J.; Estes, J.A.; Schmitz, O.J.; Constant, V.; Kaylor, M.J.; Lenz, A.; Motley, J.L.; Self, K.E.; Taylor, D.S.; Wolf, C. What is a trophic cascade? *Trends Ecol. Evol.* **2016**, *31*, 842–849. [CrossRef] [PubMed]
37. Blair, J.M.; Collins, S.L.; Knapp, A.K. Ecosystems as functional units in nature. *Nat. Resour. Env't.* **1999**, *14*, 150.
38. Jax, K. Ecological Units: Definitions and Application. *Q. Rev. Biol.* **2006**, *81*, 237–258. [CrossRef] [PubMed]
39. Niner, H.J.; Randalls, S. Good enough for governance? Audit and marine biodiversity offsetting in Australia. *Geoforum* **2021**, *120*, 38–45. [CrossRef]
40. Resilient Islands. Valuing the Benefits of Mangroves and Coral Reefs in the Caribbean. Coastal Resilience. 2024. Available online: https://media.coastalresilience.org/Resilient_Islands/BenefitsOfMangrovesAndCorals_TechReport.pdf (accessed on 16 April 2024).
41. UNEP-WCMC. WorldFish Centre, WRI, TNC. Global Distribution of Warm-Water Coral Reefs. 2024. Available online: https://doi.org/10.34892/t2wk-5t34 (accessed on 12 February 2024).
42. UNEP-WCMC. Short FT. Global Distribution of Seagrasses (Version 7.1). 2024. Available online: https://doi.org/10.34892/x6r3-d211 (accessed on 12 February 2024).
43. Chami, R.; Cosimano, T.F.; Fullenkamp, C.; Oztosun, S. Nature's Solution to Climate Change: A strategy to protect whales can limit greenhouse gases and global warming. *Financ. Dev.* **2019**, *56*. Available online: https://www.actu-environnement.com/media/pdf/news-34050-article-finances-developpement.pdf (accessed on 17 April 2023).
44. Howard, J.; Hoyt, S.; Isensee, K.; Telszewski, M.; Pidgeon, E. *Coastal Blue Carbon: Methods for Assessing Carbon Stocks and Emissions Factors in Mangroves, Tidal Salt Marshes, and Seagrasses*; IUCN: Gland, Switzerland, 2014.
45. Bumpus, A.G. The Matter of Carbon: Understanding the Materiality of tCO2e in Carbon Offsets. *Antipode* **2011**, *43*, 612–638. [CrossRef]
46. Trading Economics. EU Carbon Permits—Trading Economics. 2023. Available online: https://tradingeconomics.com/commodity/carbon (accessed on 12 April 2023).
47. Vandana, S. *Voluntary Carbon Market—S&P Global*; Agnihotri, A., Ed.; S&P Global: New York, NY, USA, 2022. Available online: https://www.spglobal.com/commodity-insights/en/market-insights/latest-news/energy-transition/010622-voluntary-carbon-market-rally-set-to-stretch-into-2022-on-demand-optimism (accessed on 13 February 2023).
48. Fleischman, F.; Basant, S.; Fischer, H.; Gupta, D.; Lopez, G.G.; Kashwan, P.; Powers, J.S.; Ramprasad, V.; Rana, P.; Rastogi, A.; et al. How politics shapes the outcomes of forest carbon finance. *Curr. Opin. Environ. Sustain.* **2021**, *51*, 7–14. [CrossRef]
49. IEG. Intrinsic Exchange Group. 2023. Available online: https://www.intrinsicexchange.com/home-1 (accessed on 12 February 2023).
50. OECD. *Lessons Learnt from Experience with Debt-for-Environment Swaps in Economies in Transition*; OECD: Paris, France, 2007; p. 39.
51. The Commonwealth. Case Study: Innovative Financing—Debt for Conservation Swap, Seychelles' Conservation and Climate Adaptation Trust and the Blue Bonds Plan, Seychelles (On-Going). 2020. Available online: https://thecommonwealth.org/case-study/case-study-innovative-financing-debt-conservation-swap-seychelles-conservation-and (accessed on 14 September 2022).
52. Fuller, F.; Zamarioli, L.; Kretschmer, B.; Thomas, A.; De Marez, L. Debt for climate swaps: Caribbean outlook. In *Impact: Science Based Implementation of 1.5 C Compatible Action for LDCs and SIDS*; Climate Analytics: Berlin, Germany, 2018; pp. 1–18. Available online: https://ca1-clm.edcdn.com/assets/debt_for_climate_swap_impact_briefing.pdf (accessed on 27 September 2022).

53. Bigger, P.; Christiansen, J.; Dempsey, J.; DiSilvestro, D.; Irvine-Broque, A.; Nelson, S.; Rojas-Marchini, F.; Schuldt, A.; Shapiro-Garza, E. *Beyond the Gap: Placing Biodiversity Finance in the Global Economy*; Third World Network: Penang, Malaysia, 2021. [CrossRef]
54. Victurine, R.; Meyers, D.; Bohorquez, J.; Box, S.; Blythe, J.; Callow, M.; Jupiter, S.; Schweigart, K.; Walsh, M.; Bieri, T. *Conservation Finance for Coral Reefs. Vibrant Oceans Initiative Whitepaper*; Wildlife Conservation Society: New York, NY, USA, 2022; p. 10. Available online: https://doi.org/10.19121/2022.Report.43864 (accessed on 18 September 2022).
55. McGowan, J.; Weary, R.; Carriere, L.; Game, E.T.; Smith, J.L.; Garvey, M.; Possingham, H.P. Prioritizing debt conversion opportunities for marine conservation. *Conserv. Biol.* **2020**, *34*, 1065–1075. [CrossRef]
56. Munevar, D. Making Sense of Belize's Blue Bond Proposal. EURODAD: European Network of Debt and Development. 2021. Available online: https://www.eurodad.org/making_sense_of_belizes_blue_bond_proposal (accessed on 20 February 2022).
57. Perry, K.K. From the plantation to the deep blue sea: Naturalising debt, ordinary disasters, and postplantation ecologies in the Caribbean. *Geogr. J.* **2022**, *189*, 562–574. [CrossRef]
58. World Bank. Sovereign Blue Bond Issuance: Frequently Asked Questions. 2018. Available online: https://www.worldbank.org/en/news/feature/2018/10/29/sovereign-blue-bond-issuance-frequently-asked-questions (accessed on 27 August 2022).
59. Winters, R. The Nature Conservancy Announces Its Third Global Debt Conversion in Barbados. The Nature Conservancy. 2022. Available online: https://www.nature.org/en-us/newsroom/tnc-announces-barbados-blue-bonds-debt-conversion/#:~:text=The%20Blue%20Bonds%20strategy%20combines,of%20their%20communities%20and%20economies (accessed on 26 September 2022).
60. Sumaila, U.R.; Walsh, M.; Hoareau, K.; Cox, A.; Teh, L.; Abdallah, P.; Akpalu, W.; Anna, Z.; Benzaken, D.; Crona, B.; et al. Financing a sustainable ocean economy. *Nat. Commun.* **2021**, *12*, 3259. [CrossRef] [PubMed]
61. March, A.; Failler, P.; Bennett, M. Challenges when designing blue bond financing for Small Island Developing States. *ICES J. Mar. Sci.* **2023**, *80*, 2244–2251. [CrossRef]
62. Adams, D.M.; Benzaken, D.; McCarron, B.; Taylor, L.; Thiele, T.; Tok, X. *Analysis & Development of a Pacific Ocean Bond*; Asia Research & Engagement: Singapore, 2020.
63. Mehta, A.; Tirumala, R.D.; Andrich, M.L. *Financing the Ocean Back to Health in Southeast Asia: Approaches for Mainstreaming Blue Finance*; Asian Development Bank: Metro Manila, Philippines, 2021.
64. Christiansen, J. Securing the sea: Ecosystem-based adaptation and the biopolitics of insuring nature's rents. *J. Political Ecol.* **2021**, *28*, 337–357. [CrossRef]
65. Cisneros-Montemayor, A.M.; Moreno-Báez, M.; Reygondeau, G.; Cheung, W.W.L.; Crosman, K.M.; González-Espinosa, P.C.; Lam, V.W.Y.; Oyinlola, M.A.; Singh, G.G.; Swartz, W.; et al. Enabling conditions for an equitable and sustainable blue economy. *Nature* **2021**, *591*, 396–401. [CrossRef] [PubMed]
66. Quak, E. *How Losing Concessional Finance Affects Small Island Developing States (SIDS)*; K4D Helpdesk Report No 626; Institute of Development Studies: Brighton, UK, 2019.
67. Bishop, M.; Argudin Violante, C.; Bouhia, R.; Carter, G.; Corbett, J.; Lindsay, C.; Scobie, M.; Wilkinson, E. *Just Transitions in Small Island Developing States (SIDS)*; Electronic Source; British Academy: London, UK, 2021.
68. UNEPFI. The Principles—Sustainable Blue Finance. United Nations Environment Programme Finance Initiative. 2018. Available online: https://www.unepfi.org/blue-finance/the-principles/ (accessed on 23 September 2023).
69. Thiele, T.; Gerber, L.R. Innovative financing for the high seas. *Aquat. Conserv. Mar. Freshw. Ecosyst.* **2017**, *27*, 89–99. [CrossRef]
70. Gbigbi, T.M.; Achoja, F.O. Cooperative Financing and the Growth of Catfish Aquaculture Value Chain in Nigeria. *Croat. J. Fish.* **2019**, *77*, 263–270. [CrossRef]
71. Syddall, V.M.; Fisher, K.; Thrush, S. Collaboration a solution for small island developing states to address food security and economic development in the face of climate change. *Ocean Coast. Manag.* **2022**, *221*, 106132. [CrossRef]
72. UNCTAD. Barbados Bets on Tuna to Boost the Value of Fish Exports. 2020. Available online: https://unctad.org/news/barbados-bets-tuna-boost-value-fish-exports (accessed on 16 February 2024).
73. OECS Member States. Organization of Eastern Caribbean States. 2024. Available online: https://www.oecs.org/en/who-we-are/member-states (accessed on 6 February 2024).
74. OECD. *Beyond GDP—Measuring What Counts for Economic and Social Performance*; Organization for Economic Co-Operation and Development (OECD): Paris, France, 2018. Available online: https://www.oecd.org/social/beyond-gdp-9789264307292-en.htm (accessed on 16 May 2023).
75. International Oceanographic Commission. Marine Spatial Planning. 2022. Available online: https://www.ioc.unesco.org/en/marine-spatial-planning (accessed on 18 May 2023).
76. ADB. *Financing the Blue Economy: Investments in Sustainable Blue Small-Medium Enterprises and Projects in Asia and the Pacific*; Asian Development Bank (ADB): Metro Manila, Philippines, 2022; ISBN 978-92-9269-603-0. Available online: https://doi.org/10.22617/TCS220281-2 (accessed on 17 September 2022).
77. ADB. SME BlueImpact Asia: Financing Small and Medium-Sized Enterprises that Dominate the Blue Economy. Online Webinar. Asian Development Bank (ADB) Knowledge Events. 2022. Available online: https://events.development.asia/learning-events/sme-blueimpact-asia-financing-small-and-medium-sized-enterprises-dominate-blue (accessed on 16 February 2024).
78. Abalobi. Home Page. 2023. Available online: https://abalobi.org (accessed on 16 February 2024).

79. Weirowski, F.; Hall, S.J. *Public-Private Partnerships for Fisheries and Aquaculture: Getting Started*; WorldFish Center Manual Number 1875; The WorldFish Center: Penang, Malaysia, 2008.
80. Allam, Z.; Jones, D. Climate Change and Economic Resilience through Urban and Cultural Heritage: The Case of Emerging Small Island Developing States Economies. *Economies* **2019**, *7*, 62. [CrossRef]
81. Anderson, T.L.; Leal, D.R. *Free Market Environmentalism*; Palgrave Macmillan: New York, NY, USA, 2001; ISBN 978-0-312-23503-1. [CrossRef]
82. Coates, R. What is Free-Market Environmentalism? Lean Liberty. 2023. Available online: https://www.learnliberty.org/blog/what-is-free-market-environmentalism/ (accessed on 2 February 2024).

fishes

MDPI

Article

From Co-Operation to Coercion in Fisheries Management: The Effects of Military Intervention on the Nile Perch Fishery on Lake Victoria in Uganda

Veronica Mpomwenda [1,2,3,*], Tumi Tómasson [4], Jón Geir Pétursson [2] and Daði Mar Kristófersson [3]

1 National Fisheries Resources Research Institute, Plot 45 | 49 Nile Crescent, Jinja P.O. Box 543, Uganda
2 Faculties of Life and Environmental Sciences, and Sociology, Anthropology and Folkloristics, University of Iceland, 102 Reykjavik, Iceland; jgp@hi.is
3 Faculty of Economics, School of Social Sciences, University of Iceland, Sæmundargötu, 102 Reykjavik, Iceland; dmk@hi.is
4 Marine Research Institute, Fornubúðir 5, 220 Hafnarfjörður, Iceland; tumi@hafro.is
* Correspondence: mpomwendav@gmail.com; Tel.: +354-793-1667 or +256-788922056

Abstract: In 2017, Uganda's small-scale inland fisheries underwent a significant transformation, shifting from local co-management to state military enforcement owing to ineffective enforcement of regulations and declining exports. Employing a mixed-methods approach and blending qualitative and quantitative data, we assessed the impact of military intervention on Lake Victoria's Nile perch fishery, focusing on fishing effort, catch, and exports. Our findings indicate that military operations adhered to regulations, gaining primary support from key stakeholders, specifically motorized fishing operators. Consequently, between 2016 and 2020, legal fishing activities experienced substantial growth. By 2021, approximately 90% of Nile perch catches were made by motorized vessels using longlines and gillnets, despite a declining trend in catch-per-unit effort. Between 2015 and 2021, the Nile perch fishery saw changes: boat seines made up about 5% of motorized fleet catches in 2021, while catches in paddled vessels increased from 20% to over 50%, suggesting a potential role in the growing longline fishery. Therefore, the current management approach does not increase catches or exports compared with the co-management period. The observed decline in catch-per-unit effort among motorized gillnets suggests overcapacity. Further research is needed to comprehend the broader sociological and ecological impacts of the present enforcement strategy for sustainable fishery management.

Keywords: rule compliance; law enforcement; fisheries management; Nile perch; Lake Victoria; Uganda

Key Contribution: This study assesses the effectiveness of military enforcement of fishery management rules in improving fishery performance and outcomes in the Ugandan part of Lake Victoria, replacing the existing co-management approach. However, this military enforcement approach did not result in increased catches or exports compared with the co-management period. The observed decline in catch-per-unit effort among the motorized gillnet fleet suggests overcapacity. Further research is needed to fully understand the broader sociological and ecological impacts of the current enforcement strategy for achieving sustainable fishery management.

Citation: Mpomwenda, V.; Tómasson, T.; Pétursson, J.G.; Kristófersson, D.M. From Co-Operation to Coercion in Fisheries Management: The Effects of Military Intervention on the Nile Perch Fishery on Lake Victoria in Uganda. *Fishes* **2023**, *8*, 563. https://doi.org/10.3390/fishes8110563

Academic Editor: Ben Drakeford

Received: 3 October 2023
Revised: 15 November 2023
Accepted: 16 November 2023
Published: 20 November 2023

1. Introduction

African small-scale inland fisheries are in transition. Fisheries co-management as a cooperative instrument to secure sustainable fisheries has not proven as successful as hoped, and some African countries have been exploring more coercive enforcement measures to ensure rule compliance and regulate access in their inland fisheries, which have traditionally been open access [1–4].

Further, there is an increased emphasis on export-oriented, commercial fisheries and less focus on artisanal fisheries as a source of local livelihoods [5–8]. These two policy directions are often mutually exclusive and have important political and economic implications [9]. This is amplified by the growing and globally integrated African national economies that impact both fisheries and other natural resource sectors [10]. Other common-pool natural resources on the continent, such as forestry and wildlife, have also seen a drive from cooperation towards more high-handed, coercive approaches to enforce rule compliance [11,12]. This has resulted in multiple paramilitary approaches and, in more extreme cases, the deployment of full-scale military operations.

Uganda conducted an interesting study of such a transition in small-scale inland fisheries. In November 2015, bold reforms were introduced to fishery management. Local-level fisheries co-management systems, the Beach Management Units (BMUs), were dissolved through a presidential directive, as well as the military, in the form of the Fisheries Protection Unit (FPU). This was installed to enforce rule compliance on Ugandan lakes, including the territorial waters of Lake Victoria [13]. The suspension of local-level BMUs was due to the alleged prevalence of mismanagement under the BMUs, resulting in widespread illegal fishing, bribery, and corruption [13,14]. Fishery landing site committees were established to enforce fishery regulations alongside the military, contrary to the mandate of the co-management structure, disrupting the involvement of government officials, such as district and sub-county officers who had earlier been involved in the co-management structure [13,15,16]. An extensive military operation commenced in Lake Victoria in February 2017 and is still being implemented in 2022. When the FPU was established, it aimed to eliminate all forms of illegal fishing gear and practices in the hope of reversing the trend of declining catches and exports [15]. Peak Nile perch catch and export values were achieved in 2006, 3 years after the installation of the BMUs. However, by 2015, catches had decreased by 20% to 117,600 MT, and exports had dropped by 50% to 18,408 MT [17]. Some authors argue that the political agenda is also to consolidate access to valuable fishery resources to the ruling elites [11,12,18,19].

Fishing effort on Lake Victoria has been growing steadily through new entrants, technological improvements, and increased use of illegal gear, such as small mesh monofilament nets, small hooks, and beach seines, in response to the reduced availability of large fish and growing demand for fish in national, regional, and global markets [8,20]. Nile perch fishers, which predominantly target global export markets, must invest in larger vessels and engines to access deeper waters where the likelihood of catching legal-sized fish is better, while those using smaller paddled vessels have traditionally been fishing for domestic consumption, and local or regional markets fish close to shore using small mesh gillnets and monofilament nets. Similar trends in the development of fishing efforts have been observed in other small-scale fisheries (SSFs), classified as commercial or artisanal, respectively [21,22]. The FAO voluntary guidelines for small-scale fisheries, the FAO code of conduct for responsible fisheries, and the United Nations' Sustainable Development Goal 14 highlight the need to increase economic benefits while protecting the access rights of artisanal small-scale fishers [23]. However, it is difficult to reconcile increased economic benefits while preserving access rights in some inland fishing nations, such as Uganda, where economic growth and development is a key agenda in the productive sectors.

The objective of this study is to analyze whether the strong enforcement of fisheries management rules has enhanced the fisheries performance and outcomes in the case of a productive African inland fishery. This study examines the effects of a shift from cooperation to coercion in managing the Nile perch fishery in Uganda's Lake Victoria. It assesses the impacts on fishing effort and catches after five years of military involvement. Our research questions are as follows.

- ○ How was the military intervention on Lake Victoria organized, and what rules were enforced?
- ○ To examine how coercive rule enforcement has impacted the fishing effort and fish stocks of the Nile perch on Lake Victoria over time;

- To understand the fishery policy implications of these interventions.

2. The Challenge of Rule Compliance in African Inland Fisheries

Institutions matter, and to avoid the tragic outcomes of open access to common-pool natural resources, an institutional framework involving formal and informal rules, norms, and conventions related to a particular resource is required [24,25]. Resource users are generally driven by individual interests and without definitive rules on resource use and accountability, resources typically suffer from degradation [26]. This is a key challenge with common-pool resources, such as inland Small-Scale Fisheries (SSF), where excluding access is difficult, and there is a high degree of competition among users. A key aspect of the effective governance of common-pool resources is that users adhere to the rules and institutions enforcing them are perceived as being effective and legitimate [27–29]. Two major challenges concerning compliance in inland fisheries exist: the inadequacy of rules and weak law enforcement mechanisms to guide sustainable management [30–32].

Rules governing resource use in African inland fisheries are mostly centred on managing fishing effort through gear limitations, such as mesh regulations and gear bans [33–35]. These are perceived as easier to implement in fisheries characterized by many users with varying fishing effort. However, non-compliance remains high among resource users [30,36]. Fisheries management faces insufficient financial and personnel capacity to enforce rules. In addition, there is also a growing uncertainty being expressed in the literature about whether these types of effort limitations are indeed effective in ensuring the sustainable management of SSF [34,35]. Most of the existing rules in African inland fisheries are dated and based on scientific principles perceived to address the political and economic interests of the state, commonly ignoring the needs of the fishing communities [30,34,37]. However, successful governance of common pool resources requires institutional frameworks to evolve, change, and adapt [28]. Over time, fishermen tend to diversify and evolve their activities, and with rigid rules, resources are susceptible to degradation in the long term.

Successfully co-managed, common-pool resources are mostly found when communities are characterised by well-defined boundaries, shared norms and rules, low levels of mobility, appropriate leadership, and accountability of users [24,38]. Inland fisheries are often transboundary, fishers are multicultural and highly mobile, and the nature of imposed co-management is instructive, as fishers are expected to adhere to the rules rather than participate in their formulation. Fisheries face challenges, such as inadequate resources to manage the fishery, bribery, corruption, and lack of political will and genuine enforcement to sustain the fisheries [4,14,39]. Thus, co-management to govern fisheries has found difficulties in ensuring compliance with the rules [40]. Consequently, nations such as Uganda have resorted to coercive powers to enforce rule compliance. The shift to coercive enforcement, coming in the form of the military in Uganda, also depicts the drive of individual states to control the fisheries sector. The co-management structures had a weak law enforcement mechanism characterised by inadequate resources, poor coordination among the BMU structures, and a lack of political will to empower enforcement activities [40–42]. Consequently, fishing practices remained contrary to government regulations, as BMU operations were mainly based on social ties rather than active law enforcement [14,42]. The challenges of the co-management structures to enforce fishery rules among fishers were used as arguments for the military to enforce rule compliance in Uganda. While cooperation failed to foster rule compliance among fishers, it is also argued that reliance on military approaches indicates a lack of social control since the state fails to facilitate collective control over common resources [43]. Military coercion, being inherently costly and prone to triggering violent conflicts, is not a sustainable approach, especially for low-income countries. On the other hand, cooperation relies on voluntary compliance, introducing the risk of free-rider problems. Therefore, there is a need to examine both coercive and cooperative approaches, as they present different but interesting opportunities for effective rule compliance in inland African fisheries.

3. Materials and Methods

3.1. Study Area and Context

Lake Victoria is a 68,600 km^2 transboundary water body, of which 29,584 km^2 (43%) belongs to Uganda. The country's share has an extensive shoreline of 1750 km characterised by numerous islands, making it favourable for fishing [44]. Fishing is primarily based on three key commercial species: the small pelagic silver cyprinid, locally known as dagaa (*Rastrineobola argentea* Pellegrin, 1904), Nile perch (*Lates niloticus*; Linnaeus, 1758), and Nile tilapia (*Oreochromis niloticus*; Linnaeus, 1758). Nile perch is processed as a chilled and frozen product for export markets, Nile tilapia is consumed locally, and dagaa is consumed in both local and regional markets. In recent years, a lucrative trade in Nile perch maws has developed where the value of the maw of the largest fish exceeds the value of the flesh [45].

Fisheries management in Lake Victoria evolved from self-governing communal arrangements before colonial times (<1890) to centralised top-down management for over a century [1890–2002]. A bottom-up co-management structure was introduced on the lake in 2003 [13]. The co-management system was established in the early 1980s as part of the global movement to decentralise governments. The shift was prompted by challenges faced by the centralised management system in effectively regulating fisheries [4,42]. In 2015, the co-management system was suspended in Uganda and replaced by military operations as the FPU was effective in 2017. At the time of data collection, the FPU had established eight operational areas around the lake, commonly known as sectors (Figure 1). Each sector includes a minimum of two adjacent districts. The operations in each sector are headed by a senior military commander commanding a cadre of 8–17 military personnel.

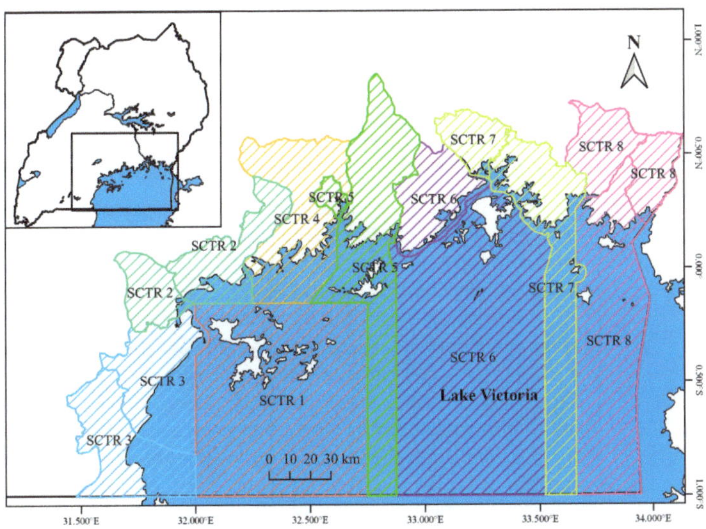

Figure 1. Boundary demarcation among the eight areas (sectors) of FPU operations along the Lake Victoria riparian districts in Uganda. Different colour patterns represent where each of the eight sector commandants conducts law enforcement on the lake.

3.2. Data Sources

To analyse the structure and effects of the current military fisheries management in the Ugandan waters of Lake Victoria, a sequential exploratory and explanatory mixed-method approach were used, using qualitative and quantitative data. The data were gathered and analyzed sequentially. The findings from the initial exploratory/qualitative dataset are substantiated by a subsequent dataset, which comprises explanatory or quantitative data in the study [46,47]. Qualitative data were obtained from the FPU personnel, while quantitative data were obtained from the National Fisheries Research Resources Institute

(NaFIRRI) and the Lake Victoria Fisheries Organisation (LVFO). Qualitative data were used to provide background information, the context of the study, and to discover the structure of the law enforcement operations, while quantitative data were used to explain the effects of the operations on fisheries effort and catch. We chose this approach to mitigate the comparative weakness of qualitative research, add richer detail to their conclusions, and make the results more credible by using different methods to collect data on the same subject as has been carried out in several other studies [46,48,49].

3.2.1. Data on the FPU Operation

Eight senior military officials, each in charge of a distinct FPU sector, were purposefully contacted and selected for interviews. They were highly ranked and had pivotal roles in law enforcement. Selected for their extensive expertise, these individuals were to provide a comprehensive understanding of military intervention in law enforcement and articulate the regulatory challenges, operational intricacies, and practical insights into maintaining law enforcement activities on the lake. The decision to engage with these key stakeholders was driven by the goal of capturing a holistic and authoritative perspective on the law enforcement dynamics in the lake. A checklist of interview questions was used; however, depending on the situation, divergences were allowed from one respondent to another to allow respondents to talk freely. Interviews were conducted from December 2019 to August 2020, recorded, and each lasted 45–60 min. Permission to collect data was sought from the FPU head office. During the fieldwork, the purpose of the study and emphasis on ethical principles of anonymity and confidentiality guiding the research was explained to the respondents, with their consent to participate confirmed before the interviews were taken [47]. In addition, records of confiscations of illegal vessels and gear covering the period from February 2017 to August 2020 were obtained from each sector and compiled for further analysis. FPU officials are tasked with recording every confiscation made per operation, which is sent to the Ministry of Agriculture, Animal Industry, and Fisheries (MAAIF) for recording purposes.

3.2.2. Fishery Effort, Catch, and Export Data

Catch and effort data were sourced from technical reports and databases at the National Fisheries Resources Research Institute (NaFIRRI) and the Lake Victoria Fisheries Organisation (LVFO). Since 2004, LVFO, in collaboration with partner fisheries research and management institutions in Uganda, Kenya, and Tanzania, has been monitoring fishing effort and catches on Lake Victoria through periodic frame surveys and continuous catch assessment surveys. Frame surveys (FS) are conducted biennially, with logistics permitting, and involve a complete census of all fishing effort variables, such as the enumeration of active landing sites, fishing vessels, gears, and fishers. Catch Assessment Surveys (CAS) involve the collection of catch data from a sample of vessels and landing sites throughout the year. Through the FS, an effective sampling frame is used to determine a representative sample of landing sites within the districts to be sampled [50,51]. Landing sites were selected randomly with a probability proportional to their size (PPS) measured in terms of the total number of vessels landing at the site, and catch data were obtained from vessels at random. The final data are representative of approximately 10% of all landing sites on the lake. These datasets are collected by trained enumerators, following the LVFO harmonized Standard Operating Procedures and data forms [50,51]. For this study, the time-series data were segregated into three discrete periods: the BMU management system era (2004 to 2015), the year of transition (2016), and the FPU period (2017 to 2021). The transition period in the study describes the period during which the BMU structures were disbanded until the commencement of FPU operations in 2017. Catch assessment data were sorted, and trips where either legal or illegal gillnets or/hooks were deployed, but not both, were selected for further analysis. This was carried out to eliminate any bias due to the uncertainty in identifying which catches or fish sizes were caught using legal or illegal gear by a specific vessel.

Fish export data were sourced from the statistical databases of the Bank of Uganda and the Uganda Bureau of Statistics [17,52], to cross-reference and validate the information retrieved from these sources. Data verification was conducted using an additional dataset provided by the Directorate of Fisheries Resources (DiFR).

3.3. Data Analysis

Interviews with the FPU senior commanders were transcribed and, in the analysis, the authors sought to find and analyse direct quotes from the interview respondents that discussed aspects related to the study objectives on how the military intervention was organised and the rules imposed. Quantitative data obtained from the sectors included records of confiscations, which were collated, summarised, and presented by year. The catch and effort data were analysed according to different types of gear and vessels.

Yearly average total catches were computed from the LVFO catch assessment survey data, whereas CPUE was computed based on vessel-level catch assessment data. For the latter, a proxy variable, the average weight of fish landed (kg fish^{-1}), was computed as a fraction of the total weight of the catch and the number of fish in a vessel for each observation. The CPUE was then presented as the kg per gear unit hours fished with gear units and hours fished as the effort variables over the study period from 2005 to 2021. The analysis and presentation of findings aimed at comparing the BMU (2005–2015) with the FPU period (2017–2021), where changes in effort, catch, CPUE and the average size of fish caught were used to explain changes related to the change in management. We compared the CPUE between vessel propulsion types and years using a two-way Analysis of Variance (ANOVA) and a post hoc Tukey test. All statistical and graphical analyses were performed using R software (4.0.3) [53] Packages dplyr and ggplot2 were used for data manipulation and visualization, respectively [54].

3.4. Limitations

Despite the study's strengths, it is essential to acknowledge its limitations, including its relatively small sample size. Qualitative data were collected from eight officials, and the quantitative data covered 10% of the total landing sites. The non-random, purposive selection of the qualitative sample limits generalizability and caution is warranted when interpreting the broader impact of military activities on the lake. Additionally, missing data in the quantitative analysis were non-random, affecting 8 out of the 17 years due to limited funding. Therefore, the findings should be cautiously interpreted, avoiding generalizations of catch trends for the entire 17-year period.

4. Results and Discussion

4.1. The Military Operation on the Lake

In 2014, 409 BMUs were recorded in the Ugandan part of Lake Victoria. In November 2015, the BMUs were suspended, and the FPU was installed to enforce rule compliance among fishers, starting its operations in February 2017. Most of the fisheries management rules that have been implemented are based on the fisheries legislation of Uganda, the most fundamental one being the Fish Act Cap 197 of 2000 and its amendments in the fish (fishing) rules of 2010, which did not change with the transition from the BMU system to the FPU. Sector commanders indicated that they had prior experience in operating in a marine environment under the Uganda Peoples Defence Force (UPDF) and this was part of the criteria for deployment in the FPU. In 2016, members of the FPUs received a 2-week orientation by the Ministry of Agriculture, Animal Industry and Fisheries (MAAIF) officials on the fisheries rules to be enforced, as noted by one respondent: "*Most of us were or are part of the UPDF marine. . .we undertook training with the ministry officials at the fisheries training institute and taught on what is acceptable on the lake, so whatever is prohibited on the lake according to the rule of law we abolish it (. . .)*".

Some of the equipment used for enforcement such as vessels and engines were provided by the MAAIF through the Directorate of Fisheries Resources (DiFR) and in some cases, additional equipment was supplied by prominent commercial fishers who supported strict enforcement. Under the FPU, fishers and vessels on the lake have to be registered and licensed, Fishers targeting Nile perch, Nile tilapia, and other larger species are prohibited from using gillnets < 5″, cast nets, beach/boat seines, monofilament nets, hooks size $\geq$ 10 (the naming of hook sizes is such that the hooks become smaller as the size number increases), and indiscriminate fishing methods and the use of vessels smaller than 28 ft/8.5 m. Dagaa fishers are allowed to fish with a maximum of 8 small seine net panels and with vessels $\geq$ 28 ft/8.5 m (Table 1). The use of illegal fishing gear and practices had been prevalent among smaller fishing vessels operating in the inshore areas of the lake [55–57]. Military officials also observed this, and hence, the eradication of small fishing vessels has been a major target throughout the enforcement work of the FPU.

Table 1. Rules enforced by the Fisheries Protection Unit (FPU).

Legislation	Rules
Fish Act Cap 197 of 2000 The fish (fishing) rules of 2010	• Vessel registration and licensing. • Prohibit indiscriminate methods of fishing, e.g., poisoning, tycoon (forcing fish into the nets by beating the water), etc., cast nets and beach seines. • Prohibit the use of gillnets < 5 inches, hook size > 10, and monofilament nets. • Acceptable fish slot size 50→85 cm TL for Nile perch and ≥25 cm TL for Nile tilapia. • Vessel size < 28 ft (8.5 m) prohibited for fishing. • Maximum of 8 panels for the seine nets for dagaa fishers.

Although the BMUs had been suspended, military officials consulted and engaged stakeholders at the local level in their activities to ensure the success of the enforcement operations. In one case, a respondent said: *"We work closely with the AFALU* (Association of Fishers and Lake Users) *and some former BMU chairmen help, we collaborated with them from the start, they gave us people to work with, in navigating the lake from the start. . .they gave us their boats and engines. . .they also feed us with information on the illegal activity hotspots"*. Engaging local stakeholders enhanced law enforcement, as the stakeholders ensured the acceptability of the regime shifts from the BMU to the FPU at the landing sites. With the suspension of the BMUs, the AFALU, whose membership is dominated by commercial Nile perch fishers, nominated two persons at every major landing site to report illegal fishing activities to the military. It is, therefore, no surprise that the AFALU has been identified as one of the major actors behind the suspension of the BMUs. Such a swift change in management in Uganda was possible because transition powers remained with government actors rather than resource users [15,16]. Mpomwenda et al. [13] indicate that commercial Nile perch fishers were dissatisfied with the local-level BMU structure and local fisheries officials. The enforcement of the rules is perceived to favour commercial Nile perch fishers over artisanal fishers using small vessels.

4.2. Enforcement Activities of the FPU and the Effect on Fishing Effort

4.2.1. Effect on Fishing Effort

At the beginning of the 21st century, most fishing vessels on Lake Victoria were unmotorised and mainly paddled (80%), while some used sails. Over the first two decades of the century, there was no growth in paddled or sailed vessels from 2006, while an exponential growth in the number of motorized vessels was recorded, with an average annual growth rate of 11% from 2000 to 2020 (Table 2).

Table 2. Fishing effort indicators on Lake Victoria in Uganda during periods of different management regimes 2000–2020. Source frame survey reports.

Management Regime	Pre-BMU	BMU							Transition Period	Military Enforcement	Average Yearly Change			
Variable	**2000**	**2002**	**2004**	**2006**	**2008**	**2010**	**2012**	**2014**	**2016**	**2020**	**2002–2004**	**2014–2016**	**2016–2020**	**2000–2020**
Landing sites	597	552	554	481	435	503	555	567	556	446	0%	−1%	−6%	−1%
Fishers	34,889	41,674	37,721	54,148	51,916	56,957	63,921	64,617	66,869	60,552	−5%	2%	−2%	3%
Motorised vessels	2031	3250	3173	5047	5156	6334	9351	9955	11,495	17,075	−1%	7%	10%	11%
Paddled vessels	12,848	14,262	12,506	17,475	15,602	16,389	17,111	17,260	17,260	8460	−7%	0%	−18%	−2%
Sailed vessels	665	1074	1096	1466	1078	682	1125	857	864	260	1%	0%	−30%	−5%
Towed vessels							17	51	28	18		−30%	−11%	1%
Foot fishers						50	367	100	463	350		77%	−7%	19%
Legal gears														
Multifilament gillnets ≥ 5″	243,209	374,642	402,351	498,037	327,098	307,052	423,155	384,849	355,348	556,767	4%	−4%	11%	4%
Hand line hooks [1]	4585	6547	8335	15,860	19,629	17,071	27,780	27,004	37,785	20,669	12%	17%	−15%	8%
Longline hooks < 10				1,681,048	1,657,458	1,389,548	1,525,810	850,493	479,767	3,178,446		−29%	47%	5%
Illegal gears														
Multifilament gillnets < 5″	54,454	52,846	56,246	91,740	76,908	66,532	59,585	78,571	79,473	8676	3%	1%	−55%	−9%
Beach/boat seine	811	880	954	1425	1649	1451	1233	1819	1968	1093	4%	4%	−15%	1%
Cast net	1276	858	659	631	1000	1095	1372	1359	1342	873	−13%	−1%	−11%	−2%
Monofilament gillnets			845		11,203	12,115	15,148	21,793	31,876	15,204		19%	−19%	18%
Basket traps	11,349	5781	5361	499	7615	10,331	7082	9000	6144	3341	−4%	−19%	−15%	−6%
Longline hooks ≥ 10				604,561	1,106,341	1,169,807	2,892,575	3,737,273	3,998,352	1,057,646		3%	−33%	4%

[1] The size of handline hooks was not specified in the frame survey reports.

The use of legal gillnets ≥ 5″ and longline hooks size < 10, which are mostly confined to motorised vessels, did not reflect the increase in the number of motorised vessels. The number of legal gillnets increased steadily from the beginning of the century until reaching a peak of about 500,000 in the 2006 census. The number of legal gillnets then fluctuated around 360,000 but increased again sharply to about 560,000 in 2020 after 4 years of military enforcement. The use of illegal multifilament gillnets fluctuated between 50,000 and 90,000 until 2016 and reduced to less than 9000 by 2020. The relative importance of multifilament gill nets < 5″ in the Nile perch fishery decreased over time, and thus military enforcement was effective in reducing the use of illegal multifilament gill nets.

Monofilament nets and basket traps are highly efficient gears on Lake Victoria and are mainly associated with the artisanal fishery targeting juvenile Nile perch near the shore. Monofilament nets were first recorded in the 2004 frame survey when less than 1000 nets were recorded. No monofilament nets were recorded in the 2006 frame survey, but in 2008 about 11,000 nets were recorded. Their numbers then grew in subsequent frame surveys and reached almost 32,000 in 2016. In 2020, they were down to about 15,000, a reduction of 52%. Over the same period, the number of paddled vessels had been reduced to 8500, a reduction of 51% from the survey in 2016. Reductions in effort variables such as landing sites, monofilament nets and beach seines are consistent with results given in [58], which indicated a general decline in these effort variables.

Longlines were first recorded in the 2006 frame survey, although longlining had already been practised to some extent earlier, both by commercial and artisanal fishers. In the beginning, the majority of longlines had legal-sized hooks, but with time, illegal smaller-sized hooks became increasingly dominant as the number of hooks increased (Table 2). The use of illegal hooks was especially prominent in commercial fisheries (Figure S1). In 2020, there had been a large drop in the use of illegal hooks from about 4 million in 2016 to 1 million in 2020. At the same time, the number of legal hooks increased from about 500,000 in 2016 to 3.3 million in 2020.

The increased use of small hook sizes (>10) and monofilament nets during the BMU period had been attributed to the decline in large-sized Nile perch, prompting fishers to shift to gear with greater fishing efficiency [55]. Illegal fishing gear usually targets immature fish, which is then thought to result in the reduction in larger fish. We note that before enforcement there was no increase in legal gear but rather illegal hooks and monofilament nets, which can be seen as an adaptation for fishers to effectively harvest the available sizes of the perch.

4.2.2. Confiscation of Illegal Gear and Vessels

During the first 3.5 years of operations, the FPU confiscated a total of 27,880 vessels (Table 3). These were mostly of the parachute type (80%) (Table 3). Unlike other vessel types, the parachute vessels are paddled and less than 6 m in length, and with their small size, can only access shallow inshore areas where they target juvenile and spawning fish [55,59]. The Ssesse-type vessels can be fitted with an outboard motor and range from 6 m to 15 m, thus including illegal ones of < 8.5 m, which were confiscated, and vessels larger than 8.5 m operating with illegal fishing gear were also confiscated, albeit with the military officials noting that the latter were given back to the owners after a fine. Despite these massive confiscations, the number of paddled vessels recorded in the frame survey in 2020 had only dropped by 8800 from the 2016 survey (Table 2).

The reduction in most illegal fishing gear from the frame surveys in 2016 to 2020 was proportional to the reduction in small fishing vessels, at around 50%, irrespective of the amount of gear confiscated. Thus, a total of about 540,000 monofilament gillnets were confiscated, which is 35 times the reduction recorded between the two last frame surveys in 2016 and 2020. The number of confiscated seines was 7 times, cast nets 3 times, and basket traps 0.6 times. The easiest and most effective gear to replace is the monofilament gillnets and this could be attributed to the low cost of doing so, which respondents also affirmed, "*these manyala [monofilament] nets and little hooks are the worst to eliminate, they are cheap to obtain...one can just get three nets and use his small boat at the shores, so they incur less capital and then within a few days, they have recovered their money back"*.

Table 3. Vessels and fishing gear were confiscated by the FPU during law enforcement operations on Lake Victoria in Uganda. Data provided by sectoral commanders for the period February 2017 to August 2020.

Effort Variable	Year	February 2017	2018	2019	August 2020	Total
Fishing vessels	Parachute ($\leq$6 m)	6136	3622	7154	5434	22,346
	Ssesse vessels (6–12 m)	924	2374	649	649	4596
	Unspecified		318	555	65	938
	Total of vessels	**7060**	**6314**	**8358**	**6148**	**27,880**
Fishing gears	Hooks size > 10	1,123,863	2,413,174	3,400,299	775,111	7,712,447
	Multifilament gillnets < 5″	22,400	15,322	44,630	2500	84,852
	Monofilament gillnets	147,331	244,949	99,992	47,391	539,663
	Basket traps	290	631	474	538	1933
	Beach/boat seines	2014	2377	1862	1185	7438
	Cast nets	1278	967	324	382	2951

Overall, the elimination of illegal gear used both by paddled and motorized vessels, such as illegal multifilament nets and small hooks, appears to have been quite effective. Confiscated multifilament nets do not appear to have been replaced and their use dropped by almost 90% between frame surveys. The use of small hooks was reduced by 73% between surveys and while these are readily replaced, the large increase in the use of legal hooks has reduced the incentive to replace confiscated small-size hooks (Table 2).

To the military officials, the confiscations have been successful, and the need for surveillance has dropped with time. A challenge, though, could be the cost of surveillance, which was mentioned by officials, in addition to the reduced intensity in conducting operations; *"Initially operations were held frequently and randomly with more support, however, this has changed because the rate at which crime is detected, and arrests made have also reduced...operations are now conducted based on information received"*. There was however no detectable reduction in the confiscation of small vessels during the first 4 years of military intervention.

4.3. Changes in Catches during the BMU and FPU Management Regimes

During the BMU period, there was a consistent decline in overall Nile perch catches from ~90,000 t in 2005 to <50,000 t in 2015, and fish exports declined by 50% from 2005 to 2008, after which they remained relatively constant until increasing in 2018 (Figure 2). Fisheries on Lake Victoria supply more than 90% of Uganda's total exports [60]. Exports continued to increase in 2019 when they were 60% higher than at the end of the BMU period. The observed decline in exports in 2020 and 2021 could be attributed to the stringent COVID-19 pandemic restrictions, where few fishery actors were required to work thus affecting the quality and quantity of Nile perch available for processing and export [61].

With the FPU in charge of law enforcement, it is reported that six more factories for processing Nile perch had opened by 2019, adding to the five existing in 2015. This has been reported as the major achievement of the FPU by the Uganda Fish Processors and Exporters Association (UFPEA) [60]. Moreover, UFPEA and AFALU have been reported to support military activities on the lake, the latter being confirmed by the FPU officials in this study [15]. Rightly or wrongly, the implication is that the shift in management enhanced the survival of the commercial Nile perch trade's value chain by imposing strict measures on illegal fishers. The surge in mechanized fishing operations, concurrent with the opening of six new factories, aimed to yield higher catch volumes. This rise in motorized vessels was driven by the demand to supply fish to the newly established processing facilities. However, the study findings indicate a non-significant increase in catch volume post-enforcement, prompting inquiries into the efficiency of vessels in meeting industry supply. Moreover, there is a deliberation on the cost-effectiveness of the enforcement measures, considering the observed outcomes.

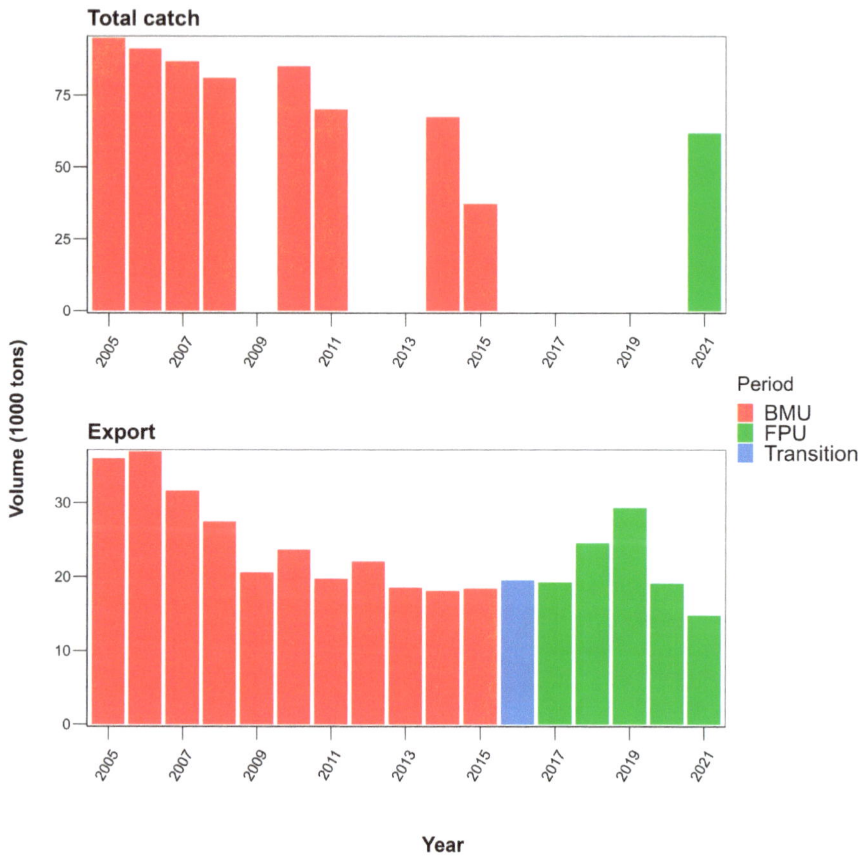

Figure 2. Annual Nile perch catch and exports as volume (1000 tons) during the BMU and FPU period.

In 2005, paddled vessels accounted for just over half of the total Nile perch catch at about 48.5 thousand tons. Thereafter, there was a gradual decline in the importance of paddled vessels, and by 2015 they accounted for about 13,000 t, equivalent to 35% of the total catch. By 2021, their catch of Nile perch was about 8000 t, around 13% of the total (Figure 3A). These changes correspond to the changes in the fleet, with rapid growth in the number of motorised vessels since 2016 and a halving of the number of paddled vessels due to the confiscation of vessels (Table 2).

In 2005, about 80% of the Nile perch caught in the commercial fishery was taken in gillnets and the rest by longlines. The proportion of Nile perch caught by longlines increased to about 30% by 2008 and remained about that figure until 2015. In 2021, longlines accounted for about 40% of the catch, gillnets for 52%, and the rest was caught using mostly illegal gear including beach seines, cast nets, and basket traps. Paddled vessels used a greater variety of gear to catch Nile perch, but gillnets and hooks (longlines and handlines) accounted for around 60–80% from 2005 to 2015. In 2021, the use of gillnets by paddled vessels had dwindled to a mere 10%, while catches were dominated (51%) by beach and boat seines followed by hooks (long and handline) at 30% (Figure 3B).

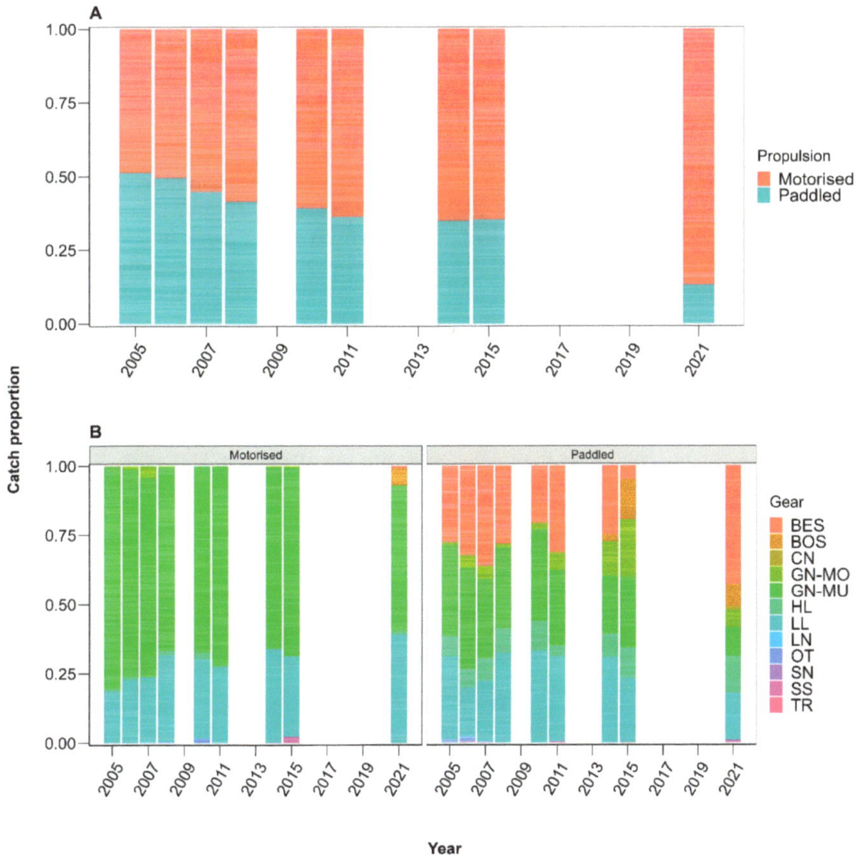

Figure 3. The proportion of Nile perch catches from vessels sampled in the study period by (**A**) vessel propulsion as paddled and motorised and (**B**) and gears; BES—Beach Seines, BOS—Boat Seines, CN—Cast nets, GN–MU—Multifilament gillnets, GN–MO—Monofilament gillnets, HL—Hand line hook, LL—Longlines, LN—Lift nets, MF—Monofilament, and OT—Other unidentified gears, SN—Scoop net, SS—Seine nets, TR—Basket traps.

Since 2015 there has been a marked reduction in the use of illegal gear and small unmotorised vessels, while the number of motorized vessels has continued to grow. There has been some increase in the use of legal gillnets, but the main increase has been in the use of legal-size hooks in both the artisanal and commercial fishery. The increased use of larger hooks has led to an increase in catches of large-sized Nile perch (Figure S2), as has also been reported in other studies [45,62,63]. This shift in the fishery has increased the demand for bait fish, which is likely the reason for the increased importance of beach seines in the artisanal fisheries (Figure 3B, [45]).

Increased effort targeting large Nile perch reflects the emergence of a lucrative fish maw business, where the economic returns are in favour of capturing the largest Nile perch individuals [45,64,65]. The Nile perch fishery, which is now mostly composed of motorised fishing vessels, does not only aim at targeting the high-value factory-sized Nile perch (≥50–85 cm TL), but much larger specimens for the fish maw trade which also appears to be attractive to the artisanal fishers with 30% of them operation long-lines and hand lines. Larger fish sizes are reported to have larger maws for instance the length of fresh maws for the 50 cm and 80 cm TL Nile perch individuals was estimated at 17 and 28.5 cm respectively, price per kg was also higher for larger maws at USD 210–270 while smaller maws were valued at USD 40–55 [45,64]. Thus, fishing for maws is economically attractive for fishers and other actors along the Nile perch value chain.

Nile Perch CPUE

A significant decline in the CPUE for the motorised gillnet Nile perch vessels was observed throughout the study period, from 4.5 kg panel^{-1} h fished at the beginning of the study period, to 3 kg panel^{-1} h fished in 2015, and further down to 1.8 kg panel^{-1} h fished in 2021. The CPUE for the artisanal gillnet fishers, who primarily use monofilament gillnets [56], declined from 4.2 kg panel h^{-1} fished in 2015 to 3.5 kg panel h^{-1} fished in 2021, but there has, however, not been a significant decline throughout the study (Figure 4, Tables S1 and S2), indicating that there has not been a detectable change in the density of small Nile perch in coastal waters. Small Nile perch of 10–50 cm TL are the most abundant (>80% of the sampled individuals) in the inshore areas of Lake Victoria, as measured in acoustic surveys [66]. A high abundance of small-sized Nile perch has long been reported on the lake [58,66,67]. Although the CPUE of monofilament gillnets has remained high, the importance of this gear in the Nile perch fisheries has dwindled as it only accounts for 1% of the total catches among all gear in 2021.

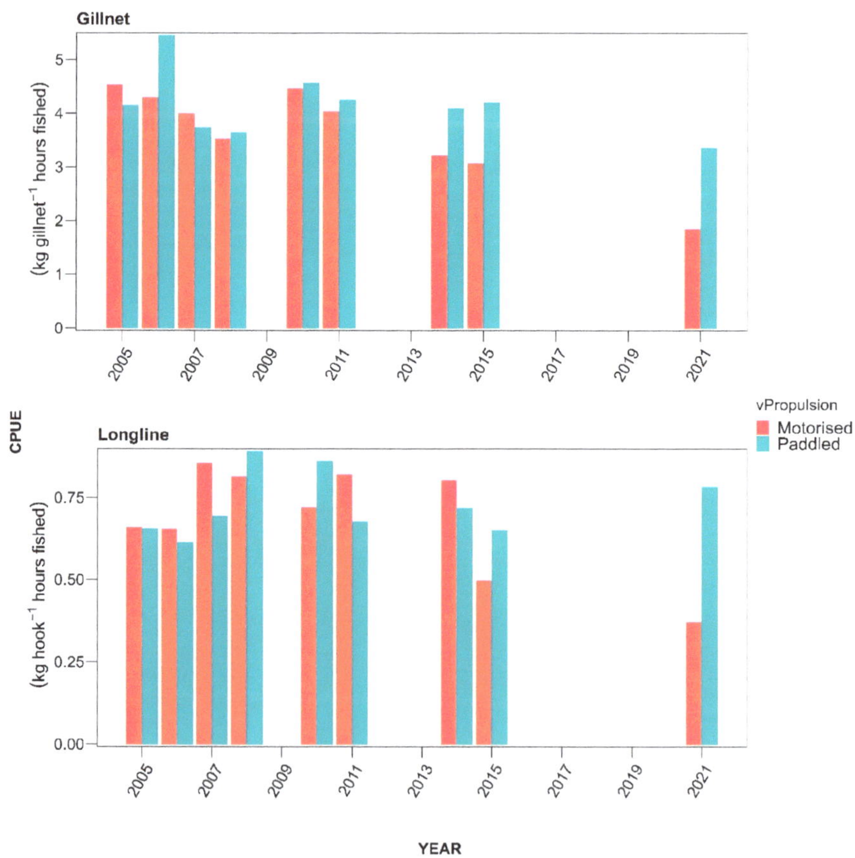

Figure 4. CPUE for Nile perch gears in kg per unit hours fished for motorised and paddled vessels for the catch assessment survey period 2005–2007, 2010–2011, 2014–2015, and 2021 [56].

Legal hooks dominated both paddled and motorised vessel groups at the beginning of the study period, with a shift to illegal hooks by the end of the BMU period. With the FPU setting in, a dominance of legal-size longline hooks was observed in the frame survey in 2020 and the catch assessment survey in 2021. (Table 2, Figures S1 and S2). A slight increase in the average CPUE was observed in the artisanal fishery from 2015 to 2021, and a reduction in the commercial fishery was not found to constitute a statistically significant difference ($p = 0.113$) (Table S2).

The CPUE in the longline fishery did not indicate a significant difference between vessel groups from the beginning of the study period and throughout the BMU period (Figure 4, Table S2). Changes over time in the use of illegal longline hooks were quite similar for both motorised and paddled vessels, which is also reflected by the similar size distribution of Nile perch harvested by motorised and paddled vessels (Figures S1 and S2).

5. Conclusions and Implications for Policy

The study examined the structure and effects of military law enforcement on the Nile perch fisheries on Lake Victoria in Uganda. The results show that the Nile perch fishery on Lake Victoria in Uganda is highly dynamic and responds rapidly to changes in the biological, social, and economic environment in which it operates.

After the first 5 years of military intervention, there has been an increased emphasis on the long-line fishery and catches of large Nile perch have increased substantially, both in absolute terms and regarding the proportion of larger fish caught. Exports of Nile perch increased rapidly from 2016 to 2019 but declined significantly in 2020 and then reached the lowest point in 2021. Although this can be attributed to the COVID-19 pandemic, it is also possible that national and regional markets have become more important, and that a substantial part of the catch is comprised of individuals that are too large for the factories.

The fishing rules that have been enforced were specified in Uganda's fisheries legislation; however, these rules foster more commercial fishery operations than subsistence ones. In the early 20th century, regulations, including the imposition of a minimum gillnet mesh size, were implemented to promote sustainable harvesting of commercial tilapia fisheries on Lake Victoria. However, these measures proved ineffective in preserving the native Oreochromis esculentus, the targeted tilapia species at that time. [67]. The Nile perch fleet alone makes up almost half of the fishing fleet on Lake Victoria in Uganda, and thus, to the commercial fleet owners and stakeholders along the value chain, the rules could be perceived as effective and legitimate. The institution and operation of the FPU in Uganda addressed the perceived "laxity" of co-management in enforcing rules, and the military has confiscated large numbers of small vessels and illegal gear since 2017. However, the results indicate that the goals to increase catches and exports relative to the BMU period may not be achieved in the long term. An increase in catch and exports was also realised in the early years of the BMU regime. Based on the findings of this study, there are no indications that the current regime has achieved the stated goals of increasing catch and exports relative to the preceding co-management period. A continued decline in catch-per-unit effort was observed for the motorized gillnet fleet, indicative of the overcapacity of the fleet. Driven by individual interests, the Nile perch stocks could suffer from resource degradation, as evidenced by the proportion of larger specimens being sought for economic benefits. It remains to be seen how the increased fishing pressure on the largest Nile perch may change the population structure, but most likely lead to reduced size at maturity and the maximum size attained [67]. Such a scenario would have serious implications for both the export of Nile perch and the maw trade.

On the side of the paddled vessels, we notice that in the first 4 years of military intervention, around 8000 small vessels were confiscated annually. Meanwhile, the use of illegal monofilament gillnets decreased, and most surveyed paddled vessels in 2021 were engaged in longlines and handlines to target large Nile perch or beach seines. This shift is likely due to the increased demand for bait fish in the expanding longline fishery. By 2020, the number of paddled vessels declined from approximately 17,000 in 2016 to 8500 in 2020, while almost 28,000 vessels were confiscated, indicating a persistent incentive to operate them. This is expected as no alternative incentive was given to the "illegal" artisanal fishers before the commencement of law enforcement.

The research provides valuable insights into the complex interactions involving military law enforcement on Lake Victoria, Uganda. It reveals the challenges associated with enforcing regulations, particularly in developing countries where the local socio-economic fabric is deeply intertwined with these vital natural resources. Two significant findings of the study are worth highlighting. Firstly, the study raises concerns about regulations inadvertently promoting excessive growth in fishing capacity in the commercial Nile perch sector. This, along with size-selective fishing, contributes to ongoing declines in catch levels of the commercial fleet. Secondly, the removal of smaller vessels implies a risk of social disruption in fishing communities, potentially causing the breakdown of social structures. This disruption may impact fisheries management, employment, and livelihoods. Hence, an assessment of the effectiveness of the current fishery laws concerning fish population dynamics and deliberate economic diversification may provide viable solutions to address these issues.

While our study has limitations, we acknowledge the need for future research to address these concerns. Diversifying interviewees to include fishers, local government actors, and fishing communities can provide a more comprehensive understanding of the social implications of military rule. Further research should explore the potential ecological effects of size-selective fishing and the necessity for revising fishery regulations considering the shifting fish population structure. Lake Victoria, a eutrophic lake, faces challenges exacerbated by a rapid increase in the human population in its catchment. Algal blooms in coastal areas lead to reduced dissolved oxygen, making these zones less suitable for larger fish [68–70]. Consequently, decreasing fishing effort in these areas may not necessarily increase larger fish. Although the study did not explore fish–environment interactions, such dynamics should be considered in formulating fisheries management plans. These efforts contribute to sustainable fishery management practices, which are crucial in addressing the complex issues related to law enforcement and fisheries.

Supplementary Materials: The following supporting information can be downloaded at: https://www.mdpi.com/article/10.3390/fishes8110563/s1, Figure S1: Legal/illegal hook size use between paddled and motorised gillnets for the study period based on catch assessment surveys in 2005–2008; 2010–2011; 2014–2015 and 2021, Figure S2: Comparison of the fish size distribution for the vessel groups; motorised and paddled and gears; gillnets and longlines; Table S1: ANOVA results for CPUE of the gillnets and longline vessel trips, Table S2: CPUE post hoc tests for gillnet and longline vessel trip comparisons.

Author Contributions: V.M.; conceptualization, methodology, investigation, visualization, writing—original draft: T.T.; supervision, funding acquisition, methodology, writing—original draft: D.M.K.; data curation, supervision, and writing—original draft: J.G.P.; conceptualization, methodology, data curation, supervision writing—original draft. All authors have read and agreed to the published version of the manuscript.

Funding: This research received no external funding.

Institutional Review Board Statement: The study did not involve any animal or human experimentation, and thus the declaration statement, approval code and date from the University of Iceland and Helsinki declaration are not applicable to this article.

Informed Consent Statement: Informed consent was obtained from all subjects involved in the study.

Data Availability Statement: Data for the study have restricted access; however, considerable explanation could be made available upon request from the corresponding author.

Acknowledgments: All staff of the FPU who participated in the interviews, staff of the National Fisheries Resources Research Institute, Directorate of Fisheries Resources and Lake Victoria Fisheries Organization that contributed to the collection of data used in the study are hereby acknowledged. Special thanks to Anthony Taabu Munyaho for the invaluable support rendered concerning data collection with the FPU, Nakiyende Herbert for providing us with guidance and feedback on the paper, and Bairon Mugenyi and Basooma Anthony. Lastly, we would like to express our gratitude to the UNESCO-GRO Fisheries Training Program that provided funding to the corresponding author for PhD studies at the University of Iceland.

Conflicts of Interest: The authors declare no conflict of interest.

References

1. Andrew, N.L.; Béné, C.; Hall, S.J.; Allison, E.H.; Heck, S.; Ratner, B.D. Diagnosis and management of small-scale fisheries in developing countries. *Fish Fish.* **2007**, *8*, 227–240. [CrossRef]
2. Nielsen, J.R.; Degnbol, P.; Viswanathan, K.K.; Ahmed, M.; Hara, M.; Abdullah, N.M.R. Fisheries co-management—An institutional innovation? Lessons from South East Asia and Southern Africa. *Mar. Policy* **2004**, *28*, 151–160. [CrossRef]
3. Nunan, F.; Hara, M.; Onyango, P. Institutions and Co-Management in East African Inland and Malawi Fisheries: A Critical Perspective. *World Dev.* **2015**, *70*, 203–214. [CrossRef]
4. Béné, C.; Belal, E.; Baba, M.O.; Ovie, S.; Raji, A.; Malasha, I.; Njaya, F.; Na Andi, M.; Russell, A.; Neiland, A. Power Struggle, Dispute and Alliance Over Local Resources: Analyzing 'Democratic' Decentralization of Natural Resources through the Lenses of Africa Inland Fisheries. *World Dev.* **2009**, *37*, 1935–1950. [CrossRef]
5. Smith, H.; Basurto, X. Defining Small-Scale Fisheries and Examining the Role of Science in Shaping Perceptions of Who and What Counts: A Systematic Review. *Front. Mar. Sci.* **2019**, *6*, 236. [CrossRef]
6. Branch, G.M.; Hauck, M.; Siqwana-Ndulo, N.; Dye, A.H. Defining fishers in the South African context: Subsistence, artisanal and small-scale commercial sectors. *Afr. J. Mar. Sci.* **2002**, *24*, 475–487. [CrossRef]
7. Fulgencio, K. Globalisation of the Nile perch: Assessing the socio-cultural implications of the Lake Victoria fishery in Uganda. *Afr. J. Political Sci. Int. Relat.* **2009**, *3*, 433–442.
8. Beuving, J. Spatial diversity in small scale fishing: A socio-cultural interpretation of the Nile perch sector on Lake Victoria, Uganda. *Tijdschr. Voor Econ. En. Soc. Geogr.* **2015**, *106*, 1–16. [CrossRef]
9. Cohen, P.J.; Allison, E.H.; Andrew, N.L.; Cinner, J.; Evans, L.S.; Fabinyi, M.; Garces, L.R.; Hall, S.J.; Hicks, C.C.; Hughes, T.P.; et al. Securing a just space for small-scale fisheries in the blue economy. *Front. Mar. Sci.* **2019**, *6*, 171. [CrossRef]
10. Calabrese, L.; Tang, X. Africa's Economic Transformation: The Role of Chinese Investment. Available online: https://degrp.odi.org/publication/africas-economic-transformation-the-role-of-chinese-investment/ (accessed on 23 January 2022).
11. Duffy, R.; Massé, F.; Smidt, E.; Marijnen, E.; Büscher, B.; Verweijen, J.; Ramutsindela, M.; Simlai, T.; Joanny, L.; Lunstrum, E. Why we must question the militarization of conservation. *Biol. Conserv.* **2018**, *232*, 66–73. [CrossRef]
12. Massé, F.; Lunstrum, E.; Holterman, D. Linking green militarization and critical military studies. *Crit. Mil. Stud.* **2018**, *4*, 201–221. [CrossRef]
13. Mpomwenda, V.; Kristófersson, D.M.; Taabu-Munyaho, A.; Tómasson, T.; Pétursson, J.G. Fisheries management on Lake Victoria at a cross-roads: Assessing fishers' perceptions on future management options in Uganda. *Fish. Manag. Ecol.* **2021**, *2*, 196–211. [CrossRef]
14. Nunan, F.; Cepić, D.; Yongo, E.; Salehe, M.; Mbilingi, B.; Odongkara, K.; Onyango, P.; Mlahagwa, E.; Owili, M. Compliance, corruption and co-management: How corruption fuels illegalities and undermines the legitimacy of fisheries co-management. *Int. J. Commons* **2018**, *12*, 58–79. [CrossRef]
15. Kantel, A.J. Fishing for Power: Incursions of the Ugandan Authoritarian State. *Ann. Am. Assoc. Geogr.* **2019**, *109*, 443–455. [CrossRef]
16. Nunan, F. The political economy of fisheries co-management: Challenging the potential for success on Lake Victoria. *Glob. Environ. Chang.* **2020**, *63*, 102101. [CrossRef]
17. BOU. Composition of Exports. Available online: https://www.bou.or.ug/bouwebsite/PaymentSystems/dataandstat.html (accessed on 22 March 2022).
18. Kjær, A.M. Political Settlements and Productive Sector Policies: Understanding Sector Differences in Uganda. *World Dev.* **2015**, *68*, 230–241. [CrossRef]
19. Kjær, A.M.; Muhumuza, F.; Mwebaze, T.; Katusiimeh, M. *The Political Economy of the Fisheries Sector in Uganda: Ruling Elites, Implementation Costs and Industry Interest*; DIIS Working Paper: Copenhagen, Denmark, 2012. Available online: https://nru.uncst.go.ug/handle/123456789/7612 (accessed on 15 February 2023).
20. Mkumbo, O.C.; Marshall, B.E. The Nile perch fishery of Lake Victoria: Current status and management challenges. *Fish. Manag. Ecol.* **2015**, *22*, 56–63. [CrossRef]
21. Arthur, R.I. Small-scale fisheries management and the problem of open access. *Mar. Policy* **2020**, *115*, 103867. [CrossRef]
22. Funge-Smith, S. *Review of the State of the World Fishery Resources: Inland*; FAO (Fish and Aquaculture Organisation): Rome, Italy, 2018.
23. FAO. *Voluntary Guidelines for Securing Sustainable Small-Scale Fisheries in the Context of Food Security and Poverty Eradication*; FAO (Fisheries and Aquaculture Organization): Rome, Italy, 2015; p. 34.
24. Ostrom, E. *Governing the Commons: The Evolution of Institutions for Collective Action*; Cambridge University Press: New York, NY, USA, 1990.
25. North, D.C.; Wallis, J.J. Integrating Institutional Change and Technical Change in Economic History A Transaction Cost Approach. *J. Institutional Theor. Econ. (JITE)/Z. Die Gesamte Staatswiss* **1994**, *150*, 609–624.

26. Hanna, S.; Munasinghe, M. *Property Rights and the Environment: Social and Ecological Issues*; Beijer Institute of Ecological Economics and World Bank: Washington, DC, USA, 1995.
27. Dletz, T.; Ostrom, E.; Stern, P.C. The struggle to govern the commons. *Science* **2003**, *302*, 1907–1918. [CrossRef]
28. Nielsen, R.J. An analytical framework for studying: Compliance and legitimacy in fisheries management. *Mar. Policy* **2003**, *27*, 425–432. [CrossRef]
29. Hauck, M.; Kroese, M. Fisheries compliance in South Africa: A decade of challenges and reform 1994–2004. *Mar Policy* **2006**, *30*, 74–83. [CrossRef]
30. Hauck, M. Rethinking small-scale fisheries compliance. *Mar. Policy* **2008**, *32*, 635–642. [CrossRef]
31. Chuenpagdee, R.; Jentoft, S. Exploring challenges in small-scale fisheries governance. In *Interactive Governance for Small-Scale Fisheries: Global Reflections*; Jentoft, S., Chuenpagdee, R., Eds.; Springer: Berlin/Heidelberg, Germany, 2015; Volume 3, pp. 3–16.
32. Hardin, G. The tragedy of the commons. *Science* **1968**, *162*, 1243–1247. [CrossRef]
33. Purcell, S.W.; Pomeroy, R.S. Driving small-scale fisheries in developing countries. *Front. Mar. Sci.* **2015**, *2*, 44. [CrossRef]
34. Kolding, J.; Van Zwieten, P.A.M. The Tragedy of Our Legacy: How do Global Management Discourses Affect Small Scale Fisheries in the South? *Forum Dev. Stud.* **2011**, *38*, 267–297. [CrossRef]
35. Jul-Larsen, E. Analysis of effort dynamics in the Zambian inshore fisheries of Lake Kariba. In *Management, Co-Management or No Management ? Major Dilemmas in Southern African Freshwater Fisheries*; Jul-Larsen, E., Kolding, J., Overå, R., Nielsen, R.J., van Zwieten, P.A.A.M., Eds.; Food & Agriculture Orgnization: Rome, Italy, 2003; p. 271.
36. Ogutu-Ohwayo, R.; Wandera, S.B.; Kamanyi, I.R. Fishing gear selectivity for Lates niloticus L., Oreochromis niloticus L. and Rastrineobola argentea P. in Lakes Victoria, Kyoga and Nabugabo. *Conserv. Biol.* **1998**, *7*, 33–38.
37. Malasha, I. Colonial and postcolonial fisheries regulations: The cases of Zambia and Zimbabwe. In *Management, Co-Management, or No Management? Major Dilemmas in Southern African Freshwater Fisheries*; Fisheries Technical paper 2; Jul-Larsen, E., Kolding, J., Overå, R., Nielsen, R.J., van Zwieten, P.A.M., Eds.; Food & Agriculture Orgnization: Rome, Italy, 2003; pp. 253–266.
38. Agrawal, A. Common property institutions and sustainable governance of resources. *World Dev.* **2001**, *29*, 1649–1672. [CrossRef]
39. Cepić, D.; Nunan, F. Justifying non-compliance: The morality of illegalities in small scale-fisheries of Lake Victoria, East Africa. *Mar. Policy* **2017**, *86*, 104–110. [CrossRef]
40. Van Hoof, L. Co-management: An alternative to enforcement? *ICES J. Mar. Sci.* **2010**, *67*, 395–401. [CrossRef]
41. Nunan, F. Wealth and welfare? Can fisheries management succeed in achieving multiple objectives? A case study of Lake Victoria, East Africa. *Fish Fish.* **2014**, *15*, 134–150. [CrossRef]
42. Lawrence, T.J.; Watkins, C. It takes more than a village: The challenges of co-management in Uganda's fishery and forestry sectors. *Int. J. Sustain. Dev. World Ecol.* **2012**, *19*, 144–154. [CrossRef]
43. Sjöstedt, M.; Linell, A. Cooperation and coercion: The quest for quasi-voluntary compliance in the governance of African commons. *World Dev.* **2021**, *139*, 105333. [CrossRef]
44. Mafuta, C.; Levi, W.; Lake Victoria Basin Commission. *Lake Victoria Basin: Atlas of our Changing Environment*; GRID-Arendal: Arendal, Norway, 2017. Available online: https://policycommons.net/artifacts/2390339/lake-victoria-basin/3411606/ (accessed on 17 July 2023).
45. Bagumire, A.; Muyanja, C.K.; Kiboneka, F.W. Report for The Value Chain Analysis of Nile Perch Maw Trade in East Africa. In *The Responsible Fisheries Business Chains Project of Deutsche Gesellschaft für Internationale Zusammenarbeit (GIZ) under Contract, 83285575*; LVFO (Lake Victoria Fisheries Organisation): Jinja, Uganda, 2018; pp. 1–52. Available online: www.foodsafetyltd.com (accessed on 6 March 2023).
46. Creswell, J.W. *Research Design; Qualitative, Quantitative and Mixed Methods Approaches*, 3rd ed; SAGE Publications: London, UK, 2009.
47. Newing, H.; Eagle, C.M.; Puri, R.K.; Watson, C.W. *Conducting Research in Conservation Social Science: A Social Science Perspective*, 1st ed.; Taylor & Francis Group: New York, NY, USA, 2011.
48. Léopold, M.; Guillemot, N.; Rocklin, D.; Chen, C. A framework for mapping small-scale coastal fisheries using fishers' knowledge. *ICES J. Mar. Sci.* **2014**, *71*, 1781–1792. [CrossRef]
49. Samy-Kamal, M.; Teixeira, C.M. Diagnosis and Management of Small-Scale and Data-Limited Fisheries. *Fishes* **2023**, *8*, 1. [CrossRef]
50. LVFO. *The Standard Operating Procedures for Catch Assessment Surveys*; LVFO (Lake Victoria Fisheries Organisation): Jinja, Uganda, 2021.
51. LVFO. *Standard Operating Procedures (SOPs) for Catch Assessment Surveys on Lake Victoria*; LVFO Standard Operating Procedures: Jinja, Uganda, 2007; No. 3.
52. UBOS. *2022 Uganda Statistical Abstract*; UBOS (Uganda Bureau of Statistics): Kampala, Uganda, 2022. Available online: https://www.ubos.org (accessed on 21 March 2023).
53. RStudio Team. *RStudio: Integrated Development Environment for R. Rstudio*; PBC: Boston, MA, USA, 2020. Available online: http://www.rstudio.com/ (accessed on 19 May 2021).
54. Wickham, H.; Wickham, H. *Data Analysis*; Springer: Berlin/Heidelberg, Germany, 2016.
55. Njiru, M.; Getabu, A.; Taabu, A.M.; Mlaponi, E.; Muhoozi, L.; Mkumbo, O.C. Managing Nile perch using slot size: Is it possible? *Afr. J. Trop. Hydrobiol. Fish.* **2010**, *12*, 9–14. [CrossRef]

56. Mpomwenda, V.; Tomasson, T.; Pétursson, J.G.; Taabu-munyaho, A.; Nakiyende, H.; Kristófersson, D.M. Adaptation Strategies to a Changing Resource Base: Case of the Gillnet Nile Perch Fishery on Lake Victoria in Uganda. *Sustainability* **2022**, *14*, 2376. [CrossRef]
57. Kayanda, R.; Taabu, A.; Tumwebaze, R.; Muhoozi, L.; Jembe, T.; Mlaponi, E.; Nzungi, P. Status of the Major Commercial Fish Stocks and Proposed Species-specific Management Plans for Lake Victoria. *Afr. J. Hydrobiol. Fish.* **2008**, *12*, 15–21. [CrossRef]
58. Nyamweya, C.S.; Natugonza, V.; Kashindye, B.B.; Mangeni-Sande, R.; Kagoya, E.; Mpomwenda, V.; Mziri, V.; Elison, M.; Mlaponi, E.; Ongore, C.; et al. Response of fish stocks in Lake Victoria to enforcement of the ban on illegal fishing: Are there lessons for management? *J. Great Lakes Res.* **2023**, *49*, 531–544. [CrossRef]
59. Njiru, M.; Van der Knaap, M.; Taabu-Munyaho, A.; Nyamweya, C.S.; Kayanda, R.J.; Marshall, B.E. Management of Lake Victoria fishery: Are we looking for easy solutions ? *Aquat. Ecosyst. Health Manag.* **2014**, *17*, 70–79. [CrossRef]
60. UFPEA. *Uganda Fish Processors and Exporters Association: E-Newsletter; UFPEA*; no. August; Uganda Fish Processors and Exporters Association: Kampala, Uganda, 2019; p. 17. Available online: https://drive.google.com/file/d/1l44CVnOs03hnIF8eIeFgtv0ijjdpbuCV/view (accessed on 6 July 2023).
61. Diyamett, L.D.; Mbilingi, B.; Mutambala, M. *COVID-19 Pandemic in Developing Economies: Lockdown versus No Lockdown Scenarios for the Fisheries*; Southern Voice: Dar es Salaam, Tanzania, 2021.
62. Msuku, B.S.; Mrosso, H.D.J.; Nsinda, P.E. A critical look at the current gillnet regulations meant to protect the Nile Perch stocks in Lake Victoria. *Aquat. Ecosyst. Health Manag.* **2011**, *14*, 252–259. [CrossRef]
63. Mkumbo, O.C.; Mlaponi, E. Impact of the baited hook fishery on the recovering endemic fish species in Lake Victoria. *Aquat. Ecosyst. Health Manag.* **2007**, *10*, 458–466. [CrossRef]
64. Ben-Hasan, A.; de Mitcheson, Y.S.; Cisneros-Mata, M.A.; Jimenez, A.; Daliri, M.; Cisneros-Montemayor, A.M.; Nair, R.J.; Thankappan, S.A.; Walters, C.J.; Christensen, V. China's fish maw demand and its implications for fisheries in source countries. *Mar. Policy* **2021**, *132*, 104696. [CrossRef]
65. de Mitcheson, Y.S.; To, A.W.L.; Wong, N.W.; Kwan, H.Y.; Bud, W.S. Emerging from the murk: Threats, challenges and opportunities for the global swim bladder trade. *Rev. Fish. Biol. Fish.* **2019**, *29*, 809–835. [CrossRef]
66. LVFO. *A Report of the Lake-Wide Hydroacoustic Survey on Lake Victoria 15th September–13th October 2019*; LVFO (Lake Victoria Fisheries Organisation): Jinja, Uganda, 2019.
67. Natugonza, V.; Nyamweya, C.; Sturludóttir, E.; Musinguzi, L.; Ogutu-Ohwayo, R.; Bassa, S.; Mlaponi, E.; Tomasson, T.; Stefansson, G. Spatiotemporal variation in fishing patterns and fishing pressure in Lake Victoria (East Africa) in relation to balanced harvest. *Fish. Res.* **2022**, *252*, 106355. [CrossRef]
68. Downing, A.S.; van Nes, E.H.; Balirwa, J.S.; Beuving, J.; Bwathondi, P.; Chapman, L.J.; Cornelissen, I.J.M.; Cowx, I.G.; Goudswaard, K.P.C.; Hecky, R.E.; et al. Coupled human and natural system dynamics as key to the sustainability of Lake Victoria's Ecosystem services. *Ecol. Soc.* **2014**, *19*, 31–49. [CrossRef]
69. Nyamweya, C.S.; Natugonza, V.; Taabu-Munyaho, A.; Aura, C.M.; Njiru, J.M.; Ongore, C.; Mangeni-Sande, R.; Kashindye, B.B.; Odoli, C.O.; Ogari, Z.; et al. A century of drastic change: Human-induced changes of Lake Victoria fisheries and ecology. *Fish. Res.* **2020**, *230*, 105564. [CrossRef]
70. Aura, C.M.; Musa, S.; Yongo, E.; Okechi, J.K.; Njiru, J.M.; Ogari, Z.; Wanyama, R.; Charo-Karisa, H.; Mbugua, H.; Kidera, S.; et al. Integration of mapping and socio-economic status of cage culture: Towards balancing lake-use and culture fisheries in Lake Victoria, Kenya. *Aquac. Res.* **2018**, *49*, 532–545. [CrossRef]

fishes

Article

Research on Legal Risk Identification, Causes and Remedies for Prevention and Control in China's Aquaculture Industry

Chang Xu [1], Yang Liu [2] and Zhaobin Pei [2,*]

[1] Graduate School, Korea Maritime & Ocean University, Busan 49112, Republic of Korea; xuchangdlou@163.com
[2] School of Marine Law and Humanities, Dalian Ocean University, Dalian 116023, China; liuyang2255@dlou.edu.cn
* Correspondence: pzb@dlou.edu.cn; Tel.: +86-0411-84762119

Abstract: Aquatic products fulfill the protein needs of people and play an important role in food safety. And aquaculture is prized for its high productivity, sustainability and environmental friendliness. Considering the importance of aquaculture, the legal risks exposed during the aquaculture process deserve attention in order to prevent them from hindering the development of the aquaculture industry. Through online research, literature analysis and practical communication, it is shown that the current legal risks with commonalities include land use violations, lack of legal documents, failure to meet tailing water criteria, unquarantined fry and misuse of prohibited agricultural pharmaceuticals through online research, literature analysis and practical communication. By analyzing the reasons for the formation of legal risks and combining the experiences in sustainable development of three major aquaculture countries, which are Korea, Norway and Chile, this paper provides targeted preventive remedies and suggestions for aquaculture operators, administrative parties, legislators and other parties on legal risks. It includes promoting the improvement of the rule of law in multiple aspects, clarifying the positioning of the aquatic breeding certificates, improving and propagating the standards for wastewater discharge, increasing the self-sufficiency rate of aquatic fry and fingerlings, as well as making use of the synergy of soft law and hard law.

Keywords: aquaculture; legal risks; land use regulation; the right to use sea areas; wastewater discharge

Key Contribution: First, this paper emphasizes the impact of the law on the aquaculture economy. Focusing on legal risks that are of low concern and providing remedies for prevention and control can help prevent aquaculture operators from undermining their production and development dynamics due to the loss of financial interests, which would consequently affect the market and the industry. Second, based on the analysis of common legal risks in practice, this paper responds to the needs of foreign scholars who want to learn about the current situation in China and provides warnings and references for other countries that are developing aquaculture industries. Third, this paper focuses on the transition between former and new legal provisions and the separation between law and practice and uncovers a phenomenon that has received limited attention but is in urgent need of remedy. It includes the legal connection in terms of property rights, the update of regulatory provisions for the discharge of wastewater and the illegal use of non-pharmaceuticals. It also provides doctrinal explanations of disputes over aquatic breeding certificates and offers complementary recommendations for the issue of quarantine of aquatic fry or fingerlings beyond the general perspective. This helps highlight the important role of the law in the green development of the aquaculture industry.

Citation: Xu, C.; Liu, Y.; Pei, Z. Research on Legal Risk Identification, Causes and Remedies for Prevention and Control in China's Aquaculture Industry. *Fishes* **2023**, *8*, 537. https://doi.org/10.3390/fishes8110537

Academic Editor: Ben Drakeford

Received: 19 September 2023
Revised: 24 October 2023
Accepted: 27 October 2023
Published: 29 October 2023

1. Introduction

Fish provide 17% of animal protein and 7% of total protein consumed globally. Moreover, based on an analysis of potential mariculture production, it is projected that oceans can provide almost two-thirds of the total protein demand of the world population [1]. It is predicted that production from the oceans could provide nearly two-thirds of the total protein needs of the world's population, based on an analysis of potential mariculture production. The fisheries and aquaculture sectors have been increasingly recognized for their essential contributions to global food security and nutrition in the twenty-first century. However, the FAO's long-term monitoring of assessed marine fishery stocks confirms that marine fishery resources have continued to decline [2]. Consequently, aquaculture, both mariculture and freshwater aquaculture, has been suggested as a possible alternative for fisheries towards mitigating food security problems and preventing the loss of wild fish supplies [3]. According to the data provided by the FAO for the period 2017–2020, it is known that aquaculture accounts for about half of the world's fish supply and is expected to grow further, which makes it an important part of the supply of high-quality protein for the global population [4]. Moreover, aquatic foods have a lower environmental footprint than other animal-sourced proteins [5]. Joseph Poore et al. suggest that farmed fish have been estimated to have 87% smaller carbon footprints than beef, use 49% less land than poultry, and require 84% less stress-weighted fresh water than pigs [6]. In conclusion, aquaculture is prized for its high productivity, sustainability and environmental friendliness.

However, there are legal risks associated with the production stage of aquaculture, which directly affect the production and development viability of aquaculture operators and consequently affect the expansion of the industry. At the production stage, aquaculture operators are inclined to focus on technical issues directly related to actual output and trends towards market prices, while paying insufficient attention to legal issues. Currently, increasing legal provisions for administrative orders have diluted their close relationship with morality. It is difficult for aquaculture operators to fulfil legal obligations beyond their perception based on general public morality in the absence of an objective and comprehensive knowledge of the law. And that is where the legal risk exists. Whereas the perception of illegality is the basis for the establishment of a crime in the liability doctrine, the establishment of an administrative penalty does not require illegality or subjective intent as a necessary condition. In other words, aquaculture operators are responsible for legal risks that they cannot identify.

The resulting significant losses, including the clearance of fishponds and high fines, have affected operators' ability to continue their production and development, with further impacts on the market and industry. Legal risks, on the other hand, arise internally and are, therefore, controllable. Therefore, there is an important practical significance on how to identify, analyze and prevent the legal risks in the aquaculture field for the economic interests of the operators, the stability of the market, and the expansion of the industry. This article's theoretical significance also lies in the fact that it takes into account the inability of many aquaculture and food policy scholars to access literature in the Chinese language [7]. By researching the basis of the current status of China, this paper is able to increase scholars' understanding of the aquaculture aspect of Chinese aquacultural practices. And it also takes into account the current state of affairs in China, where 57% of the total aquaculture volume and 59% of the global value of its output is produced, and where changes in its policies can have a significant impact on the state of aquaculture in the world [2]. This paper analyzes the legal risks prevalent throughout China to be able to provide a warning to other countries interested in developing aquaculture and may even be able to provide some new legal insights for other countries in terms of preventive and control measures.

Based on the literature, it is clear that the above impacts of law on the aquaculture economy have not received sufficient attention in the academic field. As a whole, up to September 2023, 8185 articles can be searched by using Web of Science as the search platform and "aquaculture" and "risk" as the keywords. Its content focuses on natural disciplines, such as environmental sciences, fisheries, marine freshwater biology, and so on, with less literature on the social sciences. In particular, there are only five articles under the law classification. In the following, the research will adopt the methods of literature analysis and comparative study on the basis of analyzing the relevant results of domestic and foreign scholars. On a specific level, respectively, legal risks in aquaculture mainly exist in property rights, licensing, discharge, fry or fingerlings and pharmaceuticals.

First, as a matter of consensus, stable property rights are the legal basis for developing long-term aquaculture and achieving economic growth [8–12]. And the reason for the instability of property rights is that the value of aquaculture is in conflict with other values. As noted by the FAO in its review of aquaculture development in the Near East and North Africa, conflicts of interest can occur among different authorities involved in governance and regulation, which may lead to poor management, strategies and policies [13]. In Norway, this conflict also exists between coastal zones and aquaculture management [14,15]. In this regard, policy integration can reduce the incidence of conflict [16]. And in the face of national issues in food security, the value of aquaculture (mainly freshwater aquaculture on land) takes a back seat and cannot be safeguarded. For example, Myanmar, which now ranks among the top 10 global aquaculture producers, has strictly enforced laws against converting rice fields to fishponds in smallholder areas, despite potential employment and income gains [17–19].

With regard to the risk of unsecured property rights resulting from low-value parity in aquaculture, the existing literature has focused on providing advice on site selection to avoid negative policy-induced impacts [20–24]. Little attention has been paid to the interface between the former and new laws or policies. Therefore, this paper will provide additional research in a targeted manner.

Second, access permits are the legal threshold for engaging in aquaculture. Due to the overlap in geographic space, the right to use waters and mudflats for aquaculture (including mariculture and freshwater aquaculture) is manifested as the simultaneous utilization of the right to use the sea areas and the right to practice aquaculture in the case of mariculture and as the simultaneous utilization of the right to use the lands contracted for management and the right to practice aquaculture in the case of freshwater aquaculture. Acquiring the right to use non-private sea areas or lands becomes, in essence, a license for access to aquaculture. The particularity of mariculture is that the operator is required by law to obtain both a certificate for the right to use the sea area and a certificate for aquaculture.

Some Chinese scholars have raised objections to such provisions. Jianyuan believes that different rights existing in the same sea area are prone to conflict and that the system of rights to use the sea areas should be abolished [25]. Hui believes that "double certificates" are duplicated contrary to the principle of efficiency, and it should be legal aquaculture even if the operator only has the certificate of the right to use the sea area [26]. Wanzhong believes that the conflict between the two rights raises another conflict of interest between the subject of the right to use the sea area and the subject of the right to participate in aquaculture, which results in a chaotic status quo of the management system of sea use for aquaculture [27].

Some scholars have raised objections. Ying believes that the certificates of the rights to use the sea areas confer the right to use the sea areas on the users, while the aquaculture certificate is a license for the industry, and the two certificates could co-exist at the same time in practice [28]. And in the face of the issues in practice, the abolition of certificates conveying rights to use the sea areas is not an effective way. Lifeng suggested that the current role of aquaculture certificates has not been effectively played, mainly because of the unclear relationship between the nature of aquaculture certificates and other certificates, which needs to be clarified in the legislation and issuance of certificates [29]. Shiling and others proposed that the violations should be distinguished into administrative violations and civil violations, and the right to use the sea areas for aquaculture and the property rights of aquaculture should be distinguished in the trial process and reflected in the judgment results [30].

The relevant literature focuses on the domestic context, as the system of the right to use sea areas is unique to China. However, there is also literature abroad that addresses the problem of the unclear expression of provisions. Engle argues that if regulations are complicated and lack clarity, aquaculture producers must spend more time and financial resources identifying what rules apply and how to comply with them [31]. It was a common view that complex provisions would undermine the effectiveness of implementation [32–35]. In support of this view, this paper focuses on the specific system in China and clarifies the law's dual-certification requirements for aquaculture, which will help operators to carry out their activities and help scholars from other countries conduct comparative research in the case of China.

Third, the strength of the regulation of wastewater is related to the sustainability of the aquaculture industry. The law, as a regulation of behavior, should update the regulatory measures in concert with the development of practice. For example, for the Contaminants of Emerging Concern (CECs), a number of countries have taken proactive measures. The EPA developed a Contaminant Candidate List (CCL) of chemicals that may contaminate public drinking water [36,37]. The European Environment Agency (EEA) assists with relevant assessments to improve the environment in European Union (EU) member states [38]. Chinese scholars also call for a new response based on existing provisions. The internal structure of the mariculture sector in China is changing fast [39], and the laws should be implemented and updated in line with the changes in the structure of mariculture [40]. Lack of effective monitoring and legislation on effluent discharge are the main bottlenecks that are currently limiting appropriate aquaculture site selection and carrying capacity management in China [41].

To update the regulations, Wang suggested that the relevant authorities need to establish standards, including emission limit values (ELVs) and environmental quality standards (EQSs), in the mixing zone for important contaminants present in the effluent, which apply to the water quality at the end of point-source and within the receiving environment, respectively [42]. For monitoring after standards have been established, Ottinger et al. proposes expanding the monitoring area by using remote sensing techniques to acquire and analyze environmental data at various spatial and temporal scales [43]. This paper will also put forward institutional proposals in response to this issue, so as to alleviate legal lags.

Fourth, aquatic fry and fingerlings may carry viruses or pathogens, which may adversely affect production safety, personal health and the environment.

It is important to detect the ability of the host to carry the virus at an early stage so that the continued spread of the virus can be prevented [44–46]. Existing perspectives from abroad focus on strengthening import quarantine. Ortega proposed in his study to develop stricter prevention methods to prevent the introduction of diseases, even while acknowledging the dependence of Mexican rainbow trout aquaculture on imported trout eggs [47]. Based on the importance of aquatic fry and fingerlings, the Animal Health Protection Act (AHPA) in the U.S. gives the United States Department of Agriculture (USDA) a great deal of discretion in dealing with disease [48]. And using its discretionary power provided by the AHPA, the USDA could limit or ban imports, exports and interstate movement; impose importation quarantines; or order the destruction of certain imported aquatic species, their parts, and articles [49]. The environmental threats of non-native aquatic animals motivated the European Commission to pass. Council Regulation No. 708/2007, which requires that approval to introduce non-native species be supported by a risk assessment [50]. However, quarantine cannot offer a high likelihood of disease detection. Whittington et al. even suggest limiting the number of ornamental species and their country of origin [51].

In China, the self-sufficiency of some imported aquatic fry and fingerlings, such as South American white shrimp, is low [52,53]. One of the incentives for operators to avoid quarantine procedures is the mismatch between the quantity of imported aquatic fry and fingerlings brought in and the efficiency of quarantine. Therefore, this paper starts with the origin of the issue and offers suggestions beyond mainstream viewpoints to provide an adjunctive role.

Fifth, aquaculture pharmaceuticals are an unavoidable means of production in the aquaculture process. Although antibiotics are illegal, other uses are common in the aquaculture industry [54]. For example, antibiotics banned for use in livestock feed in the United States are misused as growth promoters and prophylactics to avoid disease in fish mangrove production in Vietnam [55–58]. Albert Tacon suggested that animal feed contamination (including veterinary drug residues) could be passed along the food chain to consumers through contaminated aquaculture products [59]. In response, the *Code of Practice for Fish and Fishery Products* issued by WHO contains provisions such as that products should be registered with the appropriate national authority and that those should only be prescribed or distributed by authorized personnel authorized [60]. However, in China, the illegal use of aquaculture pharmaceuticals can also be seen in their illegal use under the name of non-pharmaceuticals [61]. This results in massive mortality of aquatic creatures [62,63], and even safety hazards [64].

2. Legal Risk Identification in Aquaculture

2.1. Status of Administrative Penalties in Aquaculture

Article 329 of the *Civil Code of the People's Republic of China* provides that mineral exploration rights, mining rights, water intake rights and the right to use water areas or intertidal zones for aquaculture or fishery, which are obtained in accordance with the law, shall be protected by law [65]. However, the civil right to practice aquaculture is also subject to administrative law. If an aquaculture operator violates the relevant legal provisions, it would be liable, even if the violation was not subjectively intentional, but rather due to ignorance of the law. Considering that aquaculture operators may not possess complete legal knowledge and professional capacity, coupled with minimal legal literacy campaigns and policy changes, there is a possibility that aquaculture operators are aware of the legal provisions only after they have been held legally liable. The current situation is analyzed below based on the number of administrative penalties.

As of 8 September 2023, a total of 5017 decisions on administrative penalties were displayed on the PKULaw platform by searching for settings in which the name of the target of the penalty included "aquaculture". According to the classification of penalty results, there are 2145 cases of fines, confiscation of illegal gains, and confiscation of illegal property, 1608 cases of temporarily detaining the license, lowering the qualification level, and revoking the license; and 522 cases of restricting the development of production and business operation activities, ordering suspension of production and business, ordering closure, and restricting employment. According to categorization by topics, a total of 1781 cases are related to market supervision, 120 cases are related to agriculture, forestry, animal husbandry and fisheries, and 27 cases are related to food products, medicines and medical treatments.

2.2. Legal Risks in Aquaculture

Based on the above search content and combined with communication with practitioners, the following five legal risks can be summarized for aquaculture, including freshwater aquaculture and marine aquaculture.

2.2.1. Legal Risk I: Property Rights Legal Risk

Contrary to fishing, which involves the capture of fish in a public, open location, aquaculture involves the preservation of the production infrastructure and ownership of the products produced. In order for markets to operate effectively and for economies to run efficiently, clear property rights are considered to be essential [66]. The issue of land used for aquaculture is highly influenced by legal policies (the risks associated with land-use regulation are examined here, whereas the functional zoning of the oceans is more stable; therefore, mariculture is not included in this section). In particular, the use of land, including the use of cropland for any purpose other than agriculture, is one of the focuses of government efforts to implement regulations [67]. However, policies are influenced by the reality of unstable environments and can change. Operators will bear the legal risk of being ordered to demolish the aquaculture facilities for a limited time or being charged with a crime if they fail to become aware of the relevant rules in a timely manner and fail to comply with the use of land.

There are various legal risks regarding the use of land. The first is the legal risk of non-compliance with land use in substance. Here is an illustration of the policy change of using permanent basic farmland for facility agricultural land. *Document No. 4* states that land for aquaculture facilities is allowed to use permanent basic farmland, and the notice is currently effective. On 17 December 2019, the Ministry of Natural Resources and the Ministry of Agriculture and Rural Affairs issued the *Notice on Issues Related to the Management of Land for Facility Agriculture* (Nature Resource Regulation [2019] No. 4). However, *Document No. 166* has made adjustments to the provisions of Document No.4, stating that the new occupation of permanent basic agricultural land for the construction of aquaculture facilities is absolutely forbidden. On 27 November 2021, the Ministry of Natural Resources, the Ministry of Agriculture and Rural Affairs, and the State Forestry and Grassland Administration issued a *Notice on Issues Related to Strict Control of Arable Land Use* (Natural Resources Development [2021] No. 166). Meanwhile, *Document No. 166* mentioned that the provisions of previous documents of the Ministry (Department), which are inconsistent with this notice, are not effective. It is not appropriate to presume that the operators are good law-seekers. It must be aware of the possibility that the operators only find *Document No. 4* instead of *Document No. 166*, and this causes the legal risk that the land for farming facilities occupies permanent basic agricultural land and needs to be demolished.

Second, a situation that is consistent with the use of land but against the scale of the site is also considered to not meet the requirements of land management. And this carries legal risks. The land for agricultural facilities used for the construction of farms contains two parts: land for production facilities and land for auxiliary facilities. *Document No. 4* states that various types of land for agricultural facilities are determined by the departments of natural resources and the departments of agriculture and rural areas in each province (autonomous region and municipality), with reasonableness based on the scale of production and construction standards.

Third, in addition to the substance, there are also risks of procedural illegality faced by operators. For example, *Document No. 166* states that new rural roads, livestock and poultry breeding facilities, aquaculture facilities and planting facilities that destroy the cultivation layer and other agricultural facilities are strictly controlled in using general arable land for construction. If such facilities are required, they should be approved and comply with the relevant standards. In other words, aquaculture facilities on general arable land must follow relevant procedures for review and recording. Compared to the requirements for recording in *Document No. 4*, the current procedures put forward higher requirements. However, in practice, there is a legal risk of procedural illegalities by operators who are not aware of the exact operating policies of approval.

2.2.2. Legal Risk II: Licensing Legal Risks

An aquatic breeding certificate is the statement of an administrative licensing decision indicating that the certificate holder is allowed to practice aquaculture. Whereas the interests to which the certificate points are drawn from an analysis of laws and jurisprudence, there are no explicit legal provisions in practice. As a result, non-legal professionals are prone to misunderstand the certificate, which may lead to disputes.

Specifically, the law stipulates that the term "the right to practice aquaculture" refers to the right obtained in accordance with the law to use waters and tidal flats for aquaculture. (Article 3 of *Measures for Licensing and Registration of Aquaculture in Waters and Tidal Flats)* This provision is supposed to ensure that obtaining the right to practice aquaculture requires an application. Although the original text also refers to the use of waters and tidal flats, the acquisition of the right to use sea areas or the usufruct on rural land is stated separately in other laws. The term "conventional usufruct on rural land for agricultural operation" is a legal term that approximates the right to use the land. Article 3 of the *Law of the People's Republic of China on the Administration of Sea Areas* states that the sea areas shall belong to the state and that the State Council shall exercise ownership over the sea areas on behalf of the state. No entity or individual may usurp on, buy, sell or by any other means transfer sea areas. Article 6 stipulates that the state shall establish a system for registering the right to use sea areas. The lawfully registered rights to use sea areas shall be protected by law. Article 24 of the *Law of the People's Republic of China on the Contracting of Rural Land* states that the state implements the unified registration of arable land, forest land, and grassland, among others, and a registration body shall issue a certificate of a conventional usufruct on rural land for agricultural operations or a certificate of a forest right, among others, to a grantee, and maintain a register thereof, to confirm the conventional usufructs on rural land for agricultural operations. Thus, there is no practical meaning to the expression "use waters and tidal flats" in this provision, which cannot be the legal basis for the right to use sea areas. However, it is difficult for a non-legal professional to accurately access all the relevant legal provisions and perform the above analysis. In addition to the geographical space overlap, the public is prone to be confused and wrongly believes that applying for an aquatic breeding certificate means jointly claiming the sea area or land involved [68]. In fact, both aquatic breeding certificates and the certificate of right to use sea areas or the certificate of a conventional usufruct on rural land for agricultural operations are required, especially the certificate of right to use sea areas. The law states that the right to use sea areas comes upon application and the origin of the right is the granting of public authority. Article 3 of *Measures for Licensing and Registration of Aquaculture in Waters and Tidal Flats* states that where waters and tidal flats are used for aquaculture, the aquaculture rights of waters and tidal flats shall be identified by licenses issued by the local people's government at or above the county level. Article 5 states that where state-owned waters and tidal flats are to be used for aquaculture, an application shall be filed to the fishery administration department of the local people's government at or above the country. A conventional usufruct on rural land for agricultural operation could originate from a contract, and the public authority's registration is merely an administrative confirmation without involving a grant of the right. In contrast, the lack of a certificate of the right to use sea areas carries a heavier legal responsibility.

2.2.3. Legal Risk III: Legal Risk No. 3: Wastewater Discharge Legal Risks

The abuse of inputs can lead to problems such as eutrophication in aquaculture water and drug residues [69]. These are not only related to the quality and safety of aquatic products but will also pollute the ecological environment in the form of illegal discharge of wastewater. Article 7 of the *Aquaculture Quality and Safety Management Regulations* stipulates that the inlet and outlet systems of a culture farm or pond should be separated. The discharge of wastewater products from aquaculture should reach national discharge standards. In terms of discharge standards, the relevant standards currently in effect include the Integrated Wastewater Discharge Standard [70], Freshwater Pond Culture Water Discharge Requirements [71] and Mariculture Water Discharge Requirements [72]. Moreover, there are different relevant standards based on the category of water in which such waste is discharged. If wastewater is used for agricultural irrigation, it should comply with the stricter standards for agricultural water use, which are the Water Quality Standards for Agricultural Irrigation [73]. If it is to be discharged into the surface water system, the standard of "Environmental Quality Standard for Surface Water" needs to be implemented [74]. If the site is in the vicinity of drinking water sources, the primary A standard of "Discharge Standards for Pollutants from Urban Sewage Plants" must be implemented [75]. In terms of application, there should be no conflicts in the implementation of national integrated discharge standards and natural industrial discharge standards. If a national industry standard for water pollutant discharge exists, the standard should be followed. At the same time, a distinction should be made between mandatory national standards and recommended national standards. Operators may discharge wastewater that does not satisfy particular criteria due to erroneous understanding of the rules when confronted with a sophisticated and complex set of standards. Article 85 of the *Water Pollution Prevention and Control Law of the People's Republic of China* provides that the local people's government administrative protection department at or above the county level shall order violators to stop breaking the law and adopt treatment measures within a prescribed time limit to eliminate pollution and impose a fine on it. If it fails to take treatment measures within the prescribed time limit, the administrative environmental protection unit may designate an entity capable of such treatment to do so with the required expenses borne by the violator. In summary, the stage of discharge of wastewater produced by aquaculture presents the legal risk of operators being fined for the misapplication of relevant standards.

2.2.4. Legal Risk IV: Aquatic Fry and Fingerlings Legal Risk

Aquatic fry is located at the core of the aquaculture industry chain. China is able to carry out the breeding of some of the fry independently, but current operations are mainly stuck in the promotion and demonstrate stages. Moreover, some species of fry still rely on imports. Article 17(1) of the *Fisheries Law* provides that quarantine must be executed for the import and export of aquatic fingerlings in order to prevent disease from passing into or out of the territory. Specific quarantine work shall be carried out in accordance with regulatory provisions on the quarantine of imported and exported animals and plants. Aquaculture operators must go through the process of applying for the use of the development of isolation quarantine sites and applying for a quarantine permit for imported animals and plants. In practice, however, there are countries that prohibit the export of certain aquatic fry, such as eel fry. The information is derived from interviews with people from the industry. Some aquaculture operators have turned to smuggling eel fry because they cannot obtain specific aquatic fry legally. This not only breaks the regulatory order of entry and exit but also has legal risks, as the imported fry is not quarantined and may be prone to genetic and viral risks.

2.2.5. Legal Risk V: Pharmaceutical Legal Risks

Due to the restricted policy on land use and other reasons, areas used for aquaculture are decreasing [76]. Operators have responded by adopting high-density aquaculture methods. However, aquatic animals are prone to sensitivity to high-density aquaculture [77]. Moreover, higher feed intake tends to worsen water quality and increase fish diseases. Controlling fish diseases is more difficult than livestock disease control since aquatic animals live underwater and changes in their behavior in response to disease are difficult to detect before the diseases have advanced beyond the stage at which they are most easily treated. It is also difficult for drugs to reach the disease site directly due to the complex water environment of aquaculture, the large number of aquatic animals and the difficulty in capturing them [78]. In order to avoid serious losses caused by diseases, some operators are indeed using drugs abusively and blindly abusing antibiotics and other drugs, leading to excessive drug residues in aquatic products and raising concerns among domestic and international consumers [79–82]. In order to strengthen the quality and safety supervision of aquatic products, in early 2021, the Ministry of Agriculture and Rural Affairs issued a *Notice on Strengthening the Regulation of Inputs Used in Aquaculture*, which clearly implemented a pilot white-list system for the usage of inputs for aquaculture. It also carried out three-year special rectification of irregularities related to veterinary drugs, forage and forage additives for aquaculture. However, a category of inputs used in large quantities to regulate water quality and aquaculture substrates cannot be added to the white list due to a lack of standards. These products cannot be banned because of need, in spite of long-term issues such as vague definitions of their efficacy [83]. Consequently, some producers take advantage of this regulatory vacuum to sell substances containing components of drugs that meet national standards under the guise of water adjustment products [84]. Some even add banned or restricted drugs illegally [85]. Producers purposefully omit the terms "prevention, treatment, diagnosis of diseases in aquaculture animals" and "purposeful adjustment of physiological functions of aquaculture animals" from their product instructions and give inaccurate product information in order to avoid supervision [86]. While there is a lack of guidance for operators, the improper use of such products can easily lead to excessive drug residues in aquatic products, pollution of the ecological environment and other problems. And this leads to the legal risk of the operator being fined or even held criminally liable.

3. Causes of Legal Risk in Aquaculture

3.1. Objective Reasons

In general, the above risks can be objectively attributed to the problem of industrial structure. China's aquaculture methods are still relatively mature and are mainly decentralized, small scale and lack systematic planning and not forming large-scale [87]. The disadvantages of decentralized aquaculture include low economic efficiency, as well as a lack of resilience to risk. In a competitive situation, it is easy to induce moral hazard in the production management of the operators, that is, to act detrimentally in terms of property rights, licenses, discharges, aquatic fry and fingerlings and pharmaceuticals, which appears as a violation of the law in the result.

The series of actions taken by the government also reflects that the industrial structure is the cause of the problems. The large number of violations in the aquaculture field has not only triggered heated debates in the community but has also attracted the attention of the government. For example, in November 2015, the central government made it clear in the *Overall Proposal for the Reform of the Ecological Civilization System* that the protection and environmental restoration of areas producing aquatic products will be strengthened, their aqua culture will be controlled, and mechanisms will be established for the protection of aquatic plants and animal life. After research, in January 2019, the Ministry of Agriculture and Rural Development and others issued *Several Opinions on Accelerating the Green Development of Aquaculture*. The document is a programmatic document to guide the green development of aquaculture in China in the current and future period [88]. The document pointed out that in recent years, China's aquaculture industry has been characterized by both irrational layout and industrial structure, and excessive aquaculture density in some regions and suggested strengthening the scientific layout and transforming aquaculture modes [89]. Data from the Food Safety Incident Big Data Monitoring Platform show that during the 10-year period from 2009–2018, the number of quality and safety problems of aquatic products and their products showed a trend of increasing and then decreasing, with a rapid increase from 1643 to 3943 in 2013 and then a rapid decrease to 636 in 2018 [90]. In 2018, the qualification rate of routine monitoring of aquatic products increased from 94.4 percent in 2013 to 97.1 percent, an increase of 2.7 percentage points in five years. Although the qualification rate of routine monitoring of aquatic products has been lower than the overall qualification rate of routine monitoring of agricultural products during 2013–2018, the gap between the two has been gradually narrowing, from 3.1 percentage points in 2013 to 0.4 percentage points in 2018 [91].

3.2. Subjective Reasons

The subjective awareness of aquaculture operators to understand and comply with the law is not high, and their scientific literacy and food safety awareness also need to be improved. Some scholars, after on-site investigations, have found that aquaculture operators are deficient in terms of professionalism. For example, among the 982 aquaculture operators in the port town of Zhongshan City, none graduated from aquaculture-related majors, and most of the operators were aged 50 to 60, with many of them having only primary or junior high school education [92]. Many scholars have also suggested that aquaculture operators have a reserve of specialized knowledge that needs to be improved [93], a poor ability to innovate [94], and a weak concept of the law [95]. These factors make it easy for operators to create incentives to violate the law.

4. Related Beneficial Experiences

As Korea has experienced a shift from extensive to environmentally friendly aquaculture, its experience offers valuable lessons [96]. Norway's aquaculture was its second largest export industry [97], and the government put in place a series of safeguards to ensure the sustainable development of the industry [98]. Its experience is worth learning from. As one of the more economically developed countries in Latin America, the progress of Chile's aquaculture industry is a successful case of the industrial upgrading of the fishery economy in developing countries [99]. Therefore, the relevant experiences of Korea, Norway and Chile are described below.

4.1. Korea

In August 2019, Korea's Ministry of Maritime Affairs and Fisheries consolidated the relevant issues stipulated in the *Fisheries Act* and the *Inland Waterways Fisheries Act* to form the *Aquaculture Industry Development Act* and its enforcement measures. The aim is to contribute to the sound development of the aquaculture industry and the national economy by enhancing aquacultural productivity. This reflects Korea's attitude toward strengthening the competitiveness of aquaculture. In addition, Korea has one of the world's fastest-growing aquaculture industries [100]. The problems it has encountered, and its solutions could inspire China.

Korea considers China's non-pharmaceuticals as pharmaceuticals, since they are substances that are directly applied to moving water and are absorbed into the bodies of animals to be effective. Schedule 1 of *Related provisions on the scope and designation of aquatic and veterinary non-pharmaceuticals*, which is named *Scope of Non-Pharmaceuticals for Aquatic and Veterinary (Article 2, paragraph 1 related)* listed, antibiotics, inhibitors, repellents and pesticides for aquatic animal pests. None may keep or use toxic chemicals prescribed in subparagraph 7 of Article 2 of the *Chemical Substances Control Act* for the purpose of cultivating fishery resources or removing foreign substances attached to fishing implements or fishing nets. The term "hazardous chemical substances" includes substances requiring permits, restricted substances or prohibited substances, substances requiring preparation for accidents, or other chemical substances that present or are likely to present a hazard or risk. The *Aquaculture Foster Act* was enacted in January 2002 by the Ministry of Agriculture, Food and Rural Affairs of the Republic of Korea to establish a disease treatment system for aquatic creatures, which stipulates financial supports, aquatic creature treatment, and so on.

In addition to support measures, the management of pharmaceuticals for aquatic use in Korea is characterized by diversification. The management system for aquatic pharmaceuticals in Korea is shown below (Figure 1).

The regulations for aquatic pharmaceuticals in Korea are scattered in several laws and are shown below (Table 1). Above all, with technical and financial support and the regulation of the pharmaceutical management system, the abuse of pharmaceuticals in Korean aquaculture is not common.

Regarding the discharge of aquaculture, Article 6 of the *Fishing Grounds Management Act* provides for the inspection of the fishing ground environment in Korea. The Minister of Oceans and Fisheries establishes and operates a network for regularly inspecting fishing ground environments. Whenever necessary, the Minister of Oceans and Fisheries assesses the need to urgently inspect the environment of fishing grounds in the event of serious environmental pollution. Regarding the specific contents, the Ministry of Environment in Korea issued *Guidelines for Setting and Managing Water Quality Standards for Discharge Water from Aquaculture Facilities* and set enforcement rules for the same act. The *Guide* is divided into three categories: general fish farms, eel farms, and tank-type land-based aquaculture facilities, with different water quality standards to design their programs. General fish farms with flowing water allow continuous water flow through the breeding pond, characterize by quite high water consumption, low pollutant concentrations, and a rapid rise in pollutant concentrations in discharge water due to loss of feed and sediment disturbance caused by fish activity during feeding. Therefore, the usual standard (average value), standard at feeding (maximum value), and standard at cleaning (maximum value) are used to set management standards for flowing water farms. The amount of water in and out of eel farms changes according to the seasons, and there is almost no wastewater discharge on weekdays, just intermittent discharges during the time of screening and releasing eels. Therefore, the discharge concentration is set as the instantaneous concentration to determine the optimal discharge concentration. The concentration of pollutants in the water discharged from a tank-type, land-based aquaculture facility sharply increases during feeding, and it discharges 20% to 30% of the sediment in the tank at once after supply. Therefore, the standards are set and managed using the normal standard (average value) and the standard at feeding (maximum value).

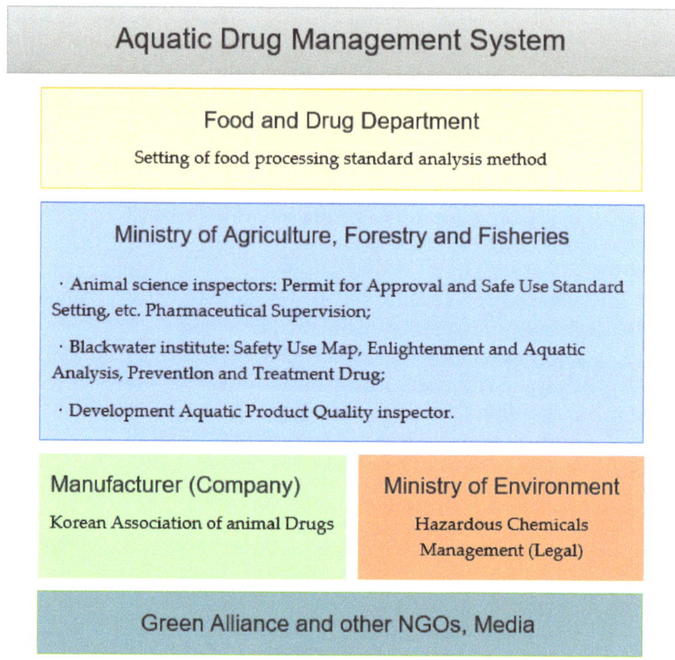

Figure 1. Drug management system for aquatic use in Korea.

Table 1. Legal system for aquatic drug management in Korea.

Law	Content
Article 85(3) of the *Pharmaceutical Affairs Law*	A person who intends to use animal drugs shall observe the standards.
Article 3 of the *Standard for the Safe Use of Veterinary Drugs*	The matters to be observed by the user, including the target animals prescribed in the declaration license or the objects and usage and dosage.
Article 40(1) of the *Aquatic Organism Disease Control Act*	No aquaculture business entity and its worker shall use animal drugs which have not been approved or reported under Article 31(2) of Pharmaceutical Affairs Act and Article 85(1) of the same Act or hazardous chemical substances under subparagraph 7 of Article 2 of Chemical Substances Control Act: Provided, That the same shall not apply where he or she has obtained approval for use under other Acts, such as approval for use under the proviso of Article 25(2) of the Fishery Resources Management Act.

Table 1. *Cont.*

Law	Content
Article 98(10) of the *Pharmaceutical Affairs Act*	Any of the following persons shall be subject to an administrative fine of not more than one million won: a person who fails to observe the standards for use of drugs, etc. for animals, in violation of Article 85(3).
Article 53(9) of the *Aquatic Animal Disease Management Act*	Those who use veterinary drugs or dangerous chemicals in violation of Article 40(1) shall be punished by imprisonment of up to three years or a fine of up to 30 million won, and their crimes shall be prosecuted by the judicial police.

4.2. Norway

One of the direct impacts of land-use regulations on the aquaculture sector is the reduction in the area for aquaculture. The development of freshwater aquaculture in China is already constrained, and near-shore mariculture is overloaded [101]. Thus, the only way to maintain the sustainable development of the aquaculture industry is to explore the deep sea. Reference can be made here to the Norwegian deep-sea netting aquaculture practices, which can be used to increase fish farming capacity and contribute to the restoration of near-shore agricultural lands or wetlands that were previously used for mariculture [102].

Like many countries, Norway has a licensing system, and fishing is a highly regulated industry with stringent licensing requirements. Both the specific and general rights and obligations associated with the license can be regarded as part of that license. Article 5(1) of the Norwegian *Aquaculture Act*, an aquaculture license, permits the production of specific species in limited geographic areas (sites) subject to any prescribed restrictions on the license scope that may apply at any given time. Norway also has specific aquaculture licenses for certain species—in particular, for salmon, trout and rainbow trout—as stipulated in Article 7 of the same *Aquaculture Act*.

Most Norwegian sea-farms are cage systems located in the deep, sheltered fjords [103]. For example, in April 2017, CIMC Raffles, an ocean technology group, signed an agreement with Norway's Ocean Aquafarms AS, and the construction of five Hex Box tanks would enable Norwegian salmon aquaculture to escape the geographical limitations of the fjord. The Hex Box platform is the most recent development in deep-sea fish aquaculture complexes, with a platform diameter of 90 m, a total height of 35.5 m, an empty vessel weight of 5400 t, and a farming capacity of 2 million fish. This method of aquaculture can clearly increase farming capacity. As deep-sea netting has the advantages of being free from land and near-shore restrictions and improving aquaculture efficiency, aquaculture operators could consider switching to deep-sea netting for their aquaculture practices [104–106].

Norway has also designated mariculture genetic breeding as a top priority research area. The Norwegian government launched a selection program for Atlantic salmon and rainbow trout in the early 1970s and invested in a genetic research center [107]. With government support and the involvement of private breeding companies, Norway's social and commercial breeding industry system is becoming increasingly mature, with good salmon and trout breeding stock for aquaculture, which decisively supports those industries in Norway and worldwide [108]. According to a 2012 study, farmed Atlantic salmon showed a 115% increase in growth rate and a 23% decrease in bait coefficient (feed conversion rate) compared with farmed stocks in the 1970s [109]. This reflects the government's active role in the industry.

4.3. Chile

Chile attaches great importance to the development of the marine economy, and its aquaculture production reached a record high of 1.48 million tons in 2020. Globally, Chile is the second largest producer of salmon and mussel products, although salmon is not a native fish in the Southern Hemisphere [110]. Chile's success in aquaculture stems from its strong support for the introduction of aquatic fry and fingerlings and the protection of property rights, among other aspects. In the late 1960s, commercial fresh fish farming in Chile began with experimental activities in the lakes, rivers and fjords of southern Chile through a national and international cooperative effort [111]. At the beginning of the development of the Atlantic salmon aquaculture industry, the Chilean government took the initiative to fund the introduction of foreign fry and encouraged operators to experiment while also promoting imports of net tanks through financial assistance. In its *Penal Code*, Chile declared that all waters are national assets and available for public use. If aquaculture involves the installation of aquaculture centers in marine, lake or river waters, exclusive use rights are required. Aquaculture concessions also apply to public beaches within 80 m of the highest tide line on the coast [112]. The license holder has the right to use the water surface and the space under it. The aquaculture operator can fully exercise its right to possession of farmed salmon throughout the entire farming process, from hatching to the final catch [113]. After the development of the industry, the government privatized the relevant public enterprises, allowing them to play their full role in the market economy, while the government focused on emergency responses to large-scale epidemics, anti-dumping, and other issues [114]. After the government withdrew its control of the aquaculture industry, the industrial structure shifted toward private capital, with a high degree of commercial organization. Risk management is an explicit goal of aquaculture companies and the salmon aquaculture association called Salmon Chile [115]. In addition to ecological and economic risks, the companies not only implement technological improvements and safety measures but also mobilize their social technologies—that is, adaptive political strategies to shape social and cultural life according to the company's objectives [116].

5. Preventing and Controlling Legal Risks in Aquaculture

5.1. Promote an Improved Rule of Law

5.1.1. Raise Legal Awareness among Operators

Agricultural land involves the country's food security, and the regulation of such land is a legal system in which public power intervenes to ensure the efficient allocation of land resources. Article 4 of the Land Administration Law of the People's Republic of China provides that the State is to place strict control on the usages of land. Given this strict regulation, operators must clarify the properties and usage of the land in question. They need to be aware of land classifications and naming to avoid violating land use regulations, according to which agricultural land includes paddy fields, reservoir water surfaces, pond water surfaces, and ditches, among others [117]. It is worth noting that, influenced by land-centrism, watersheds fall into the category of land [118]. Classification of agricultural land also includes land for facilities and ancillary facilities for the production of aquaculture, such as manure disposal, inspection and quarantine, although excluding land for processing sites. Aquaculture operators must also pay attention when transferring land, as some provinces are carrying out the work of "retiring ponds for cereal". If the price of land for transfer is low, operators have to pay attention to land use regulations to avoid losses caused by illegal acts.

Operators must retrieve the relevant legal provisions and apply them with references based on an awareness of land classifications and naming practices. There are several key regulations relevant to aquaculture in the laws on land use regulation, depending on the nature of the land in question. First, for permanent basic agricultural land, aquaculture operators may not dig ponds for aquaculture, build aquaculture facilities or otherwise destroy the cultivation layer. Article 37(3) of the Land Administration Law provides that it is forbidden to occupy permanent basic farmland to develop horticulture or dig ponds to breed fish. Document No. 166 stipulates that it is strictly forbidden to newly occupy permanent basic agricultural land to build livestock and poultry breeding facilities, aquaculture facilities and planting facilities that destroy the cultivation layer. Second, if general arable land—which encompasses the arable land outside the permanent basic agricultural land—belongs to the overall plan for balancing occupation and supplement, operators must be approved and accepted by the village collective, as well as changing the contract for land management rights and the certificate of ownership in a timely manner [119]. If the land does not belong to the plan, the aquaculture operators must obtain approval to build aquaculture facilities that meet the relevant standards. Third, for integrated three-dimensional farming—this would include, for example, rice and a fishery, rice and shrimp, or rice and crab integrated farming—the aquaculture operators cannot destroy permanent basic farmland if that is the land subject to use. Where general cropland is used, more than 10% of land cannot be used for ditch pits. It is stated in the Technical Specification for Integrated Rice-Fishery Farming: 4. Technical indicators, the proportion of ditch-pits: the proportion of ditch-pits shall not exceed 10% (referring to general arable land).

Operators also need to fulfill the necessary procedures in advance instead of starting work without authorization; otherwise, they may be liable for breaching the relevant plans. In general, the rules differ slightly from place to place, but they can be broadly divided into three aspects.

First, the location and scope of the land sought for the construction of agricultural facilities by the operators should be in line with spatial, agricultural development and rural planning modalities, as well as the provisions for the scale of land for the construction of agricultural facilities issued by the local natural resources departments. Use of inefficient, idle or unused land is recommended, such as barren hills, land and beaches, to prevent the occupation of arable land.

Second, operators should sign land-use agreements with collective economic organizations to clarify the conditions for land use. In some places, construction plans and land-use agreements must be announced on township governments and village public affairs boards, and agreements can only be signed after the announcement period has expired without objection.

Third, the operators or rural collective economic organizations should file the land-use agreements in a timely manner with the township government after signing. After completing the filing, the township government must remit the filing information to the county-level natural resources departments and the agricultural and rural departments in charge [120]. In addition to the value of the procedure itself, the operator can also learn about the conditions related to the use of the land, such as its length, usage, time limits for land reclamation requirements, the surrender of the land and the liability for breach of contract. This may help operators avoid the risk of substantial violations of land-related laws.

5.1.2. Follow the Principle of Proportionality and the Principle of Reliance Interests

Administrative subjects should follow the principle of proportionality in the enforcement of law and consider the circumstances of the violations related to land use to make the appropriate administrative decision based on those facts. For example, in the case of a project involving agricultural facilities for which no relevant formalities have been completed, the person concerned should be ordered to complete the formalities within a certain period and not be subject to heavier penalties. In practice, however, an administrative subject may order the cancelation of the deadline due to procedural violations that violate the principle of proportionality [121]. Administrative subjects should also follow the principle of proportionality in operating public affairs. For example, aquaculture plants in some rivers or lakes were all outlawed by the relevant authorities because of the requirements to construct an ecological civilization and protect the environment. A case in point is the right to use Taihu Lake for aquaculture, which was withdrawn by Jiangsu Province in April 2018 [122]. The penalties clearly did not match the severity of the circumstances and instead increased the legal liability of the aquaculture operators.

The suggestion to enhance legal awareness among operators is a call for operators to comply with the law consciously, but it is not meant to eliminate the obligation of the administrative organs to promote the law and the relevant approval and filing procedures.

The administrative subject is the organization that has administrative functions and powers, as well as the independent responsibility for doing so; it should abide by the spirit of the law when enforcing that law. For example, an aquaculture operator is generally required to complete several steps to build an aquatic breeding farm. This process includes an application to the township government where the construction is taking place, a site survey of the fish farm by the relevant departments (e.g., the Department of Land, Agriculture, and Environmental Protection) in the district or county, approval by the district or county government, and a review of the site design by the relevant departments organized by the land department. The operators trust the administrative subject as their authority. After fulfilling the necessary procedures, the operators act based on policy guidelines or administrative guidance due to this trust. The legitimate interests arising from such behavior should be protected.

5.1.3. Clear Guidance from Legislators

In addition to compliance and enforcement efforts, improvements should also be made at the legislative level. Regulatory documents are universally and repeatedly applied within a certain period [123], and their reform could improve administrative efficiency, as well as support enforcing and improving laws, regulations and higher-level policies. It is important to maintain consistency between laws and policies to maintain the unity of the national legal system and the smooth flow of government orders. Regulatory documents that cannot be applied should be cleaned up or organized in a timely manner. Consider the case of *Document No. 4,* which was mentioned earlier in this paper. It is a regulatory document based on a more liberal land policy that was current at the time of its promulgation. The later *Document No. 166* indicated a change in policy as a result of a change in circumstances. It obviously does not align with the spirit of *Document No. 4,* yet both documents are currently in effect and there has been no sound transition. *Document No. 166* states, of course, that the provisions of previous documents that are inconsistent with the current document are no longer in force, but the contradiction remains. Lawmakers (including policymakers) should therefore clean up the content of regulatory documents that cannot be applied because there is no guarantee that operators have a high level of legal literacy necessary to face such complicated changing land regulations. The authorities should consolidate effective regulatory content and provide clear guidance to all other parties in the administrative legal relationship.

5.1.4. Upgrade of Government-Supported Aquaculture

Land use controls are based on considerations of food security. The related policy advocacy is thus dictated by the reality of the situation. Food harvests are difficult to predict due to a combination of market, climate and other factors, so it is difficult to maintain a stable land policy, and it is therefore difficult to maintain the area for aquaculture within a stable range on a certain amount of land. High-density aquaculture does not increase capacity but is counterproductive. To escape this dilemma, authorities should encourage aquaculture operators to make full use of the sea area and adopt new modes of aquaculture, such as deep-water nets, marine pastures, and aquaculture work boats, which are indeed encouraged by the *2023 Central Document No. 1.*

This promotion of new aquaculture modes will naturally be constrained by a variety of factors, including production philosophy, technology level, and production costs. There are currently many small enterprises and aquaculture operators in China's mariculture industry that can hardly afford the costs and risks inherent in such transformation and upgrading. It would thus be useful to refer to Chile's experience in providing financial subsidies. The government should consider introducing similar preferential policies to support aquaculture operators in upgrading their methods based on the comprehensive local situation.

5.2. Clarify the Positioning of Aquatic Breeding Certificates

Some aquaculture operators do not understand the relationship between an aquaculture license and other certificates, such as the sea area use right certificate. They may believe that there is an inclusive or overlapping relationship between such documents. They may, therefore, fail to complete the necessary documentation and are thus penalized. The root of the above legal risks lies in the unclear positioning of the breeding certificate. This is essentially a problem of overlap between administrative licensing and administrative confirmation, which may seem to strengthen its legal effect, but its scope is limited to aquaculture and may not be broadly construed as the right to use the land or sea. The relevant regulations state that the aquaculture certificate confirms the right to engage in aquaculture in waters and tidal flats. However, the aquaculture certificate is subject to an application and issuance process, and it is also within the scope of authorization by the public power to engage in aquaculture production. Article 11 of the *Fisheries Law* provides that where a unit or an individual uses a water area or beach with ownership by the whole people, which is determined by the State programming to be used for aquatic breeding industry, the user shall apply to the department in charge of fishery administration of the local people's government at the county level or above for the aquatic breeding certificate, which shall be checked and issued by the people's government at the same level.

Up to this point, it is possible to distinguish the difference between the aquaculture license and other tenure certificates based on content. This can perhaps best be explained in the judicial case. The *Fisheries Law* sets up an administrative license for aquaculture, so the right to engage in aquaculture is divorced from the right to use water and mudflats. It is thus a special right that can only be obtained through an administrative license. The legal aquaculture license and the certificate indicating legal land (or sea) use rights involve different rights and thus do not conflict. The same aquaculture water surface can be issued both a land (sea) use right certificate and an aquaculture license. However, the doctrinal clarification is opaque to the public and requires legislative recognition. Policymakers should thus further clarify the positioning of the aquaculture license in the *Fisheries Law*.

5.3. Improve and Popularize Wastewater Discharge Standards

The discharge standards for aquaculture wastewater are complicated and obsolete, while failing to fit local conditions. If the local authorities set standards for the discharge of aquaculture wastewater, it would help provide clear and scientific guidelines for operators and prevent the above-mentioned legal risks. The development of such local standards is currently underway. The *Implementation Plan for Agricultural Non-point Source Pollution Control and Supervision Guidance (Trial)*, jointly issued by the Ministry of Ecology and Environment and the Ministry of Agriculture and Rural Development, proposes to guide localities in setting standards and regulations for the discharge of waste water from aquaculture. *Opinions of Strengthening the Supervision of Ecological Conditions of Marine Culture*, also jointly issued by the same, require coastal provinces (autonomous regions and municipalities) to set relevant standards for the discharge of waste water from aquaculture by the end of 2023 in accordance with the relevant requirements of the *Technical Guidelines for the Formulation of Local Standards on Controlling the Discharge of Waste Water from Aquaculture*. *Technical Guidelines for the Formulation of Local Standards on Controlling the Discharge of Waste Water from Aquaculture* (HJ 1217-2023, hereinafter referred to as Technical Guidelines), issued by the Ministry of Ecology and Environment in February 2023, is used to guide and regulate the formulation of relevant local pollutant discharge standards in a more scientific, precise and standardized manner.

To guarantee that local standards are scientific, formulators should increase basic research to provide effective direction for practical implementation. In this formulation, they should clarify the standards for calculating emissions. The condition of aquaculture wastewater discharged from ponds under different species varies, so the water quality index measured by water samples collected at a time point within the whole discharge cycle does not accurately reflect water quality as a whole. Regulators should set the time point for collecting wastewater at a suitable point in the discharge cycle close to the average mass concentration of water quality indicators in the total discharge [124]. The Korean experience mentioned above can be combined to set up different plans to supervise the quality of wastewater based on the different aquaculture methods for the various fish species. This is in line with 6.2.4 of the *National Technical Guidelines for the Formulation of Standards for the Discharge of Water Pollutants* (HJ 945.2), which is cited in the *Technical Guidelines*. It contains that emission monitoring data should be collected, mainly including online monitoring, law enforcement monitoring, emission unit self-monitoring, environmental protection acceptance monitoring data at the completion of construction projects, which includes instantaneous, hourly and daily average emission concentrations of pollutants, discharge volume, as well as the capacity designed by the enterprise, actual capacity, production load, and so on. It also contains an analysis of the level of pollutant emissions, the proportion of compliance, and the emission characteristics of various production processes.

Before the launch of the local rules, operators should also raise awareness and understand the conditions for applying the standards, such as being able to distinguish between mandatory (GB) and voluntary national standards (GB/T). Article 2 of the *Standardization Law of the People's Republic of China* stipulates that compulsory standards must be implemented while the state merely encourages the adoption of voluntary standards. The Integrated Wastewater Discharge Standard (GB 8978–1996) is a mandatory national standard to control water pollution and protect the quality of surface water, such as rivers, lakes, canals, channels, reservoirs and oceans, as well as groundwater. The "Freshwater Pond Culture Water Discharge Requirements" (SC/T 9101–2007) and "Marine Culture Water Discharge Requirements" (SC/T 9103-2007), meanwhile, are voluntary standards, and their application is encouraged but not mandatory [125].

5.4. Increase the Self-Sufficiency Rate of Aquatic Fry and Fingerlings

Individual aquaculture is currently common in China. Without formulating a scalable situation, it is difficult to supervise the quarantine of aquatic fry comprehensively and rigorously. Instead of monitoring after the event, it is better to solve the problem at the root in advance. The lack of quality fry at present restricts the sound development of mariculture in China. By only emphasizing the legal introduction of fry through regulation without taking into account the various operator costs, positive enforcement results remain elusive. China should therefore consider the experience of diversification of industrial inputs in Norway. The government, joint research institutes, large aquaculture enterprises and other related parties should thus form a consortium to collectively solve the problem of aquaculture fry breeding. The government should take the lead in organizing and defining research and development needs, incentive mechanisms, supervision and management. The institutes could provide scientific and talent support, while the enterprises could provide financial support and promote the transformation of scientific and technological achievements through the power of the market.

5.5. Make Use of the Synergy of Soft Law and Hard Law

The lack of product standards is the main reason why water adjustment products are in a gray area and cause many legal risks, which makes such products difficult to supervise. The National Fisheries Technology Extension Center and the China Society of Fisheries issued a notice in response to this problem. According to this notice, they will focus on water adjustment products in need of regulation and develop a group standard for the China Society of Fisheries. The work will be based on surveys, and experts will determine the recommended group standard catalog of water adjustment products for aquaculture through research. A multi-sector group will then draft the standard and seek the views of the relevant parties. They will then submit the draft to the China Society of Fisheries for review and approval. During the development of the group standard, the relevant departments must adjust the white list scientifically. Indeed, one scholar noted that the formulation and dosage of some national standard fishery pharmaceuticals have not changed for a long time. Considering the resistance of disease-causing organisms, the pharmaceuticals currently in use may not effectively inhibit or destroy disease-causing organisms in the future [126]. To ensure the operability of the white list system, the departments concerned should take into account the reasonable needs of the operators to detect bacterial resistance, in addition to strengthening the supervision of water regulation products. They can then adjust the white list to achieve scientific use. At the legal level, the standard must make a distinction between non-pharmaceuticals and water adjustment products that use the name of non-pharmaceuticals but are essentially pharmaceuticals; for example, it is possible to absorb the provisions of Korea and determine that substances absorbed into the animal and effective are not non-pharmaceuticals.

Operators should also avoid buying inputs that lack critical product information, such as manufacturers, production licenses, product labels and product quality inspection certificates. They must follow Article 5 of the *Provisions on the Quality Safety Management of Aquaculture*, which provides that aquaculture water should comply with the Ministry of Agriculture regulations for "Pollution-free Food Seawater Aquaculture Water Quality" (NY5052-2001), "Pollution-free Food Freshwater Aquaculture Water Quality" (NY5051-2001), and other standards. The state prohibits the use of water sources that do not meet the water quality standards for aquaculture.

6. Discussion

Although the above issues are focused on China, much of the information is relevant to other countries as well. Property rights, licensing of access, food safety, environmental pollution and fry quarantine are all unavoidable issues in the aquaculture industry. Such common problems are not only concerns in China but are also worthy of attention in other countries. In addition to the interests of the operators, aquaculture is also related to human health, the environment and other social issues that are closely related to public life. In the 1990s, the public began to become aware of the potential problems of aquaculture, and acceptance of aquaculture practices and its products declined [127]. In the present era of advanced information and globalized trade, the situation of the aquaculture industry in a single country will positively and negatively influence multiple countries that do frequent trade transactions with it.

The following insights can be drawn from this study. First, from the perspective of aquaculture operators, pursuing only short-term economic benefits is not desirable in the context of increasing public attention to social issues. They should therefore pay attention to long-term benefits and comply with legal requirements on issues such as inputs and fry to avoid legal risks. They must make even greater efforts, above the legal minimum, to assume social responsibility by developing environmentally friendly aquaculture and joint efforts with the government to promote the regional economy.

Second, from the perspective of the government, urging aquatic operators to comply by strengthening enforcement and inspection is not the only possible solution. In addition to post-event monitoring, it is important to solve the problem at the root by, for example, developing production standards for inputs based on pharmaceutical resistance testing and introducing and breeding aquatic fry adapted to the local environment. The government should also pay attention to the role of private subjects, such as joint self-regulatory associations, and use favorable policies (among other tactics) to guide the participation of multiple parties in social governance.

Third, from the perspective of the legislator, the legal protection system for aquaculture should be improved. This could include removing and integrating the legal codification of relevant policies or clarifying the positioning of aquaculture licenses at the legal level. Moreover, regarding the preventive measures for related risks, China can learn from the experiences of other countries, and likewise, other countries can also implement countermeasures used by China or draw inspiration from Chinese practices or transformed experiences. In other words, this research on the legal risks of and preventive measures in aquaculture in China is also beneficial to other countries.

Understanding the legal risks and prevention of aquaculture in China is valuable not only in the proactive approach described above, in which other countries can learn from or draw inspiration from China, but also in a reactive sense, in which the trade dealings of other countries with China can be affected. Although China is a major aquaculture producer, it is also an important consumer. One scholar has pointed out that China's huge imports are related to Chile's "blue transition", defined as the shift from a reduction in fish biomass caused by aquaculture to the exploitation of aquatic resources for recovery [128]. As the transformation and promise of Chilean aquaculture depend greatly on China, the question of how China will develop its aquaculture industry and whether this development will relieve pressure on wild fisheries is a key question for the blue transition and the future of the ocean [129]. There is also a space for the theory discussed here to be applied to countries with similar trade relations. China's perceptions and standards for aquaculture will not only influence its domestic industry but will also influence other countries through the power of the market based on the public's evaluation of aquaculture in each country. Thus, the development of aquaculture in China is relevant to the blue transition of all countries.

7. Conclusions

Land use violations, lack of legal documents, sub-standard wastewater, unquarantined aquatic fry and fingerlings, and misuse of prohibited aquatic pharmaceuticals are five of the common current legal risks. Their occurrence is linked not only to the economic loss incurred by aquaculture operators but is also closely related to food safety, the environment, and even the economic development of China and other countries, which will counteract aquaculture operators through market forces. Operators, as stakeholders and participants in risk management, must take this into account. A complete legal framework, improved legal awareness among aquaculture operators, reasonable enforcement in accordance with the law, and resistance to and supervision of unlawful elements in the aquaculture industry by other parties are all necessary to prevent these risks. Currently, China has abandoned its once-expensive approach to aquaculture and is actively transforming and upgrading toward ecologically friendly and healthy aquaculture. The green development of aquaculture will make significant progress through the collaboration of public and private parties, which will ultimately result in the high-quality, sustainable development of the aquaculture economy.

Author Contributions: Supervision, Z.P.; resources, Z.P.; writing—original draft, C.X. and Y.L.; writing—review and editing, C.X. and Y.L. All authors have read and agreed to the published version of the manuscript.

Funding: This research was funded by Economic and Social Development Research Project of Liaoning Province in 2024 "Research on the improvement path of comprehensive law-enforcement capacity in marine fishery" (2024lsljdybkt-028), Liaoning Education Department's 2022 Basic Research Project for Higher Education Institutions "Research on Rectification and Legal Application of Fishing-related Vessels with no Vessel Number and Name, no Vessel Certificate and no Port of Registry in Liaoning Province" (LJKZR20220092), Major Projects of National Social Science Foundation of China's "Study on the Development Strategy of China's 'Dark Blue Fisheries' under the Background of Accelerating the Construction of a Marine Power" (Grant No.21 and ZD100), Economic and Social Development Research Project of Liaoning's "Research on intangible cultural heritage promoting rural revitalization in Liaoning" (2023lslybkt-09), Social Science Planning Fund of Liaoning's "Research on high-quality development path of marine economy in Liaoning" (L22AJL002), Project approved by Liaoning Provincial Department of Education's "Research on the digital transformation and development of Liaoning marine industry" (KJKMR20221130), and Project approved by Liaoning Province economic and social development research "Research on the high quality development path and countermeasures of Liaoning's marine economy under the new development pattern" (2024lsklybkt-037).

Institutional Review Board Statement: Not applicable.

Informed Consent Statement: Not applicable.

Data Availability Statement: No new data were created.

Conflicts of Interest: The authors declare no conflict of interest.

References

1. FAO. World Review. In *State of the World Fisheries and Aquaculture: Sustainability in Action*, 1st ed.; FAO: Rome, Italy, 2020; p. 5.
2. FAO. Executive Summary. In *The State of World Fisheries and Aquaculture 2022. Towards Blue Transformation*, 1st ed.; FAO: Rome, Italy, 2022; pp. 1–19.
3. Noor, N.; Harun, S.N. Towards Sustainable Aquaculture: A Brief Look into Management Issues. *Appl. Sci.* **2022**, *12*, 7448. [CrossRef]
4. Tacon, A.G.J. Trends in global aquaculture and aquafeed production: 2000–2017. *Rev. Fish. Sci. Aquac.* **2020**, *28*, 43–56. [CrossRef]
5. Hilborn, R.; Banobi, J.; Hall, S.J.; Pucylowski, T.; Walsworth, T.E. The environmental cost of animal source foods. *Front. Ecol. Environ.* **2018**, *16*, 329–335. [CrossRef]
6. Poore, J.; Nemecek, T. Reducing food's environmental impacts through producers and consumers. *Science* **2018**, *360*, 987–992. [CrossRef] [PubMed]
7. Naylor, R.; Fang, S.; Fanzo, J. A global view of aquaculture policy. *Food Policy* **2023**, *116*, 102422. [CrossRef]
8. Galparsoro, I.; Murillas, A.; Pinarbasi, K.; Sequeira, A.M.; Stelzenmüller, V.; Borja, A.; O' Hagan, A.M.; Boyd, A.; Bricker, S.; Garmendia, J.M. Global stakeholder vision for ecosystem-based marine aquaculture expansion from coastal to offshore areas. *Rev. Aquac.* **2020**, *12*, 2061–2079. [CrossRef]
9. Higgins, D.; Balint, T.; Liversage, H.; Winters, P. Investigating the impacts of increased rural land tenure security: A systematic review of the evidence. *J. Rural Stud.* **2018**, *61*, 34–62. [CrossRef]
10. Hishamunda, N.; Ridler, N.; Martone, E. Policy and governance in aquaculture: Lessons learned and way forward. *FAO Fish. Aquac. Tech. Pap.* **2014**, *577*, I.
11. Lawry, S.; Samii, C.; Hall, R.; Leopold, A.; Hornby, D.; Mtero, F. The impact of land property rights interventions on investment and agricultural productivity in developing countries: A systematic review. *J. Dev. Eff.* **2017**, *9*, 61–81. [CrossRef]
12. Menon, N.; Van der Meulen Rodgers, Y.; Nguyen, H. Women's land rights and children's human capital in Vietnam. *World Dev.* **2014**, *54*, 18–31. [CrossRef]
13. El-Sayed, A.F.M. Regional review on status and trends in aquaculture development in the Near East and North Africa-2015. *FAO Fish. Aquac. Circ.* **2017**, *C1135*, I.
14. Bennett, R.G. Coastal planning on the Atlantic fringe, north Norway: The power game. *Ocean Coast. Manag.* **2000**, *43*, 879–904. [CrossRef]
15. Stokke, K. Municipal planning in sea areas–increasing ambitions for sustainable fish farming. *Kart Og Plan* **2017**, *77*, 219–230.
16. Kvalvik, I.; Robertsen, R. Inter-municipal coastal zone planning and designation of areas for aquaculture in Norway: A tool for better and more coordinated planning? *Ocean Coast. Manag.* **2017**, *142*, 61–70. [CrossRef]
17. Belton, B.; Filipski, M. Rural transformation in central Myanmar: By how much, and for whom? *J. Rural Stud.* **2019**, *67*, 166–176. [CrossRef]
18. Belton, B.; Hein, A.; Htoo, K.; Kham, L.S.; Nischan, U.; Reardon, T.; Boughton, D. A Quiet Revolution Emerging in the Fish-Farming Value Chain in Myanmar: Implication for National Food Security. 2015. Available online: https://ageconsearch.umn.edu/record/259801/ (accessed on 24 October 2023).
19. Filipski, M.; Belton, B. Give a man a fishpond: Modeling the impacts of aquaculture in the rural economy. *World Dev.* **2018**, *110*, 205–223. [CrossRef]
20. Asche, F.; Eggert, H.; Oglend, A.; Roheim, C.A.; Smith, M.D. Aquaculture: Externalities and Policy Options. *Rev. Environ. Econ. Policy* **2022**, *16*, 282–305. [CrossRef]
21. Deniz, H.; Benli, A.C.K. Environmental Policies in Force and Its Effects on Aquaculture in Turkey. In Proceedings of the Workshop on the State-of-the-Art of ICM in the Mediterranean and the Black Sea, Mugla, Turkey, 14–18 October 2008; pp. 193–195.
22. Dierberg, F.E.; Kiattisimkul, W. Issues, impacts, and implications of shrimp aquaculture in Thailand. *Environ. Manag.* **1996**, *20*, 649–666. [CrossRef]
23. Galland, D.; McDaniels, T. Are new industry policies precautionary? The case of salmon aquaculture siting policy in British Columbia. *Environ. Sci. Policy* **2008**, *11*, 517–532. [CrossRef]
24. Zajicek, P.; Corbin, J.; Belle, S.; Rheault, R. Refuting Marine Aquaculture Myths, Unfounded Criticisms, and Assumptions. *Rev. Fish. Sci. Aquac.* **2023**, *31*, 1–28. [CrossRef]
25. Jianyuan, C. System of the Right to Use Maritime Space and Its Reflection. *Trib. Political Sci. Law* **2004**, *22*, 55–64.
26. Available online: http://www.maxlaw.cn/htf-dlhtjflawer-com (accessed on 24 October 2023).
27. Zhao, W. On National Legislative Perfection of Fishing Rights. *J. Guangdong Ocean Univ.* **2008**, *28*, 1–5.
28. Chen, Y. Government could issue both land use rights certificates and aquaculture certificates for the same aquaculture water surface. *People's Jurisd.* **2011**, *55*, 50–52.
29. Cui, L. Revision and improvement of the aquaculture license system to provide rule of law safeguards for the high-quality development of the fisheries industry. *China Fish.* **2021**, *64*, 50–52.

30. Study on Difficult Issues in the Trial of Cases of Illegal Use of Sea Area. Available online: http://mp.weixin.qq.com/s?__biz=MzU3NTY0ODcwNA==&mid=2247485493&idx=1&sn=8abacee8c13821539235a018ac148fd5&chksm=fd1eae9aca69278c832d42f29d61dfcf29ad12ec546335d27baf3d5529e06fd0477efd78396b&mpshare=1&scene=1&srcid=0907cDsyoBOKpAkExHnbz8F4&sharer_shareinfo=87b800e9b94bb7958bba2daed9903d39&sharer_shareinfo_first=87b800e9b94bb7958bba2daed9903d39#rd (accessed on 24 October 2023).
31. Engle, C.R.; Stone, N.M. Competitiveness of US aquaculture within the current US regulatory framework. *Aquac. Econ. Manag.* **2013**, *17*, 251–280. [CrossRef]
32. D'Agostino, J. Posner's Folly: The End Of Legal Pragmatism And Coercion's Clarity. *Br. J. Am. Leg. Stud.* **2018**, *7*, 365–400. [CrossRef]
33. Matthew, A. Trust, Social Licence and Regulation: Lessons from the Hayne Royal Commission. *J. Bank. Financ. Law Pract.* **2020**, *31*, 103–118.
34. Tafani, L. Enhancing the quality of legislation: The Italian experience. *Theory Pract. Legis.* **2022**, *10*, 5–21. [CrossRef]
35. Valles, C.; Pogoretskyy, V.; Yanguas, T. Challenging Unwritten Measures in the World Trade Organization: The Need for Clear Legal Standards. *J. Int. Econ. Law* **2019**, *22*, 459–482. [CrossRef]
36. Júnior, C.A.M.; da Costa Luchiari, N.; Gomes, P.C.F.L. Occurrence of caffeine in wastewater and sewage and applied techniques for analysis: A review. *Eclética Química*, 2019; 44, 11–26.
37. Naidu, R.; Jit, J.; Kennedy, B.; Arias, V. Emerging contaminant uncertainties and policy: The chicken or the egg conundrum. *Chemosphere* **2016**, *154*, 385–390. [CrossRef]
38. Ahmad, A.; Kurniawan, S.B.; Abdullah, S.R.S.; Othman, A.R.; Hasan, H.A. Contaminants of emerging concern (CECs) in aquaculture effluent: Insight into breeding and rearing activities, alarming impacts, regulations, performance of wastewater treatment unit and future approaches. *Chemosphere* **2022**, *290*, 133319. [CrossRef] [PubMed]
39. Lindmark, M.; Vikström, P. The determinants of structural change: Transformation pressure and structural change in Swedish manufacturing industry, 1870–1993. *Eur. Rev. Econ. Hist.* **2002**, *6*, 87–110. [CrossRef]
40. Bao, X.; Zhang, J.; Ding, J. Best management practices on non-point source pollution of aquaculture. *South China Fish. Sci.* **2012**, *8*, 79–86.
41. Aquaculture Site Selection and Carrying Capacity Management in the People's Republic of China. Available online: https://www.fao.org/fishery/docs/CDrom/P21/root/14.pdf (accessed on 24 October 2023).
42. Wang, X.; Cuthbertson, A.; Gualtieri, C.; Shao, D. A review on mariculture effluent: Characterization and management tools. *Water* **2020**, *12*, 2991. [CrossRef]
43. Ottinger, M.; Clauss, K.; Kuenzer, C. Aquaculture: Relevance, distribution, impacts and spatial assessments–A review. *Ocean Coast. Manag.* **2016**, *119*, 244–266. [CrossRef]
44. Biosecurity in Aquaculture, Part 1: An Overview. Available online: https://thefishsite.com/articles/biosecurity-in-aquaculture-part-1-an-overview (accessed on 24 October 2023).
45. Goel, A.D.; Bhardwaj, P.; Gupta, M.; Kumar, N.; Jain, V.; Misra, S.; Saurabh, S.; Garg, M.K.; Nag, V.L. Swift contact tracing can prevent transmission—Case report of an early COVID-19 positive case. *J. Infect. Public Health* **2021**, *14*, 260–262. [CrossRef] [PubMed]
46. Hasan, S.S.; Kow, C.S.; Zaidi, S.T.R. Social distancing and the use of PPE by community pharmacy personnel: Does evidence support these measures? *Res. Soc. Admin. Pharm.* **2021**, *17*, 456–459. [CrossRef] [PubMed]
47. Ortega, C.; Valladares, B. Analysis on the development and current situation of rainbow trout (Oncorhynchus mykiss) farming in Mexico. *Rev. Aquac.* **2017**, *9*, 194–202. [CrossRef]
48. 7, U.S. Code § 8303—Restriction on Importation or Entry. Available online: https://www.law.cornell.edu/uscode/text/7/8303 (accessed on 24 October 2023).
49. Palermo, D. A Ribbiting Proposal. *Vermont J. Environ. Law* **2021**, *22*, 77–99.
50. Peeler, E.J.; Oidtmann, B.C.; Midtlyng, P.J.; Miossec, L.; Gozlan, R.E. Non-native aquatic animals introductions have driven disease emergence in Europe. *Biol. Invasions* **2011**, *13*, 1291–1303. [CrossRef]
51. Whittington, R.; Chong, R. Global trade in ornamental fish from an Australian perspective: The case for revised import risk analysis and management strategies. *Prev. Vet. Med.* **2007**, *81*, 92–116. [CrossRef] [PubMed]
52. Sun, J.C.D. The innovative development path of modern aquaculture seed industry in China. *Res. Agric. Mod.* **2021**, *42*, 377–389.
53. Yang, H. Several strategies for the modernization of the construction of the aquaculture seed industry silicon valley. *Mar. Sci.* **2018**, *42*, 1–7.
54. Wilson, G.M. A Day on the Fish Farm: FDA and the Regulation of Aquaculture. *Va. Environ. Law J.* **2004**, *23*, 351–395.
55. Antimicrobial Usage and Antimicrobial Resistance in Vietnam. Available online: https://dokumen.tips/documents/antimicrobial-usage-and-antimicrobial-resistance-in-usage-and-antimicrobial-resistance.html?page=1 (accessed on 24 October 2023).
56. Mujeeb Rahiman, K.M.; Mohamed Hatha, A.A.; Gnana Selvam, A.D.; Thomas, A.P. Relative prevalence of antibiotic resistance among heterotrophic bacteria from natural and culture environments of freshwater prawn, Macrobrachium Rosenbergii. *J. World Aquac. Soc.* **2016**, *47*, 470–480. [CrossRef]

57. Rico, A.; Phu, T.M.; Satapornvanit, K.; Min, J.; Shahabuddin, A.; Henriksson, P.J.; Murray, F.J.; Little, D.C.; Dalsgaard, A.; Van den Brink, P.J. Use of veterinary medicines, feed additives and probiotics in four major internationally traded aquaculture species farmed in Asia. *Aquaculture* **2013**, *412*, 231–243. [CrossRef]
58. Rico, A.; Satapornvanit, K.; Haque, M.M.; Min, J.; Nguyen, P.T.; Telfer, T.C.; Van Den Brink, P.J. Use of chemicals and biological products in Asian aquaculture and their potential environmental risks: A critical review. *Rev. Aquac.* **2012**, *4*, 75–93.
59. Tacon, A.G.; Metian, M. Aquaculture feed and food safety: The role of the food and agriculture organization and the Codex Alimentarius. *Ann. N. Y. Acad. Sci.* **2008**, *1140*, 50–59. [CrossRef] [PubMed]
60. FAO; WHO. Aquaculture Production. In *Code of Practice for Fish and Fishery Products*, 1st ed.; WHO: Rome, Italy, 2020; p. 76.
61. Wang, Y. The history and current situation of the development of aquatic animal protection industry. *Curr. Fish.* **2019**, *44*, 84–87.
62. Li, N. Impact of fishery drugs on quality and safety of aquaculture products and countermeasures. *Heilongjiang Anim. Sci. Vet. Med.* **2017**, *60*, 283–285.
63. Li, X. Strengthening the management of aquatic non-pharmaceuticals to promote the healthy development of aquaculture industry. *Anhui Agric. Sci. Bull.* **2019**, *25*, 63–64+76.
64. Bei, Y.; Zhou, F.; Zhu, N.; Ma, W.; Ding, X. Problems and Suggestions in Production and Use of Water and Sediment Quality Improver for Aquaculture. *Mod. Agric. Sci. Technol.* **2019**, *14*, 212–214.
65. Ji, L.W. Research on the Eligibility of Natural Resource Concession Guarantee-Based on Article 329 of the Civil Code. *Res. Real Estate Law China* **2022**, *26*, 121–136.
66. Ruan, R.; Su, J.; Zheng, F. Use Right Confirmation and Rural Finance—An example of the impact of farming certificates on the availability of loans to fishermen. *Econ. Perspect.* **2017**, *58*, 53–67.
67. Yu, F.W.; Dai, M.H.; Lin, S. Cultivated Land Protection Based on Bottom Line Thinking of Food Security: Current Situation, Difficulties and Countermeasures. *Econ. Rev.* **2022**, *38*, 9–16.
68. Huang, R.H.; Pei, Z.B. On Perfecting the Legal System and Supervision Mechanism of Fishery Administrative Licensing. *J. Shenyang Agric. Univ.* **2020**, *22*, 348–353.
69. Gu, Z.J.; Liu, X.G.; Cheng, G.F.; Wang, X.D. The effect of ecological ditches in the management of wastewater from freshwater pond culture and the technology of construction. *Technol. Innov. Appl.* **2019**, *26*, 127–132.
70. Integrated Wastewater Discharge Standard. Available online: https://www.mee.gov.cn/ywgz/fgbz/bz/bzwb/shjbh/swrwpfbz/199801/t19980101_66568.shtml (accessed on 24 October 2023).
71. Freshwater Pond Culture Water Discharge Requirements. Available online: https://www.doc88.com/p-1075432615259.html (accessed on 24 October 2023).
72. Mariculture Water Discharge Requirements. Available online: https://max.book118.com/html/2019/1106/8070000006002062.shtm (accessed on 24 October 2023).
73. Standard for Irrigation Water Quality. Available online: https://www.mee.gov.cn/ywgz/fgbz/bz/bzwb/shjbh/shjzlbz/202102/t20210209_821075.shtml (accessed on 24 October 2023).
74. Environmental Quality Standards for Surface Water. Available online: https://www.mee.gov.cn/ywgz/fgbz/bz/bzwb/shjbh/shjzlbz/200206/t20020601_66497.shtml (accessed on 24 October 2023).
75. Discharge Standard of Pollutants for Municipal Wastewater Treatment Plant. Available online: https://english.mee.gov.cn/Resources/standards/water_environment/Discharge_standard/200710/t20071024_111808.shtml (accessed on 24 October 2023).
76. Yu, X.J. Stabilizing Aquaculture: An Analysis Based on National Fisheries Statistics. *China Fish.* **2023**, *66*, 22–25.
77. Yu, J.Y.; Wang, J.; Sun, P.; Tang, B.J. Physiological response of Larimichthys crocea under different stocking densities and screening of stress sensitive biomarkers. *Mar. Fish.* **2021**, *43*, 327–339. [CrossRef]
78. Aquaculture, the Concept of Disease Prevention Focused on Treatment Should be Changed. Available online: http://www.ysfri.ac.cn/info/2048/38463.htm (accessed on 24 October 2023).
79. Guo, H.; Dong, X. Discussion on drug residues in aquaculture animals and aquatic product safety. *J. Aquac.* **2015**, *36*, 46–48.
80. Jiang, Q.; Hu, K. Driving Forces and Equilibrium Analysis of Social Responsibilityin Marine Aquatic Products Supply Chain. *Ocean Dev. Manag.* **2016**, *33*, 50–54.
81. Pan, W.; Luo, Q.; Liu, W.; Lin, Q.; Tu, J. Analysis and Countermeasure of the Problems of Drug Residues in Aquatic Product and Aquatic Feed. *Fujian J. Agric. Sci.* **2011**, *26*, 1096–1100.
82. Liu, X.; Liu, X.; Ren, X.; Zhang, J.; Nelson, R.G. Drug residue issues of aquatic products export from China. In *LISS 2014, Proceedings of 4th International Conference on Logistics, Informatics and Service Science*; Springer: Berlin/Heildelberg, Germany, 2015; pp. 1135–1141.
83. Discussion on the Significance of Setting Group Standards for Water Adjustment Products in Aquaculture for the Development of Aquatic Science and Technology. Available online: https://mp.weixin.qq.com/s/kcuDceuRX605ZlN4TR8edA (accessed on 24 October 2023).
84. Rectification of fake drugs! Yang Xianle voice: The future of aquatic animal protection + fishing medicine is "0". Available online: https://mp.weixin.qq.com/s/JEYn2nDB63y1GemmPR_lKA (accessed on 24 October 2023).
85. Chen, C.F.; Chen, H. Analysis of the Problem of Aquatic Non-pharmaceutical and Suggestions for Management in China. *China Fish.* **2018**, *61*, 99–100.
86. Liu, Y.F. Screening of Unlabeled Pesticide and Veterinary Drugs in Non-drugs Fishery Inputs and Its Impact on Aquaculture. Master's Thesis, Shanghai Ocean University, Shanghai, China, 2022.

87. Yuan, Y. Exploring the key to aquaculture development in the new era from a multi-dimensional perspective. *Mod. Agric. Mach.* **2022**, *40*, 107–108.
88. Aquaculture Industry Leaders Comprehensively Explain Aquaculture Development Trends, Cutting-Edge Technology. Available online: https://www.ixueshu.com/h5/document/03d3884ec484856c1defc7dec9f46d64318947a18e7f9386.html (accessed on 24 October 2023).
89. Several Opinions on Accelerating the Green Development of Aquaculture. Available online: https://www.gov.cn/zhengce/zhengceku/2019-10/22/content_5443445.htm (accessed on 24 October 2023).
90. Finance and Economics: Quality and Safety of Fish Products in China. Available online: https://mp.weixin.qq.com/s/Rvu_FVaSQDmmqj0SnBd5Pg (accessed on 24 October 2023).
91. Research Report on the Quality and Safety Status of Fish Products in China. Available online: https://mp.weixin.qq.com/s/1PuhwKmL5nTwiAz7o1NT8A (accessed on 24 October 2023).
92. The Aquaculture Safety Problems and Countermeasures Research of GangKou ZhongShan. Available online: http://www.cnki.net/KCMS/detail/detail.aspx?dbcode=CMFD&dbname=CMFD201701&filename=1017004568.nh&uniplatform=OVERSEA&v=kssYELNVfsT7uiCQODmbb72zkLln4t1n2yZsYEOwcCWbW2lrkao9Y4-xag6XfIoM (accessed on 24 October 2023).
93. Deng, J.; Jia, B.; Zhao, Y.; Ma, H.; Cen, J.; Chen, S.; Yang, X.; Li, L. Application status and standard management of aquaculture inputs. *J. Food Saf. Qual.* **2023**, *14*, 200–206.
94. Ding, H.; Xie, X. Jinxi:How aquaculture breaks through in the new normal. *Jiangxi Agric.* **2015**, *34*, 51.
95. Sun, C.; Yuan, S. Research of aquatic products quality and safety problem in China based on circulation channel. *Sci. Technol. Food Ind.* **2014**, *35*, 275–279.
96. Kim, I.B. Aquaculture–For Sustainable Systems of the Future. *Korean Soc. Fish. Aquat. Sci.* **2015**, *61*, 306.
97. Lu, K.; Wu, W. Main Measures for Efficient Development of Norwegian Mariculture and Lessons Learned from Experience. *World Agric.* **2016**, *38*, 190–193.
98. Liu, C.; Liu, H.; Liu, X.; Lu, S.; Xu, Y.; Long, L.; Zhang, C. Reference from sustainable development of Norwegian Atlantic salmon industry to Chinese aquaculture. *Fish. Inform. Strategy* **2021**, *36*, 208–216.
99. Xu, H.; Luo, S. Upgrading Path of Chilean Salmon Aquaculture Industry and Implications for the Development of China's Mariculture Industry. *Ocean Dev. Manag.* **2010**, *27*, 83–88.
100. Yang, B.R. Environmental status of aquaculture farms in Korea and countermeasures. *China Fish.* **2007**, *50*, 24–25.
101. Ming, L. Developing Large-Scale Deep Sea Aquaculture: Problems, Modes and Realization Ways. *Manag. World* **2022**, *38*, 39–58.
102. Yan, G.Q.; Ni, X.H.; Mo, J.S. Research Status and Development Tendency of Deep Sea Aquaculture Equipments: A Review. *J. Dalian Ocean Univ.* **2018**, *33*, 123–129.
103. Fisheries and Aquaculture Country Profiles. Available online: https://www.fao.org/fishery/zh/facp/nor?lang=en (accessed on 24 October 2023).
104. Li, Z. Study on Countermeasures for the Development of Deep-water Net-pen Aquaculture in Hainan Province. *Rural Agric. Farmer* **2023**, *38*, 22–24.
105. Yang, H.; Wang, W.; Li, Q.; Fu, Q. Discussion on the development of deep-water net-pen aquaculture in Lingshui County, Hainan, China. *Ocean Dev. Manag.* **2015**, *32*, 49–52.
106. Zhu, Y.; Ju, X.; Chen, Y. Offshore cage aquaculture of China: Current situation, problems andcountermeasures. *Chin. Fish. Econ.* **2017**, *35*, 72–78.
107. Thodesen, J.; Gjedrem, T. Breeding Programs on Atlantic Salmon in Norway: Lessons Learned. *World Fish Cent.* **2006**, *73*, 22–26.
108. Zhang, S.D.; Li, Y.X.; Jiang, Y.; Tian, J.Y.; Liu, X.Q.; Zhao, X.X.; Ma, J.; Du, W.; Hu, J.T. Pathway Analysis for Developing Modern Mariculture Under the Background of Oceanic Great Power Strategy. *Ocean Dev. Manag.* **2021**, *38*, 18–26.
109. Gjedrem, T.; Robinson, N.; Rye, M. The Importance of Selective Breeding in Aquaculture to Meet Future Demands for Animal Protein: A Review. *J. Aquac.* **2012**, *350*, 117–129. [CrossRef]
110. Fishery and Aquaculture Country Profiles-Chile. Available online: https://www.fao.org/fishery/zh/facp/chl?lang=es (accessed on 24 October 2023).
111. Basulto, S. El Largo Viaje de los Salmones. Una crónica olvidada. Propagación y cultivos de Especies Acuáticos en Chile. *Not. Mus. Nat. Hist. Mon. Newsl.* **2003**, *352*, 21.
112. Evolution of the Aquaculture Environmental Regime in Chile. Available online: https://www.scielo.cl/scielo.php?script=sci_arttext&pid=S0718-68512014000100013 (accessed on 24 October 2023).
113. Yu, H.G.; Liang, Z.L. Property rights arrangements and aquaculture development—The case of Mexico, Chile and Washington State, USA. *Ocean Dev. Manag.* **2006**, *23*, 145–149.
114. Ren, C.L.; Gao, J.J. On the Successful Experience of the Atlantic Salmon Breeding in Chile. *Ocean Dev. Manag.* **2016**, *33*, 70–74.
115. Rabellotti, R. Chile's Salmon Industry: Policy Challenges in Managing Public Goods. *Res. Policy* **2017**, *46*, 898–899. [CrossRef]
116. Benson, P.; Kirsch, S. Capitalism and the Politics of Resignation. *Curr. Anthropol.* **2010**, *51*, 459–486. [CrossRef]
117. Current Land Use Classification. Available online: http://c.gb688.cn/bzgk/gb/showGb?type=online&hcno=224BF9DA69F053DA22AC758AAAADEEAA (accessed on 24 October 2023).
118. Zhou, H.X. Research on the Right to Use Inland Waters. Ph.D's Thesis, Hunan University, Hunan, China, 2021.
119. Is It Allowed to Build Farms on General Arable Land or Not? The New Rules in 2022 Have Changed! Available online: https://mp.weixin.qq.com/s/2IkE9yMWli5q9qD3HYUkGg (accessed on 24 October 2023).

120. Zhu, G.M. Whether Land for Construction of Agricultural Facilities Requires Approval for Agricultural Land Conversion. *China Land* **2023**, *444*, 60.
121. Judgment of the Second Instance of Deng Wencai v. Shimen County Natural Resources Bureau Land Administrative Punishment. Available online: https://wenshu.court.gov.cn/website/wenshu/181107ANFZ0BXSK4/index.html?docId=rnZKb0y5gfy2RR3+pd+HjOFhe99IWbLANum/yaGfazZ1an2gITk79Z/dgBYosE2g8gUOT8PqpyXZgHiZSnIM/jTFP25au9Azif75qfqTGRD8SME6W5nnHXuIZ5YpeFXW (accessed on 24 October 2023).
122. Jiangsu Province Reclaims the Right to Use Taihu Lake for Aquaculture. Available online: https://www.gov.cn/xinwen/2018-04/16/content_5282806.htm (accessed on 24 October 2023).
123. Yuan, Y. On the Definition of the Normative Document—Centered on the Constitutive Conditions and Identification Standards. *Polit. Sci. Law* **2021**, *316*, 114–130.
124. Xu, J.B.; Liu, Y.S.; Shi, Y.H.; Yuan, X.C.; Wang, J.J.; Liu, J.Z.; Jia, C.P. Dynamic Changes of Water Quality During Wastewater Discharge Cycle of Different Aquaculture Varieties. *Fish. Mod.* **2023**, *50*, 26–33.
125. Song, L.; Zhang, W.; Sun, X.; Liu, X.; Lu, Y.; Li, N.; Wang, Y. Tail Water of Aquatic Animal Culture: Control Strategy. *J. Agric.* **2019**, *9*, 59–64.
126. Chen, C.F.; Zhou, X.J. Discussion of how aquaculture input production and sales enterprises should act to face the supervision of the Ministry of Agriculture and Rural. *Curr. Fish.* **2021**, *46*, 72–73+77.
127. Bergleiter, S.; Meisch, S. Certification Standards for Aquaculture Products: Bringing Together the Values of Producers and Consumers in Globalised Organic Food Markets. *J. Agric. Environ. Ethics* **2015**, *28*, 553–569. [CrossRef]
128. Nahuelhual, L.; Defeo, O.; Vergara, X.; Blanco, G.; Marín, S.L.; Bozzeda, F. Is There a Blue Transition Underway? *Fish Fish.* **2019**, *20*, 584–595. [CrossRef]
129. Villasante, S.; González, D.; Antelo, M.; Rodríguez, S.; de Santiago, J.A.; Macho, G. All Fish for China? *J. Ambio.* **2013**, *42*, 923–936. [CrossRef]

fishes

Article

Spatial–Temporal Characteristics and Influencing Factors of Marine Fishery Eco-Efficiency in China: Evidence from Coastal Regions

Wendong Zhu [1], Wenhui Sun [1], Dahai Li [2,*] and Limin Han [1,2]

1 School of Management, Ocean University of China, Qingdao 266100, China; zhuwendong@stu.ouc.edu.cn (W.Z.); 13206478936@163.com (W.S.)
2 Institute of Marine Development, Ocean University of China, Qingdao 266100, China
* Correspondence: sunrain1932@163.com

Abstract: Marine fishery is an important part of China's maritime power strategy. Improving the ecological efficiency of marine fishery is the inevitable way to achieve the sustainable development of fishery. Based on the perspective of industrial sustainable development, this study used the Super-SBM model to evaluate the ecological efficiency of marine fishery in 11 coastal provinces of China from 2011 to 2020. Combined with Malmquist index, Moran index and other methods, the spatial and temporal evolution characteristics were analyzed. On this basis, the Tobit panel model was used to explore the influencing factors of marine fishery eco-efficiency. The results show that: (1) From 2011 to 2020, the marine fishery eco-efficiency in the 10 coastal provinces and cities of China shows a clear trend of improvement, and the efficiency values in high-efficiency areas remain basically stable. The relative gap between efficient and inefficient regions remains significant. (2) From the perspective of spatial distribution characteristics, the ecological efficiency of marine fishery in coastal provinces and cities in China had no obvious spatial correlation, and showed a trend of cross-distribution between high-efficiency regions and low-efficiency regions. (3) The change of marine fishery eco-efficiency is the result of a variety of influencing factors. Fishery industrial structure, scientific and technological support levels and environmental regulation play a role in promoting the improvement of marine fishery eco-efficiency. Therefore, optimizing the structure of the fishery industry, improving environmental regulation and increasing investment in science and technology are all effective measures for local governments to improve the eco-efficiency of marine fisheries.

Keywords: marine fishery; ecological efficiency; Super-SBM model; spatial and temporal characteristics; Tobit regression

Key Contribution: This paper analyzes the spatial–temporal characteristics and influencing factors of the ecological efficiency of marine fisheries.

Citation: Zhu, W.; Sun, W.; Li, D.; Han, L. Spatial–Temporal Characteristics and Influencing Factors of Marine Fishery Eco-Efficiency in China: Evidence from Coastal Regions. *Fishes* **2023**, *8*, 438. https://doi.org/10.3390/fishes8090438

Academic Editor: Ben Drakeford

Received: 30 June 2023
Revised: 20 August 2023
Accepted: 22 August 2023
Published: 28 August 2023

1. Introduction

The marine economy serves as a crucial driver for regional development in the modern era and significantly contributes to enhancing national strength. The 19th National Congress of the Communist Party of China (CPC) introduced the strategy of "adhering to the overall planning of land and sea and accelerating the establishment of maritime power". Integrating the development of maritime power into the contemporary economic system, embracing the new concept of green development and enhancing marine ecological efficiency are pivotal to achieving sustainable progress in the marine economy. Modern marine fishery stands as one of the world's four primary marine industries, with China being the largest global producer of marine aquatic products. In 2020, China's marine product output reached 33.1438 million tons, accounting for 50.6% of the total global output.

Of this, marine fishing accounted for 11.7907 million tons, and mariculture accounted for 21.3531 million tons [1]. The total value of China's marine fishery production reached 87.471 billion dollars, underscoring its increasingly prominent role in bolstering the nation's marine economic strength. However, with the rapid economic development in coastal areas of China and the deepening of marine resource exploitation, the traditional extensive development mode of marine economy is no longer suitable for the current concept of conservation and intensive and recycling of resources. Problems such as the depletion of fishery resources and deterioration of marine ecological environment appear [2], which seriously restrict the sustainable development of China's marine fishery; these have become urgent problems that need to be solved in the development of China's fishery economy.

Ecological efficiency, based on efficient resource utilization and minimal environmental damage, can assess the level and quality of economic development. As a result, it has emerged as a crucial indicator for measuring the sustainable development of the economy, resources and environment. It comprehensively reflects the degree of coordinated development among regional resources, economy, and ecosystems [3]. Schaltegger et al. introduced the concept of "ecological efficiency". They defined it as the ratio between economic added value and environmental impact, emphasizing the coordination of economic and environmental benefits to create greater social benefits with reduced resource consumption and environmental impact [4]. The growing focus on sustainable development has led to an increased attention toward ecological efficiency. Currently, research on ecological efficiency primarily centers around the following aspects: (1) Research objects of ecological efficiency. Scholars have combined the concept of eco-efficiency with economic production practices, involving industry [5–8], energy [9,10], marine [11–14], agriculture [15,16], tourism [17,18], urban development [19–22] and many other fields. The economic and environmental performance was evaluated via an empirical method, and ecological efficiency was used as an optimization strategy in the actual production process. (2) Research methods of ecological efficiency. Scholars used analysis tools such as Principal Component Analysis [23], environmental resource ratio method [24], Ecological Footprint Analysis method [25], Stochastic Frontier method (SFA) [26,27] and Data Envelopment Analysis method (DEA) [28,29] to measure the ecological efficiency of different research objects. DEA and SFA are the most widely used methods, but DEA has obvious advantages: it does not need to set a production function and can handle multi-input and multi-output simultaneously. Tone et al. [30] incorporated relaxation variables into the objective function on the basis of the DEA model and built a non-radial and non-angular Slacks-Based Measure model (SBM), which has gradually become the mainstream model for measuring ecological efficiency. After a series of extensions, models such as the super-efficiency SBM model and non-expected output SBM model are formed. (3) Study on regional differences of eco-efficiency. Early studies mainly described and analyzed the results of eco-efficiency in different regions [31]. However, due to the differences of production technology in different time sections, it is difficult to reveal the complete change process and the evolution characteristics of regional differences by static evaluation of eco-efficiency only from the perspective of space. With the deepening of research and the maturity of research methods, spatial analysis methods, such as data spatial visualization, Kernel density, Theil index, spatial Markov chain, and exploratory spatial analysis, have been applied to investigate regional disparities and dynamic evolution of eco-efficiency distribution [32–34]. Research levels also involve different scales, such as country [35], province [36], special economic zone [37], and city [38]. (4) Study on influencing factors of ecological efficiency. Existing studies have covered industrial structure, economic level, government regulation and opening to the outside world, etc., and the DEA-Tobit model is usually used to investigate the impact degree of the above factors on ecological efficiency [39].

Throughout the relevant literature, ecological efficiency has been applied not only to the environmental performance evaluation of specific industries, especially high energy consumption and high-pollution industries, but also to the evaluation of overall economic and environmental coordination at different spatial scales. The concept of ecological efficiency has been well applied and developed. Specifically speaking, in the field of marine fisheries, ecological efficiency of marine fisheries refers to maximizing the benefits of marine fisheries and minimizing negative ecological benefits on the premise that the output and quality of marine fisheries meet social needs. Currently, scholars have basically reached a consensus on the resource and environmental issues facing the development of marine fisheries, but there is still no precise definition of the ecological efficiency of marine fisheries. The perspective on the ecological performance of marine fisheries is mostly limited to resource management [40], ecosystem [41], or carbon emissions [42], and few scholars have conducted systematic research on the ecological efficiency of marine fisheries from the overall perspective of industrial sustainable development. The theory of ecological efficiency has not been applied to the study of the sustainable development of marine fisheries. Based on this, from the perspective of the sustainable development of the marine fishery industry, this paper uses the super-efficiency SBM model to measure the ecological efficiency of marine fisheries in 11 coastal provinces in China from 2011 to 2020, and analyzes its spatiotemporal evolution characteristics using methods such as the Malmquist index and the Moran index. On this basis, in order to provide a feasible reference for improving the level of marine fishery ecological efficiency and realizing the sustainable development of Marine fishery economy, the Tobit panel model was used to analyze the factors influencing the ecological efficiency of Marine fishery.

2. Theoretical Analysis, Research Methods and Index Construction

2.1. Theoretical Analysis

In a broad sense, marine fishery includes not only marine fishing and mariculture, but also the value-added links of the industrial chain such as aquatic product processing, storage, transportation, circulation and service, as well as the supporting links such as seed breeding, fishery feed and fishery drug processing and fishery machinery manufacturing. It is an industrial network with horizontal and vertical systems. With the expansion of the scale of the marine fishery industry, energy resource consumption and pollutant emissions have gradually increased. At the same time, due to the decline in fishery resources and the deterioration of the offshore environment, the carbon sink function of fisheries has weakened, which has a negative impact on the ecosystem balance of coastal areas. According to previous studies, combined with the current situation of China's marine fishery development, the impact of marine fishery industry on its eco-efficiency mainly comes from four aspects: The first is the resource consumption caused by offshore fishing and fuel burning of fishing vessels. The second is the energy consumption and pollutants of mariculture, including the fuel consumption of aquaculture fishing vessels, the power consumption of oxygen supply and feeding in ponds and factories, and the eutrophication of water bodies caused by fishing drugs and feeding. The third is the three wastes (waste gas, waste water and industrial residue) produced in the processing of aquatic products due to low raw material utilization and low-level processes. Lastly, the fourth is the loss of products in the transportation process, as well as the environmental pollution caused by the consumption of related service industries. In order to show the ways or means that different links in the marine fishery industry chain may affect eco-efficiency under the constraints of resources and environment, a schematic diagram of the impact process of the marine fishery industry's eco-efficiency is constructed (Figure 1).

Based on this, the ecological efficiency of marine fishery is defined as the premise of reducing the consumption of resources and energy in the operation of the marine fishery industry, reducing the negative impact of the whole environment and maintaining a high level of income. The process of improving the ecological efficiency of marine fisheries is the process of internalizing negative externalities through a series of industrial optimization methods.

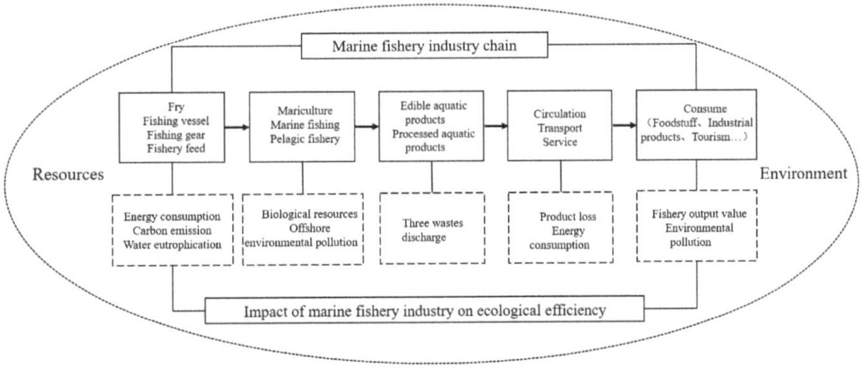

Figure 1. Impact of marine fishery industry on eco-efficiency.

2.2. Research Methods

2.2.1. Super-SBM Model

By combing various works from the existing literature, it is found that the commonly used measurement methods of ecological efficiency mainly include Stochastic Frontier Analysis (SFA), Data Envelopment Analysis (DEA), etc. Considering that the SBM model can obtain a more accurate efficiency value than the traditional DEA model and can make up for the lack of CCR model in the case of when the multiple DMU is 1. This article regards marine fisheries as a complete industrial sector and uses the super-efficiency SBM model to calculate the ecological efficiency of marine fisheries in different provinces and cities along the coast of China. The super-efficiency SBM model with constant returns to scale is expressed as:

$$\begin{aligned} \min\rho &= \frac{\frac{1}{m}\sum_{i=1}^{m}\overline{x_i}/x_{ik}}{\frac{1}{s}\sum_{r=1}^{s}\overline{y_r}/y_{rk}} \\ \text{s.t.}\overline{x_i} &\geq \sum_{j=1,j\neq k}^{n} x_{ij}\theta_j \\ \overline{y_r} &\leq \sum_{j=1,j\neq k}^{n} y_{rj}\theta_j \\ \overline{x_i} &\geq x_{ik} \\ \overline{y_r} &\leq y_{rk} \\ \theta, s^-&, s^+, \overline{y} \geq 0 \\ I = 1, 2, \ldots, m;\ r &= 1, 2, \ldots, q;\ j = 1, 2, \ldots, n\ (j \neq k) \end{aligned} \tag{1}$$

where n decision-making units (DMUs) are composed of input m and output s. The vector form is expressed as $x \in R^m$ and $y \in R^s$; x and y are matrices. s^- and s^+ represent the slack of input and output.

2.2.2. Malmquist Index

The Malmquist index is mainly used to analyze the change of production efficiency after efficiency evaluation. In order to analyze the changes of a marine fishery eco-efficiency, this paper chooses the Malmquist index used by Zhou et al. (2010) [43]:

$$M\left(x^{t+1}, y^{t+1}, x^t, y^t\right) = \left[\frac{E^t\left(x^{t+1}, y^{t+1}\right)}{E^t\left(x^t, y^t\right)} \times \frac{E^{t+1}\left(x^{t+1}, y^{t+1}\right)}{E^{t+1}\left(x^t, y^t\right)}\right]^{1/2} \tag{2}$$

The index involves two single-period distance functions $E^t\left(x^t, y^t\right)$ and $E^{t+1}\left(x^{t+1}, y^{t+1}\right)$ with constant returns to scale, and two inter-period output distance functions $E^t\left(x^{t+1}, y^{t+1}\right)$ and $E^{t+1}\left(x^t, y^t\right)$. If $M\left(x^{t+1}, y^{t+1}, x^t, y^t\right) > 1$, it indicates the technological progress. Moreover, the Malmquist index can be decomposed into technical efficiency change and technical change. Thus, (2) can be written as:

$$M\left(x^{t+1}, y^{t+1}, x^{t}, y^{t}\right) = \frac{E^{t+1}(x^{t+1}, y^{t+1})}{E^{t}(x^{t}, y^{t})} \times \left[\frac{E^{t}(x^{t+1}, y^{t+1})}{E^{t+1}(x^{t+1}, y^{t+1})} \times \frac{E^{t}(x^{t}, y^{t})}{E^{t+1}(x^{t}, y^{t})}\right]^{1/2} \tag{3}$$

2.2.3. Moran's Index

Moran's index is a coefficient [44] used to judge the correlation between entities in a certain space. The value is usually in $[-1, 1]$. When Moran's index is between $[0, 1]$, it shows that there is spatial correlation, that is, adjacent entities have similar attributes or trends; Moran's index is negatively correlated when it is less than 0; when the Moran index is equal to 0, it presents spatial randomness. The Moran index is divided into Global Moran's I and Local Moran's I, which are used to measure the existence of spatial autocorrelation and the specific location of spatial agglomeration.

2.2.4. Tobit Regression Model

The Tobit regression model is a model with limited dependent variables. When the value of the variable is cut or truncated, the Tobit regression model following the maximum likelihood method is a better choice. Since the efficiency values calculated by the data envelopment method are greater than 0, which belongs to the truncated case, the panel Tobit regression method is used to analyze the influencing factors.

2.3. Indicator Selection and Data Sources

2.3.1. Index System Construction

Combined with the connotation of marine fishery ecological efficiency, the input index is constructed from three aspects of economy, resources and environment, and the total output value of marine fishery is taken as the output index to construct the quantitative evaluation index system of marine fishery ecological efficiency (Table 1). In terms of input indicators, at the economic level, refer to existing research (Sun Kang et al., 2017) [45] selecting the number of marine fishery employees as labor input, marine fishery fixed assets stock and marine fishery intermediate consumption as capital input; at the resource level, marine aquaculture area represents natural resource input; at the environmental level, because it is difficult to measure the direct environmental investment in marine fishery, the economic loss of marine fishery caused by pollution is used to reflect the input of environmental pollution. Among them, ① there is no direct data on the investment in fixed assets of marine fisheries from 2011 to 2020 in the 'China Fisheries Statistical Yearbook'. This paper uses the investment in fixed assets of agriculture, forestry, animal husbandry and fishery to indirectly obtain the investment in fixed assets of marine fisheries. Considering the lag of the role of fixed assets, the perpetual inventory method is used to measure the stock of fixed assets over the years with 2011 as the base period. The specific formula is:

$$k_{i,t} = (1 - \delta)k_{i,t-1} + \lambda I_{i,t} \tag{4}$$

Table 1. The quantitative evaluation index system of marine fishery ecological efficiency.

Target Layer	Criterion Layer	Variable	Indicator Layer
Eco-efficiency of marine fishery	Input indicators	Labor input	Marine fishery practitioners
		Fixed asset investment	Fixed asset stock of marine fishery
		Current asset investment	Intermediate consumption of marine fishery
		Natural resources input	Area of mariculture
		Environmental pollution input	Economic losses of marine fishery caused by pollution
	Output indicators	Output value	Total economic output value of marine fishery

Among them, $K_{i,t}$ and $K_{i,t-1}$ represent the stock of marine fishery fixed assets in region i in t and $t-1$ years, respectively; $I_{i,t}$ represents the constant price fixed asset stock of region i in year t; $\delta_{i,t}$ represents the depreciation rate of fixed assets; and λ represents the capital formation rate of fixed assets. The values of the two are based on the practice of Yu et al. (2020) [46]. ② The data of intermediate consumption of marine fisheries cannot be obtained directly from the yearbook. Drawing on the ideas of Qin et al. (2018) [47], the intermediate consumption of fisheries is used for conversion. The specific conversion formula is:

$$\begin{aligned}\text{Marine fishery intermediate consumption} &\\ &= \text{Fishery intermediate consumption}\\ &\times \frac{\text{Total output value of marine fishery}}{\text{Total fishery output value}}\end{aligned}$$

The results are converted into comparable prices in 2011 according to the price index of agricultural means of production. Since the data of marine fishery pollution indicators are difficult to obtain directly, the economic loss of marine fishery caused by pollution is used instead. Based on the idea of Xu et al. (2022) [48], the specific formula is:

$$\begin{aligned}\text{Marine fishery economic losses caused by pollution} &\\ &= \text{Economic loss of aquatic products caused by pollution}\\ &\times \frac{\text{Economic value added of fishery}}{\text{GDP}} \times \frac{\text{Seawater product output}}{\text{Total output of aquatic products}}\end{aligned}$$

In terms of output indicators, the total output value of marine fisheries was selected and reduced to a comparable price in 2011 according to the agricultural producer price index. Furthermore, the super-efficiency SBM model with constant returns to scale is used to measure the ecological efficiency of marine fishery.

2.3.2. Data Source

The period of this study is from 2011 to 2020. The research objects included 10 coastal provinces of Tianjin, Hebei, Liaoning, Jiangsu, Zhejiang, Fujian, Shandong, Guangdong, Guangxi and Hainan in mainland China. Due to the serious lack of data in Shanghai, it was easy for the empirical results to deviate greatly. Therefore, this is not included in the study. The relevant data in the study are mainly derived from the "China Fisheries Statistical Yearbook (2012–2021)", "China Fisheries Yearbook (2012–2020)", "China Marine Statistical Yearbook (2012–2017)", "China Environmental Statistical Yearbook (2012–2021)", "China Rural Statistical Yearbook (2012–2021)" and "China Statistical Yearbook (2012–2021)". A small amount of missing data is supplemented by the moving average method. The statistical software used in this study includes DEA-solver, DEAP 2.1, Stata and Eviews 8.0, and the chart visualization software includes Origin2021, ArcMap10.8 and GeoDA.

3. Empirical Analysis and Results

3.1. Marine Fishery Ecological Efficiency Calculation

According to Formula (1), the DEA-solver software is used to calculate the ecological efficiency of marine fisheries in China's coastal areas from 2011 to 2020 (Table 2). According to the relevant research and the situation of this paper, it is considered that the efficiency value less than 0.4 is relatively ineffective, 0.4–0.8 is relatively inefficient, 0.8–1 is relatively effective, 1–1.2 is weakly effective, and 1.2 is highly effective.

According to the changes of marine fishery eco-efficiency values and mean values over the years, the trend chart of marine fishery eco-efficiency values in coastal provinces from 2011 to 2020 has been drawn (Figure 2).

Table 2. Ecological efficiency value of marine fishery in 2011–2020.

	2011	2012	2013	2014	2015	2016	2017	2018	2019	2020	Average Value	Ranking
Tianjin	1.0551	1.0818	1.1232	1.1789	1.1832	1.0988	1.1389	1.1375	1.0714	1.0380	1.1107	4
Hebei	0.2561	0.2810	0.3460	0.2989	0.3145	0.3034	0.2993	0.3190	0.3532	0.3812	0.3153	10
Liaoning	0.6867	0.7188	1.0878	1.0035	1.0264	1.0565	1.0571	0.6907	1.0185	1.0248	0.9371	6
Jiangsu	0.4205	0.4222	0.4325	0.4593	0.4638	1.0645	1.0454	1.0422	1.0757	1.1015	0.7528	7
Zhejiang	1.2387	1.0709	1.0128	1.0426	1.0257	1.0975	1.1403	1.1442	1.0952	1.1368	1.1005	5
Fujian	1.0873	1.1351	1.2681	1.2451	1.2832	1.2812	1.2870	1.3110	1.3232	1.4304	1.2652	2
Shandong	1.4754	1.4336	1.2280	1.2747	1.2223	1.1743	1.1534	1.1818	1.1918	1.1875	1.2523	3
Guangdong	0.5337	0.5515	0.6343	0.7364	0.6766	0.7288	0.7161	0.7623	1.0006	1.0166	0.7357	8
Guangxi	0.3715	0.4666	0.5333	0.5033	0.4719	0.5384	0.5522	0.5562	0.5612	0.5773	0.5132	9
Hainan	1.7520	1.6816	1.7141	1.6923	1.3578	1.4412	1.3371	1.3697	1.3823	1.2362	1.4964	1

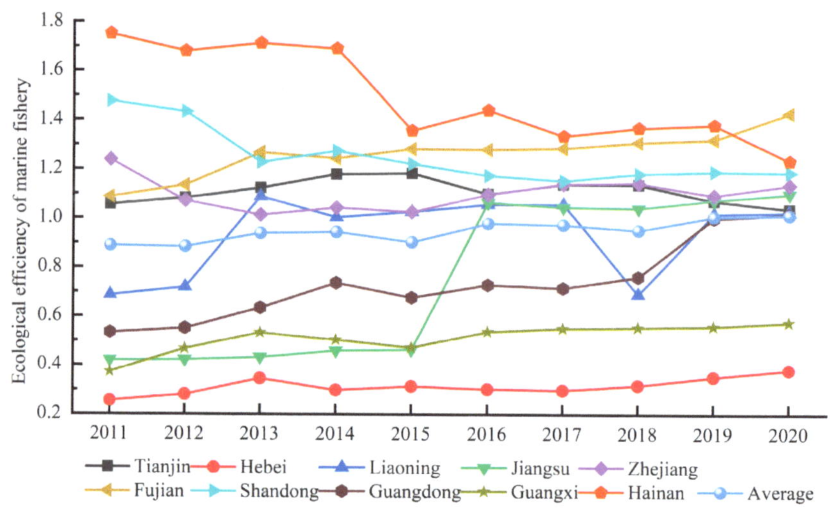

Figure 2. Trends of marine fishery eco-efficiency in coastal provinces from 2011 to 2020.

From the average calculated over the years in Figure 1, it can be seen that the eco-efficiency of marine fisheries shows a fluctuating upward trend, and the efficiency values are obviously different among regions. According to the characteristics of the data, it is divided into five categories: ① Highly effective, including Hainan, Shandong and Fujian provinces, showing that the efficiency value remains above 1.2 for a long time. The efficiency value of Hainan over the years has shown a fluctuating downward trend. The year 2014 is an important node for efficiency reduction, but the overall average is still 1.49 high. Shandong's efficiency value continued to decline steadily, with a large decline from 2012 to 2013, and began to stabilize in 2016, with an average of about 1.25 over the years. Fujian's efficiency value rises in gentle fluctuations, surpassing Hainan and Shandong in 2020 and ranking first in the country. From 2011 to 2013, the marine industry in various provinces and cities was in a period of rapid development. The degradation of offshore resources and environment has threatened the industrial sustainability in some areas. The high ecological efficiency of Hainan benefits from the superior marine environment and the tertiary industry structure. But, long-term low energy consumption and low output route will reduce the vitality of the industry, thereby reducing ecological efficiency. Shandong Province is a major marine fishery province, rich in fishery industry and marine scientific research strength. As a pioneer in the construction of marine ranching, Shandong has explored and practiced a systematic ecological farming model to restore offshore fishery resources to a certain extent, and further combined leisure tourism and modern service industries to create new economic growth points so that marine fishery ecological efficiency can be maintained at a high level for a long time. Fujian is also a traditional marine fishery province, the coastline length accounts for about 1/5 of the country, with the development of marine fishery resources, location and industrial base. In addition, Fujian's import and export trade of aquatic products is active and the output value contribution is large. The integration with international standards is conducive to forcing the green and high-quality development of the fishery industry. ② Effective efficiency refers to the state where the average efficiency is close to 1, including Tianjin, Zhejiang and Liaoning provinces. The average eco-efficiency of Tianjin from 2011 to 2020 is about 1.11, which indicates that the input and output of marine fishery are coordinated. The eco-efficiency value of marine fishery in Zhejiang Province decreased first and then increased slowly, but the efficiency value remained stable at about 1.1, reflecting a high level of marine fishery ecological performance. In addition to maintaining a good marine resource environment, Zhejiang has also formulated strict regulations on the conservation of fishery resources and habitat protection, creating a beneficial industrial development environment. Marine fisheries in Liaoning Province were relatively inefficient in 2011 and 2012, and remained effective after growing to 1.09 in 2013. Until 2018, with the reduction in the scale of aquaculture and the reduction in total output value, it entered a painful period of structural adjustment of the fishery industry, followed by a new growth trend in 2019. ③ From invalid to effective, including Guangdong and Jiangsu provinces, the efficiency value is between 0.4 and 0.8. Guangdong and Jiangsu are provinces with more developed fisheries. The ecological efficiency of marine fisheries in Guangdong has fluctuated during the study period, with an average annual growth rate of about 7.4%. In 2019, the efficiency was effective, disposing of the disparity in the environmental impact of marine fishery activities among cities, and achieving an overall improvement in ecological efficiency. The efficiency value of Jiangsu Province rose from 0.42 in 2011 to 1.1 in 2020, with an average annual growth rate of 11.29%. In 2016, it completed a leap from relatively inefficient to efficient. ④ Relatively inefficient. Guangxi's resources and environment are in good condition, but the scale of the marine fishery industry is small, the development mode is extensive, and the industrialization level is low, all of which reduce the ecological efficiency of marine fishery. ⑤ Relatively ineffective, with an efficiency value less than 0.4. Hebei is located in the Bohai Rim region, the scale of marine fisheries is not large, the marine pollution is serious and the ecological environment is fragile, causing the ecological efficiency of marine fisheries to remain at a low level for a long time.

3.2. *Time Trend of Marine Fishery Eco-Efficiency Change*

Based on the selected input and output indicators of marine fishery, DEAP 2.1 is used to measure the Malmquist index. The results obtained using the calculation Formulas (2) and (3) are shown in Tables 3 and 4. The changes of marine fishery eco-efficiency are analyzed from static and dynamic perspectives.

3.2.1. Static Analysis

In terms of regions, the technological progress of Tianjin, Zhejiang, Fujian, Shandong and Hainan is higher than the comprehensive technical efficiency, indicating that technological progress has a high contribution to the improvement of marine fishery ecological efficiency, and each province relies on technological innovation to achieve the goal in industrial ecology. At the same time, the pure technical efficiency of Tianjin, Liaoning, Zhejiang, Fujian, Shandong and Hainan is less than the technical progress, indicating that the digestion of existing technologies in the field is still insufficient, and it is necessary to use various factors efficiently to consolidate the development foundation. The scale efficiency of Hebei, Jiangsu, Guangdong and Guangxi is less than the pure technical efficiency, which is in the stage of increasing returns to scale. It can be moderately expanded to improve efficiency, while the scale advantage of the Guangdong marine fishery industry is gradually replaced, and technological innovation has become the source of industrial power.

Table 3. Total factor productivity index of marine fishery in China from 2011 to 2020.

Area	Comprehensive Technical Efficiency	Technological Progress	Pure Technical Efficiency	Scale Efficiency	Total Factor Productivity
Tianjin	1.000	1.056	1.000	1.000	1.056
Hebei	1.039	1.000	1.034	1.006	1.039
Liaoning	1.004	1.020	1.001	1.004	1.024
Jiangsu	1.079	1.035	1.074	1.005	1.117
Zhejiang	1.000	1.022	1.000	1.000	1.022
Fujian	1.000	1.062	1.000	1.000	1.062
Shandong	1.000	1.016	1.000	1.000	1.016
Guangdong	1.042	0.987	1.041	1.001	1.029
Guangxi	1.040	0.984	1.029	1.011	1.023
Hainan	1.000	1.023	1.000	1.000	1.023
Average value	1.020	1.020	1.017	1.003	1.041

Table 4. Dynamic analysis of marine fishery eco-efficiency.

Time	Comprehensive Technical Efficiency	Technological Progress	Pure Technical Efficiency	Scale Efficiency	Total Factor Productivity
2011–2012	1.0540	0.9480	1.0330	1.0200	0.9990
2012–2013	1.0530	0.9380	1.0220	1.0300	0.9880
2013–2014	0.9900	1.0510	1.0080	0.9820	1.0410
2014–2015	1.0200	1.0710	1.0220	0.9980	1.0920
2015–2016	1.0510	1.0970	1.0650	0.9870	1.1540
2016–2017	0.9940	1.0300	1.0020	0.9920	1.0240
2017–2018	1.0090	1.0160	1.0040	1.0060	1.0260
2018–2019	1.0020	0.9910	1.0020	0.9990	0.9920
2019–2020	1.0100	1.0510	1.0000	1.0100	1.0610
Average value	1.0200	1.0200	1.0170	1.0030	1.0410

3.2.2. Dynamic Analysis

From the perspective of time series, the dynamic changes of the decomposition indexes of marine fishery eco-efficiency from 2011 to 2020 are further analyzed, as shown in Table 4.

From the mean point of view, the total factor productivity of marine fisheries in 10 coastal provinces and cities in China from 2011 to 2020 was 1.041, indicating that total factor productivity has become the main driving force for the improvement of the ecological efficiency of marine fisheries in China, and the technological progress and total factor productivity in 2011–2020. The trend of change is basically consistent. It can be said that the main reason for the change of total factor productivity is technological progress. The comprehensive technical efficiency, technical progress, pure technical efficiency and scale efficiency are 1.02, 1.02, 1.017 and 1.003, respectively. It also shows that the total factor productivity of marine fishery is gradually driven by scale efficiency to technological progress, and pure technical efficiency has a significant effect on the improvement of comprehensive technical efficiency. In addition, the reason why the technological progress from 2011 to 2013 was less than 1 was that China did not pay enough attention to the ecological problems of marine fisheries during this period, and the fishery proliferation technology was relatively backward.

3.3. Spatial–Temporal Evolution Characteristics of Marine Fishery Eco-Efficiency

3.3.1. Kernel Density Estimation

In order to analyze the time evolution trend of marine fishery eco-efficiency, the four representative time nodes of 2011, 2014, 2017and 2020 were selected, and Eviews 8.0 was used to draw the corresponding kernel density curve (Figure 3).

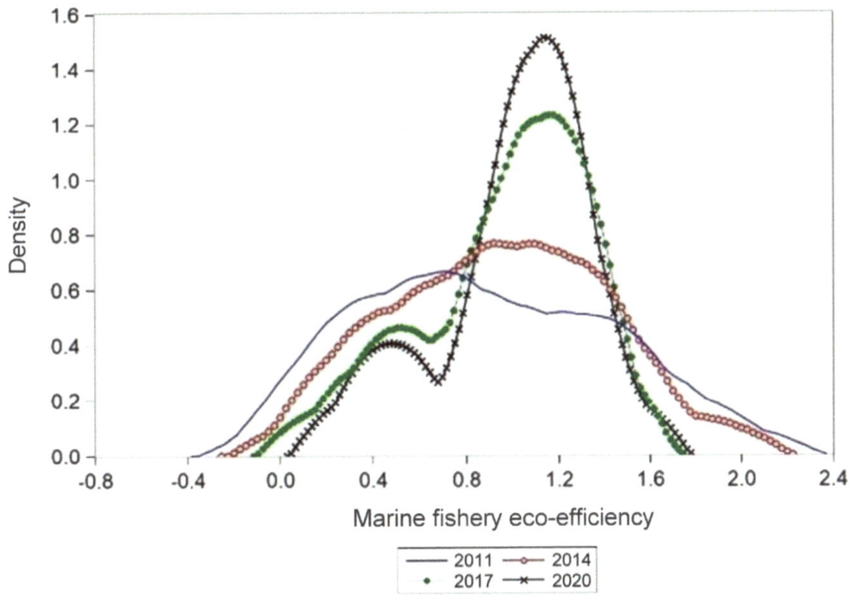

Figure 3. Temporal and spatial evolution characteristics of marine fishery eco-efficiency in China.

Firstly, from the position point of view, the density function center of the four years gradually moved to the right. Among them, 2014 has a slight right shift compared with 2011, and 2017 has a significant left shift compared with 2014. The trend in 2020 and 2017 is similar. Compared with 2014, it has a tendency to converge to the high-efficiency interval. After 2014, with the adjustment of fishery policy and the upgrading of fishery structure, the development of marine fishery industry and the ecological environment have been gradually coordinated, and the ecological efficiency has been improved.

Secondly, in terms of shape, the four sample intervals showed different distribution characteristics of high-efficiency and low-efficiency areas. The slopes in 2011 and 2014 were relatively gentle. In 2017, the density value increased, and the high density was concentrated in areas with high ecological efficiency values. In 2020, the density value will continue to rise, high density will be more concentrated in high-efficiency areas, and the density of low-efficiency areas will decrease, indicating that the relative gap of ecological efficiency of marine fisheries in China is quite pronounced.

Thirdly, from the peak point of view, the ecological efficiency of marine fisheries gradually evolved from a wide peak to a sharp peak from 2011 to 2020. The peak value of high-efficiency areas increased significantly, the vertical distance of each year widened, and the peak value of high-efficiency areas was much higher than that of low-efficiency areas, indicating that the ecological efficiency of marine fisheries has been improved to a certain extent, but there is still a large gap between provinces.

3.3.2. Moran's Index

Global Moran Index

We used the GeoDa [49] software to calculate the global Moran index of marine fishery eco-efficiency in China from 2011 to 2020 on the basis of adjacency weight, see Table 5.

From the numerical characteristics, it can be seen (Table 5) that the Moran index of China's marine fishery ecological efficiency is negative in all years except 2018, and all of them fail to pass the 5% significance test, indicating that China's marine fishery ecological efficiency presents the characteristics of cross-distribution between high-value areas and low-value areas. From the trend of change (Figure 4), the ecological efficiency of marine fishery in 2011–2020 showed a trend of segmented change. The global Moran's index decreased significantly from 2011 to 2014, indicating that the eco-efficiency value of marine fishery was more dispersed in spatial distribution during this period. To a certain extent, it reflects that as China has entered a new era of marine development, various regions have focused on the development of marine fisheries while ignoring inter-regional exchanges and cooperation, even in the development of beggar-thy-neighbors, resulting in serious border effects, forming a situation in which the ecological efficiency of marine fisheries is uneven. After 2015, the global Moran's index fluctuated and rose until 2018, from negative to positive, reaching the highest concentration state in recent years, indicating that the similarity of marine fishery eco-efficiency values in adjacent areas was rising higher and higher. In 2015, China's fishery economy stabilized at a high level. At the same time, new progress has been made in the conservation of fishery resources and the action of fishery energy conservation and emission reduction. The central government continues to increase its efforts to support the implementation and construction of proliferation and release, marine ranching, etc. At the same time, it provides guarantee for fishermen to reduce their boats and transfer their jobs, and has achieved remarkable results in the improvement of marine fishery ecology. The spillover effect of high eco-efficiency provinces expanded, so that the eco-efficiency of marine fisheries in neighboring provinces has been significantly improved. Since entering the '13th Five-Year', the development speed of the marine industry has accelerated, the fishery industry is facing the challenge of industrial transformation and upgrading, and the ecological efficiency of the marine fishery has fluctuated, but the trend is good.

Table 5. The global Moran index of marine fishery eco-efficiency in China from 2011 to 2020.

Time	Global Moran Index	Expected Value	Z-Statistic	*p*-Value
2011	−0.4122	−0.1250	−0.9182	0.1940
2012	−0.4985	−0.1250	−1.1664	0.1370
2013	−0.6146	−0.1250	−1.4424	0.0650
2014	−0.6308	−0.1250	−1.5161	0.0560
2015	−0.5777	−0.1250	−1.3476	0.0890
2016	−0.2510	−0.1250	−0.4006	0.3770
2017	−0.2583	−0.1250	−0.4250	0.3590
2018	0.0389	−0.1250	0.4770	0.3010
2019	−0.2207	−0.1250	−0.3383	0.3970
2020	−0.0868	−0.1250	0.1092	0.4430

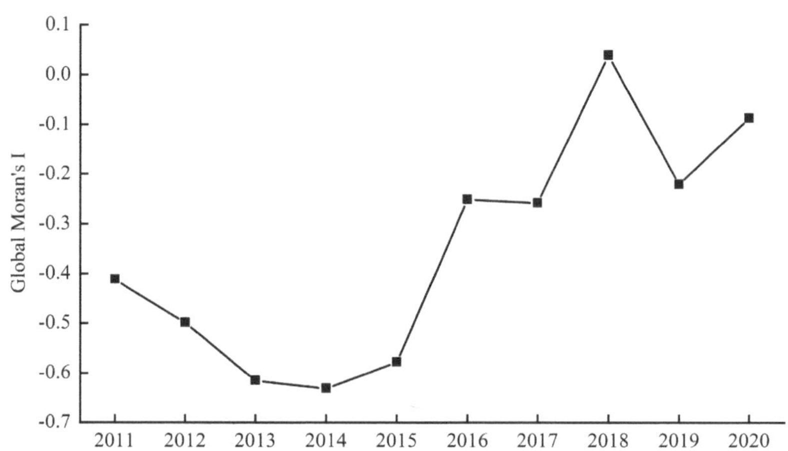

Figure 4. The change trend of the global Moreland index of marine fishery eco-efficiency.

Local Moran Index

In order to further reveal the spatial evolution characteristics of inter-regional marine fishery eco-efficiency, the GeoDa software was used to calculate the local Moran index from 2011 to 2020, and the scatter plot (Figure 5) and LISA clustering map (Figure 6) of 2011, 2014, 2017 and 2020 were generated.

(Moran's I: −0.412)

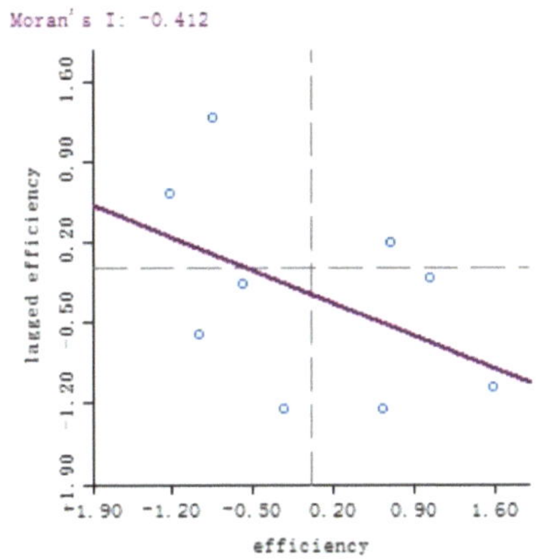

(2011)
(**a**)

(Moran's I: −0.613)

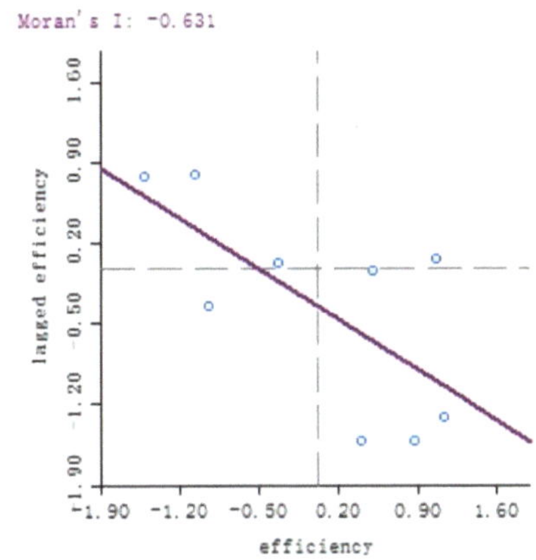

(2014)
(**b**)

(Moran's I: −0.258)

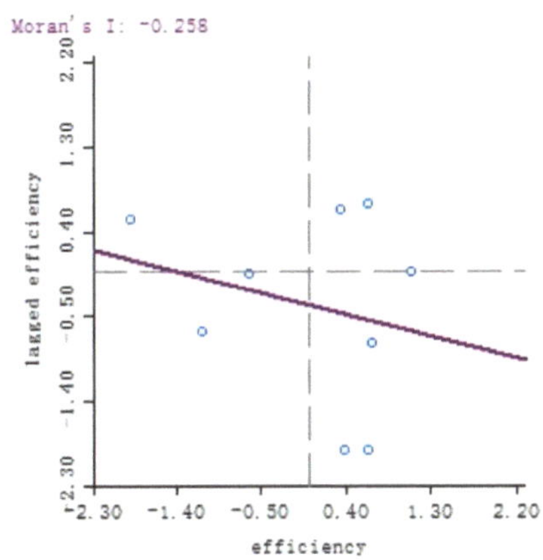

(2017)
(**c**)

(Moran's I: −0.087)

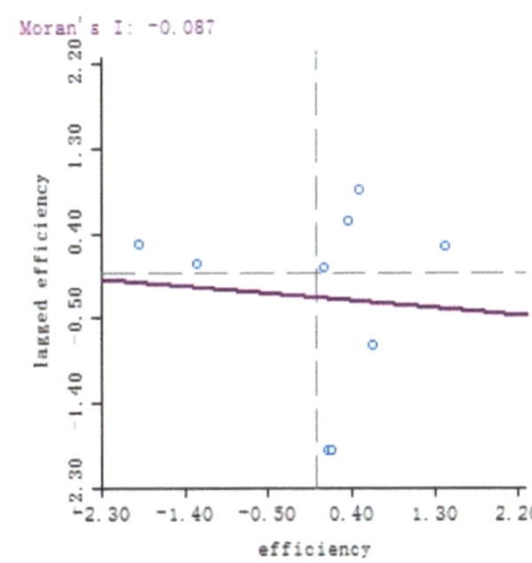

(2020)
(**d**)

Figure 5. Scatter chart of local Moran index of marine fishery eco-efficiency. (**a**) Local Moran index scatter plot for 2011; (**b**) Local Moran index scatter plot for 2014; (**c**) Local Moran index scatter plot for 2017; (**d**) Local Moran index scatter plot for 2020.

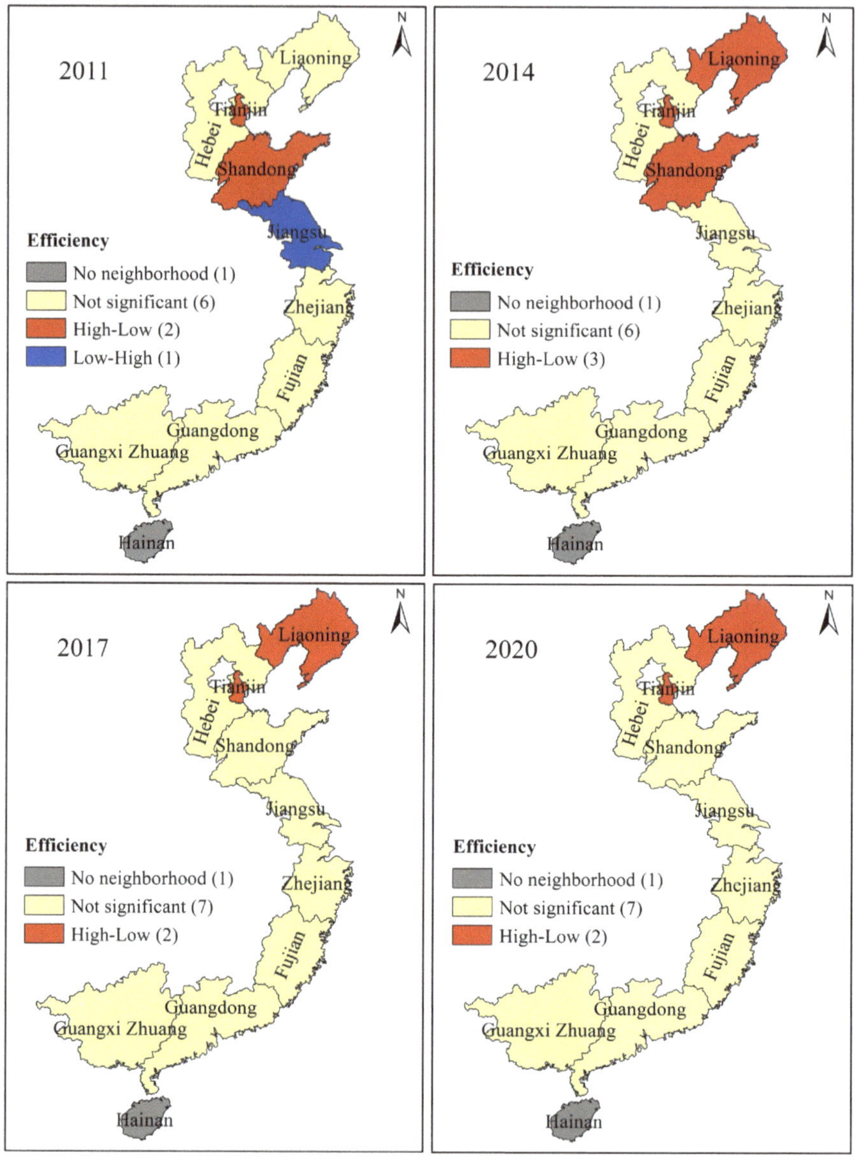

Figure 6. Spatial distribution patterns of marine fishery eco-efficiency types in 2011, 2014, 2017 and 2020.

From the analysis of Figure 5, it can be seen that the eco-efficiency of China's marine fishery has been on the rise in the past 10 years, but the spatial pattern of cross-distribution has not improved significantly. In 2011, 2014 and 2017, the provinces distributed in high–low and low–high regions accounted for a large proportion, accounting for more than 50% of the study provinces. By 2020, the provinces distributed in high–high regions have accounted for about 44% of the research provinces, indicating that the inter-provincial spatial agglomeration of marine fishery eco-efficiency has increased, and the efficiency differences between coastal provinces and cities have further narrowed. From the LISA clustering map of marine fishery eco-efficiency in Figure 6, it can be seen that Tianjin and Shandong were surrounded by low-efficiency areas in 2011 and 2014, Jiangsu was surrounded by high-efficiency areas in 2011, and the efficiency of Liaoning, Tianjin and neighboring areas was significantly different in 2014, 2017 and 2020. Combined with the distribution map, it can be seen that high-efficiency areas and low-efficiency areas are staggered from north to south. The zonal shape of China's coastal areas makes it difficult for marine fisheries to form a regional 'agglomeration' like the industrial sector. Therefore, the spatial spillover effect of high eco-efficiency areas of marine fisheries is not significant. The eco-efficiency values of Hebei and Guangxi are low for a long time, but Hebei is bordered by Tianjin and Shandong, which are high-efficiency areas, and Guangxi is close to Guangdong, which is growing rapidly. Therefore, it is necessary to explore a new model of interregional marine fishery cooperation, so that low-efficiency areas can fully accept the radiation effect from high-efficiency areas and achieve coordinated development.

4. Analysis of Influencing Factors of Marine Fishery Eco-Efficiency

4.1. Identification of Influencing Factors

Tobit regression model is used to analyze the main factors affecting the change of marine fishery eco-efficiency in China and describe its influencing mechanism and degree of action. According to the relevant research results and the principles of scientificity, generality and data availability, the factors that may affect the ecological efficiency of marine fisheries are selected to construct the index system of influencing factors, as shown in Table 6.

Table 6. Influencing factors of marine fishery eco-efficiency.

Variable	Code	Computing Method
Industrial structure	SYS	Output value of marine fishery secondary industry and tertiary industry/primary industry
Scientific support	TEC	Number of marine fishery R&D institutions/Number of fishery R&D institutions
Degree of opening-up	OPE	Total imports and exports of aquatic products/GDP of fishery economy
Environmental regulation	POL	Investment in marine environmental governance

Industrial structure determines the direction and scale of resource flows, thus affecting the level of resource consumption; the industrial structure determines the type and quantity of energy use, thus affecting the environmental situation. Some scholars have confirmed that industrial structure optimization has a positive impact on eco-efficiency [50]. Therefore, this paper expects that the optimization of marine fishery industrial structure will be conducive to the improvement of marine fishery eco-efficiency. Based on this, the proportion of the second industry, the third industry output value and the first industry output value of marine fishery in each province is selected to characterize the industrial structure of marine fishery.

Science and technology investment can promote scientific and technological innovation, thereby improving the utilization efficiency of marine fishery resources and the level of production technology, thus reducing pollutant emissions. At the same time, technological progress is also conducive to improving the level of pollutant control of related enterprises. Based on this, the proportion of the number of marine fishery research and promotion institutions in each province in the total number of national fishery research institutions is selected to characterize the scientific and technological support for marine fishery.

The implementation of fishery 'going out' is an important strategy to promote the sustainable development of fishery. Although opening to the outside world has expanded people's demand for marine aquatic products to a certain extent, the gradual expansion of import and export will lead to the excessive consumption of resources and increase in water pollution in specific sea environments; through the deep participation in international aquatic products trade, the quality of domestic fishery enterprises can be improved. Based on this, the proportion of import and export trade of aquatic products in the total output value of fishery economy in each province is selected to characterize the openness of marine fishery.

Environmental regulation refers to the direct or indirect intervention of local or central governments to related enterprises for the purpose of environmental protection and resource conservation. Under the pressure of environmental regulations, marine fishery operators will invest more resources in ecological environmental protection and carry out environmental technology innovation and upgrading in order to seek long-term development. The governments of different provinces attach different importance to the ecological environment of marine fishery, which leads to the different impacts of marine fishery industry on environmental pollution. Based on this, the investment funds of local governments in marine environmental governance are selected to represent environmental regulation.

4.2. Regression Analysis

The Tobit regression model is a model with limited dependent variables. When the value of the variable is cut or truncated, the Tobit regression model following the maximum likelihood method is a better choice. Since the efficiency values calculated using the data envelopment method are greater than 0, which belongs to the truncated case, the panel Tobit regression method is used to analyze the influencing factors.

Based on each influencing factor index, the following model is established:

$$EFF_{it} = \beta_0 + \beta_1 SYS + \beta_2 TEC + \beta_3 OPE + \beta_4 POL + \delta + \varepsilon \quad (5)$$

In the formula: EFF_{it} represents the eco-efficiency value of marine fisheries in the t year of the i province, β_0, β_1, β_2, ……, β_4 are regression coefficients of each explanatory variable, δ is the individual effect and ε is the residual term.

Panel Tobit regression of random effects of influencing factors was performed using Stata 14.0 software. The results are shown in Table 7.

Table 7. Regression results of influencing factors of marine fishery eco-efficiency.

Variable	Regression Coefficient	Z-Statistic
SYS	0.2860 ***	3.10
POL	0.0463 *	1.09
TEC	0.0040 **	2.54
OPE	−0.2212 ***	−3.03
Constant term	1.9712 ***	3.69

Note: ***, ** and *, respectively, indicate that the variables are significant at the level of 1%, 5% and 10%.

LR test results strongly reject the '$H_0 : \sigma_u = 0$'; the individual effects are existed and random effects panel Tobit regression should be used, so this model is reasonable.

According to the calculation results shown by the model:

Firstly, the industrial structure of marine fisheries has a significant positive effect on the eco-efficiency of marine fisheries, with an impact coefficient of 0.286 and a 1% significance test. This shows that the adjustment of marine fishery industrial structure in China's coastal provinces is conducive to the improvement of marine fishery ecological efficiency. The reason is that the primary industry of marine fisheries has a high degree of resource dependence, which is an industrial sector that directly interacts with the marine ecological environment, while the secondary and tertiary industries of marine fisheries have a lower resource consumption, higher technical content and added value. With the increase in the proportion of the secondary and tertiary industries, labor, capital and technology are concentrated in the fields of marine aquaculture processing, circulation and service, reducing the environmental pressure of the primary sector while obtaining more green output value, which promotes the ecological efficiency of marine fisheries.

Secondly, the level of fishery opening to the outside world has a certain negative constraint on the ecological efficiency of marine fisheries, with an impact coefficient of −0.221, and passed the 1% significance test. The higher the level of the opening up of coastal provinces and regions, the greater the demand for export and import of aquatic products. At the same time, due to the of deep processing of aquatic products, low added value of products and low processing technology, many drawbacks and structural problems are highlighted. Relying solely on marine fishery production to enhance international competitiveness has caused the deterioration of the marine fishery water environment. At present, the level of opening to the outside world is a negative indicator, but coastal provinces and regions can selectively import and export trade according to their actual situation, while paying attention to the combination of advanced technology and the protection of the ecological environment, so as to realize the benign interaction of aquatic products import and export trade, the positive effect of opening to the outside world will gradually appear.

Thirdly, science and technology support has a significant positive effect on marine fishery eco-efficiency, the influence coefficient is 0.004, and has passed the 5% significance test. The reason is that, on the one hand, the increase in R&D funding has gradually transformed the marine fishery industry from labor and capital-intensive to technology and knowledge-intensive, making full use of marine fishery resources and greatly reducing pollutant emissions. On the other hand, coastal provinces and regions have continuously improved the introduction and training mode of marine fishery talents, established scientific research institutions for marine fishery, and improved the conversion rate of achievements, which has played an important role in improving the ecological efficiency of marine fisheries.

Fourthly, environmental regulation has a positive effect on the ecological efficiency of marine fisheries, with an impact coefficient of 0.0463, and has passed the 10% significance test. Under the background of the current ecological civilization construction and green development, the coastal provincial governments pay more and more attention to the protection of the ecological environment of marine fishery waters, and attach importance to the improvement of environmental quality. Environmental regulation can directly and strongly manage and restrict the unreasonable discharge of marine fishery enterprises, and urge them to consider the cost of environmental pollution and other stakeholders while obtaining the economic benefits of marine fishery. Reasonable environmental regulation can stimulate the new demand of marine fishery-related enterprises, R&D bias green technology innovation, improve the level of cleaner production technology, upgrade pollution treatment technology and indirectly improve the ecological environment of marine fisheries.

5. Discussion

China is a maritime power, and marine fisheries are an important component of modern agriculture and the marine economy. In recent years, marine fisheries have rapidly developed, with the continuous optimization of their structure and a substantial increase in marine product output. However, the development of China's marine fisheries still follows an extensive approach, and the traditional extensive development model is no longer suitable for the current intensive and sustainable resource perspective. Problems such as excessive nearshore fishing, the depletion of fishery resources and marine environmental pollution have emerged [2]. In 2013, the Chinese government introduced "Several Opinions on Promoting the Sustainable and Healthy Development of Marine Fisheries," actively promoting the sustainable and healthy development of marine fisheries and advancing marine environmental restoration projects. The marine fisheries have gradually shifted toward ecological health development. In the future, the marine fisheries will further uphold the concept of ecological civilization construction, forging a path toward China's distinctive marine fisheries ecological development [41].

In this paper, the Super-SBM model is used to measure the ecological efficiency of marine fisheries in 10 coastal provinces and cities in China from 2011 to 2020. The

Malmquist index is used to analyze its static characteristics and dynamic changes. Kernel density estimation, GIS technology and Moran index are used to describe the temporal and spatial evolution of marine fisheries' ecological efficiency. On this basis, the panel Tobit regression model is used to analyze various factors affecting the ecological efficiency of marine fisheries.

According to the measured results of efficiency value, the marine fishery eco-efficiency value of China's 10 coastal provinces is ranked as follows: Hainan, Fujian, Shandong, Tianjin, Zhejiang, Liaoning, Jiangsu, Guangdong, Guangxi and Hebei. The ecological efficiency of marine fisheries in Hainan, Fujian and Shandong has been kept above 1.2 for a long time, which belongs to the high efficiency type. The average efficiency of Tianjin, Zhejiang and Liaoning provinces is close to 1, which belongs to the relatively efficient type. The efficiency values of Jiangsu and Guangdong are between 0.4 and 0.8, which belong to the relatively low efficiency type. Guangxi belongs to the low efficiency type. The average efficiency value of Hebei Province is less than 0.4, which belongs to the relatively invalid type.

From the static and dynamic analyses, technological progress has become an important support to enhance the ecological efficiency of marine fisheries. Each province should realize the digestion of existing technologies through the scientific allocation of elements, and explore technological innovation on this basis. On the premise of maintaining a stable ecological environment, provinces with low ecological efficiency such as Hebei and Guangxi should pay more attention to the expansion of the marine fishery scale.

From the perspective of time evolution, the marine fishery eco-efficiency in the 10 coastal provinces and cities of China shows a clear trend of improvement, and the efficiency values in high-efficiency areas remain basically stable. The efficiency values in some regions have improved rapidly, and efficiency types have been upgraded. For example, Jiangsu and Guangdong provinces have increased from low efficiency types to relatively low efficiency types, while Fujian has increased from relatively low efficiency types to high efficiency. The relative gap between efficient and inefficient regions remains significant.

From the perspective of spatial distribution, the marine fishery eco-efficiency of coastal provinces and cities in China has no obvious spatial correlation, showing a cross-distribution of high-efficiency and low-efficiency regions. Overall, the strip-shaped geographical distribution of China's coastal provinces is not conducive to the agglomeration and development of the marine fishery industry.

The change of marine fishery eco-efficiency is the result of a variety of influencing factors. The structure of the fishery industry, the level of scientific and technological support and environmental regulations have a positive effect on the improvement of marine fishery eco-efficiency, and the effect intensity is more significant. The level of opening to the outside world has a negative effect on the eco-efficiency of marine fishery at the significance level of 1%, which means that the higher the degree of opening to the outside world, the more unfavorable the improvement of the ecological efficiency of marine fisheries. Therefore, when formulating policies to improve the ecological efficiency of marine fisheries, multiple approaches must be comprehensively considered. Optimizing the structure of the marine fishery industry, enhancing the local government regulation of the marine fishery ecological environment, and increasing investment in scientific and technological funds are all effective measures to improve the ecological efficiency of marine fisheries.

6. Conclusions

To sum up, our study calculated the marine fishery eco-efficiency in China's coastal areas from 2011 to 2020, explored the rules and characteristics of its spatiotemporal changes, and analyzed the influencing factors of such changes. The results emphasize that the ecological efficiency of marine fisheries in China's coastal areas has been significantly improved, and technological progress is an important reason for the improvement of marine fishery ecological efficiency. Furthermore, the study highlights the importance of considering regional differences and acknowledges that in the future, such differentiated developments will persist for a long time. In future studies, it is necessary to continue to explore other factors that may influence the eco-efficiency of marine fisheries and to evaluate the applicability of our findings in different contexts. It is also possible to further consider the redundancy rate and deficiency rate of ecological efficiency inputs and outputs of marine fisheries.

Author Contributions: Conceptualization, W.Z. and D.L.; methodology, W.Z., W.S. and D.L.; software, W.Z. and W.S.; validation, W.S. and D.L.; formal analysis, D.L. and L.H.; investigation, W.Z.; resources, L.H.; data curation, W.Z.; writing—original draft preparation, W.Z.; writing—review and editing, W.Z. and D.L.; visualization, W.Z. and W.S.; supervision, L.H.; project administration, L.H.; funding acquisition, L.H. All authors have read and agreed to the published version of the manuscript.

Funding: This research was funded by the National Natural Science Foundation of China (42176218) and a Major Project of the National Social Science Foundation of China (21&ZD100) and Shandong Province Key R&D Plan (Soft Science Project) Key Project (2022RZB04027).

Institutional Review Board Statement: Not applicable.

Data Availability Statement: The raw data supporting the conclusions of this article will be made available by the authors, without undue reservation.

Acknowledgments: We are grateful for the funding provided by Major Project of National Social Science Foundation of China (21&ZD100), we also thank three anonymous reviewers and editors for constructive comments to improve this manuscript.

Conflicts of Interest: The authors declare no conflict of interest.

References

1. National Bureau of Statistics of the People's Republic of China. *China Fisheries Yearbook*; China Statistics Press: Beijing, China, 2021.
2. Liu, G.; Xu, Y.; Ge, W.; Yang, X.; Su, X.; Shen, B.; Ran, Q. How can marine fishery enable low carbon development in China? Based on system dynamics simulation analysis. *Ocean Coast. Manag.* **2023**, *231*, 106382. [CrossRef]
3. Guo, Z.; Tang, S.; Wang, D.; Zhang, J. Spatial distribution and factors influencing ecological efficiency of the Yellow River Basin in China. *Front. Environ. Sci.* **2022**, *10*, 235. [CrossRef]
4. Schaltegger, S.; Sturm, A. Ökologische rationalität: Ansatzpunkte zur ausgestaltung von ökologieorientierten managementinstrumenten. *Die Unternehm.* **1990**, *44*, 273–290.
5. Zhang, J.; Liu, Y.; Chang, Y.; Zhang, L. Industrial eco-efficiency in China: A provincial quantification using three-stage data envelopment analysis. *J. Clean. Prod.* **2017**, *143*, 238–249. [CrossRef]
6. Li, W.; Chen, X.; Wang, Y. Spatiotemporal patterns and influencing factors of industrial ecological efficiency in northeast China. *Sustainability* **2022**, *14*, 9691. [CrossRef]
7. Zhou, X.; Xue, Z.; Seydehmet, J. An empirical study on industrial eco-efficiency in arid resource exploitation region of northwest China. *Environ. Sci. Pollut. Res.* **2021**, *28*, 53394–53411. [CrossRef]
8. Zhang, R.L.; Liu, X.H. Evaluating ecological efficiency of Chinese industrial enterprise. *Renew. Energ.* **2021**, *178*, 679–691. [CrossRef]
9. Tan, R.; Lin, B. The influence of carbon tax on the ecological efficiency of China's energy intensive industries—A inter-fuel and inter-factor substitution perspective. *J. Environ. Manag.* **2020**, *261*, 110252. [CrossRef]
10. Stanek, W. Thermo-Ecological Cost (TEC)–comparison of energy-ecological efficiency of renewable and non-renewable energy technologies. *Energy* **2022**, *261*, 125152. [CrossRef]
11. Wang, L.; Qiu, X.; Liu, Z.; Chen, S. Ecological efficiency of China's marine economy: A convergence analysis. *J. Coast. Res.* **2019**, *94*, 983–987. [CrossRef]
12. Du, Y.W.; Jiang, J.; Li, C.H. Ecological efficiency evaluation of marine ranching based on the Super-SBM model: A case study of Shandong. *Ecol. Indic.* **2021**, *131*, 108174. [CrossRef]
13. Xu, Z.; Zhai, S.; Qian, C. The impact of green financial agglomeration on the ecological efficiency of marine economy. *J. Coast. Res.* **2019**, *94*, 988–991. [CrossRef]
14. Zhou, X.; Fang, J.; Shi, S.; Yang, L.; Xie, B.; Huang, L. Ecosystem services of ecosystem approach to mariculture: Providing an unprecedented opportunity for the reform of China's sustainable aquaculture. *Front. Mar. Sci.* **2022**, *9*, 1020. [CrossRef]
15. Liu, Y.; Cheng, X. Does agro-ecological efficiency contribute to poverty alleviation? An empirical study based on panel data regression. *Environ. Sci. Pollut. Res.* **2022**, *29*, 51892–51908. [CrossRef]
16. Chi, M.; Guo, Q.; Mi, L.; Wang, G.; Song, W. Spatial distribution of agricultural eco-efficiency and agriculture high-quality development in China. *Land* **2022**, *11*, 722. [CrossRef]

17. Lin, L. Research on the spatio-temporal evolution of tourism eco-efficiency by using the ecological footprint model. *Fresenius Environ. Bull.* **2021**, *30*, 10665–10674.
18. Ye, X.L.; Wang, L.; Zhu, Z.F.; Du, J.; Sun, R.H. Study on the ecological efficiency and its temporal dynamic evolution of China's cruise tourism from the perspective of carbon footprint. *Appl. Ecol. Environ. Res.* **2019**, *17*, 13001–13013. [CrossRef]
19. Zhang, C.; Liu, T.; Li, J.; Xu, M.; Li, X.; Wang, H. Economic growth target, government expenditure behavior, and cities' ecological efficiency—Evidence from 284 cities in China. *Land* **2023**, *12*, 182. [CrossRef]
20. Zhang, Y.; Zhang, H.; Fu, Y.; Wang, L.; Wang, T. Effects of industrial agglomeration and environmental regulation on urban ecological efficiency: Evidence from 269 cities in China. *Environ. Sci. Pollut. Res.* **2021**, *28*, 66389–66408. [CrossRef]
21. Wang, M.; Li, Y.; Liao, G. Spatial spillover and interaction between high-tech industrial agglomeration and urban ecological efficiency. *Front. Environ. Sci.* **2022**, *10*, 829851. [CrossRef]
22. Wang, D.; Wan, K.; Yang, J. Ecological efficiency of coal cities in China: Evaluation and influence factors. *Nat. Hazards.* **2019**, *95*, 363–379. [CrossRef]
23. Jollands, N.; Lermit, J.; Patterson, M. Aggregate eco-efficiency indices for New Zealand—A principal components analysis. *J. Environ. Manag.* **2004**, *73*, 293–305. [CrossRef]
24. Lin, J.; Li, Y.; Wang, W.; Cui, S.; Wei, X. An eco-efficiency-based urban sustainability assessment method and its application. *Int. J. Sustain. Dev. World Ecol.* **2010**, *17*, 356–361. [CrossRef]
25. Jin, X.; Li, X.; Feng, Z.; Wu, J.; Wu, K. Linking ecological efficiency and the economic agglomeration of China based on the ecological footprint and nighttime light data. *Ecol. Indic.* **2020**, *111*, 106035. [CrossRef]
26. Wang, L.; Zhang, F.; Fu, W.; Tan, Q.; Chen, J. Analysis of temporal and spatial differences and influencing factors of energy eco-efficiency in energy-rich area of the Yellow River Basin. *Phys. Chem. Earth* **2021**, *121*, 102976. [CrossRef]
27. Deng, X.; Gibson, J. Sustainable land use management for improving land eco-efficiency: A case study of Hebei, China. *Ann. Oper. Res.* **2020**, *290*, 265–277. [CrossRef]
28. Sun, X.; Loh, L.; Chen, Z.; Zhou, X. Factor price distortion and ecological efficiency: The role of institutional quality. *Environ. Sci. Pollut. Res.* **2020**, *27*, 5293–5304. [CrossRef] [PubMed]
29. Wang, R.; Wang, Y. Study on the measurement and Influencing factors of regional ecological efficiency in China. *Fresenius Environ. Bull.* **2018**, *27*, 4307–4317.
30. Tone, K. A slacks-based measure of super-efficiency in data envelopment analysis. *Eur. J. Oper. Res.* **2002**, *143*, 32–41. [CrossRef]
31. Guo, L.; Xu, S.Q. Industrial eco-efficiency with the integration super efficiency DEA: The central six provinces 2003–2013 date as an example. *Econ. Geogr.* **2016**, *36*, 116–122.
32. Li, C.Y.; Zhang, S.Q.; Zhang, W. Spatial distribution characteristics and influencing factors of China's inter provincial industrial eco-efficiency. *Sci. Geogr. Sin.* **2018**, *38*, 1970–1978.
33. Zhang, X.L.; Qiu, F.D.; Tan, J.T.; Wang, C.J. Spatial pattern change and influencing factors of China's industrial eco- efficiency. *Sci. Geogr. Sin.* **2020**, *40*, 335–343.
34. Kong, F.; Cui, W.; Xi, H. Spatial–temporal variation, decoupling effects and prediction of marine fishery based on modified ecological footprint model: Case study of 11 coastal provinces in China. *Ecol. Indic.* **2021**, *132*, 108271. [CrossRef]
35. Yasmeen, H.; Tan, Q.; Zameer, H.; Tan, J.; Nawaz, K. Exploring the impact of technological innovation, environmental regulations and urbanization on ecological efficiency of China in the context of COP21. *J. Environ. Manag.* **2020**, *274*, 111210. [CrossRef] [PubMed]
36. Wang, S.; Hua, G.; Yang, L. Coordinated development of economic growth and ecological efficiency in Jiangsu, China. *Environ. Sci. Pollut. Res.* **2020**, *27*, 36664–36676. [CrossRef]
37. Wang, C.; Zhang, C.; Hu, F.; Wang, Y.; Yu, L.; Liu, C. Emergy-based ecological efficiency evaluation and optimization method for logistics park. *Environ. Sci. Pollut. Res.* **2021**, *28*, 58342–58354. [CrossRef] [PubMed]
38. He, J.X.; Hu, S. Ecological efficiency and its determining factors in an urban agglomeration in China: The Chengdu-Chongqing urban agglomeration. *Urban Clim.* **2022**, *41*, 101071. [CrossRef]
39. Qin, M.; Sun, M. Effects of marine ranching policies on the ecological efficiency of marine ranching-Based on 25 marine ranching in Shandong Province. *Mar. Policy* **2021**, *134*, 104788. [CrossRef]
40. Adiga, S.M.; Ananthan, P.S.; Divya Kumari, H.V.; Mallikarjun, H. An analysis of long term catch trends in marine fishery resources of Karnataka state, India. *Appl. Ecol. Environ. Res.* **2016**, *14*, 201–213. [CrossRef]
41. Li, B.; Jin, X.M. Spatio-temporal evolution of marine fishery industry ecosystem vulnerability in the Bohai Rim Region. *China Geogr. Sci.* **2019**, *29*, 1052–1064. [CrossRef]
42. Chen, X.; Di, Q.; Hou, Z.; Yu, Z. Measurement of carbon emissions from marine fisheries and system dynamics simulation analysis: China's northern marine economic zone case. *Mar. Policy* **2022**, *145*, 105279. [CrossRef]
43. Zhou, P.; Ang, B.W.; Han, J.Y. Total factor carbon emission performance: A Malmquist index analysis. *Energy Econ.* **2010**, *32*, 194–201. [CrossRef]
44. Chen, Y.G. New approaches for calculating Moran's index of spatial autocorrelation. *PLoS ONE* **2013**, *8*, e68336. [CrossRef] [PubMed]
45. Sun, K.; Ji, J.W.; Li, L.D.; Zhang, C.; Liu, J.F.; Fu, M. Marine fishery economic efficiency and its spatio-temporal differences based on undesirable outputs in China. *Resour. Sci.* **2017**, *39*, 2040–2051.

46. Yu, X.; Hu, Q.; Shen, M. Provincial differences and dynamic changes in mariculture efficiency in China: Based on Super-SBM model and global Malmquist index. *Biology* **2020**, *9*, 18. [CrossRef]
47. Qin, H.; Zhang, Y.; Lu, Y.Y. Measurement and analysis of China's mariculture eco-economic efficiency: Based on SBM model. *J. Agrotech. Econ.* **2018**, *9*, 67–79.
48. Xu, J.; Han, L.; Yin, W. Research on the ecologicalization efficiency of mariculture industry in China and its influencing factors. *Mar. Policy* **2022**, *137*, 104935. [CrossRef]
49. Fu, Q.; Zhu, Y.; Kong, Y.; Sun, J. Spatial analysis and districting of the livestock and poultry breeding in China. *J. Geogr. Sci.* **2012**, *22*, 1079–1100. [CrossRef]
50. Cheng, J.H.; Sun, Q.; Guo, M.J.; Xu, W.Y. Research on regional disparity and dynamic evolution of eco-efficiency in China. *China Popul. Resour. Environ.* **2014**, *24*, 47–54.

fishes MDPI

Article

Governing Distant-Water Fishing within the Blue Economy in Madagascar: Policy Frameworks, Challenges and Pathways

Mialy Andriamahefazafy [1,2]

[1] Geneva-Science Policy Interface, University of Geneva, Pont d'Arve 28, 1204 Geneva, Switzerland; mialy.rann@unige.ch

[2] Centre for Blue Governance, University of Portsmouth, Winston Churchill Avenue, Portsmouth PO1 2UP, UK

Abstract: Madagascar's vast oceanic space hosts distant-water fishing (DWF) fleets from Taiwan, Japan, South Korea, Spain, France and others since the 1960s, making DWF a substantial component of the blue economy. Considering this extensive experience of managing DWF activities for more than 60 years, this paper explores the existing policy frameworks and challenges regarding managing DWF. The results show while it is well equipped legally, the country is struggling to implement its national policies and laws while continuing to adopt new management frameworks. This is due to a limited coherence on long-term policy making and policy implementation, resulting in a mismatch between the two, and a paradoxical vision that promotes DWF without the means to monitor fishing activities and their impacts. The existing institutional settings and governance frameworks make change possible in Madagascar's approach to DWF. To improve the management of DWF, this paper outlines four pathways. These include (i) a greater attention on the implementation, harmonisation and evaluation of existing policies and projects; (ii) continuing efforts on transparency for DWF operations and contract negotiations; (iii) realigning aspirations and policies with local needs; and (iv) taking better advantage of measures adopted at the regional Indian Ocean level to improve national management of DWF. An efficient implementation of an improved policy and legal framework could contribute to strengthening the governance of DWF activities toward sustaining national benefits while preserving coastal livelihoods.

Keywords: fishery governance; national legal framework; tuna fishing; distant-water fishing nations; transparency

Key Contribution: Despite the challenging socio-economic and political context in Madagascar, the management of distant-water fishing is based on a rich and evolving policy framework. Managing distant-water fishing in developing coastal states like Madagascar requires addressing challenges of institutional misfits, transparency and capacity towards monitoring, control and surveillance.

Citation: Andriamahefazafy, M. Governing Distant-Water Fishing within the Blue Economy in Madagascar: Policy Frameworks, Challenges and Pathways. *Fishes* **2023**, *8*, 361. https://doi.org/10.3390/fishes8070361

Academic Editor: Ben Drakeford

Received: 1 June 2023
Revised: 29 June 2023
Accepted: 6 July 2023
Published: 11 July 2023

1. Introduction

With more than 1.2 million km^2 of an exclusive economic zone (EEZ), Madagascar hosts important marine resources. Due to its large land mass and vast population, fisheries represent a smaller part of GDP compared to agriculture [1]. However, it is still an important source of revenue for the national budget, and it is a key pillar for Madagascar's blue economy policy [2]. Moreover, coastal populations depend on fisheries for livelihoods and protein intake. They are highly vulnerable and often have no alternative source of revenue. A little over 50% of the total Malagasy population of 26 million live within 13 coastal regions [3,4]. State revenues from fisheries are generated through the issuance of various fishing licenses and permits to fish, transport, sell and export marine products. A large part of state revenues come from fees associated with tuna fishing access agreements, which represent the Ministry of Fisheries' primary income source. In 2018, when all DWF fleets were active, fishing access agreements generated around EUR 3 million in revenue [5].

Tuna caught from distant-water fishing (DWF) is also the second most exported fishery product, representing 17% of exports in 2016, just after shrimp [6].

Distant-water fishing in the waters of Madagascar started with exploratory fishing by Japanese and Taiwanese ships in 1955 looking for yellowfin and albacore tuna around Madagascar [7]. In the 1980s, the European fleet entered Madagascar's EEZ and started fishing tuna and associated species under fishing access agreements, followed by agreements with Asian countries and companies from Japan and Taiwan. The largest DWF stakeholders are Asian fleets, mainly Taiwanese, Japanese and South Korean, with 121 licensed vessels in 2018 [5]. They account for the largest part of the tuna catch in the EEZ [8]. However, they do not land in Madagascar so there is limited information available regarding their operation. The Asian fleets are composed of longliners that catch tuna and associated species like swordfish. In 2016, their catch was estimated at around 6000 tons by the former tuna statistic unit [9]. Species of sharks are also caught as bycatch, although there is no updated data available to establish the proportion of such catch. The European Union (EU) fleet is the second largest stakeholder in Madagascar's EEZ and the most well-known by the Malagasy public. The EU fleet is the only fleet that lands in Madagascar. While it is not a legal obligation, it is incentivised in the access agreement by a reduction in licence fees if there is landing in a Malagasy port or processing company. In 2018, 61 European fishing vessels, mainly Spanish and French, were authorised to fish in the Malagasy EEZ [5]. While it records the highest catch of all fleets in the Indian Ocean, estimated at more than 210,000 tons in 2016 [10], the EU fleet has a lower catch in the waters of Madagascar, with 2600 tons in 2016, less than half of the Asian fleet's catch [9]. The EU fleet consists mainly of Spanish and French vessels [11,12]. The EU fleet is composed mainly of purse seiners but also has longliners that operate on the eastern side of Madagascar's EEZ from the island of La Réunion. Available reports from the former statistic unit estimated that the annual catch of tuna by foreign industrial fleets in Madagascar's EEZ is around 10,000 tons per year [9].

DWF activities are managed at the national level by the ministry in charge of fisheries. As of 2022, this is the Ministry of Fisheries and blue economy. The ministry oversees the adoption and implementation of legal and policy frameworks while also concluding fishing access agreements. Two units are particularly relevant for distant-water fishing: the Tuna Fisheries Unit and the Centre for Fisheries Surveillance. The Tuna Fisheries Unit oversees industrial and semi-industrial activities and administers fishing access agreements and licenses. The surveillance centre undertakes the monitoring control and surveillance of fishing activities within the EEZ. Various civil society organisations and NGOs are also indirectly involved in DWF management. The WWF, for example, funded policy initiatives including two strategies for the management of tuna fisheries; CSOs also take part in consultations of stakeholders including during the 2022 negotiation of the EU agreement renewal (pers. obs.).

With the long-term involvement of DWF fleets in the waters of Madagascar, the government also adopted various frameworks to establish the modalities of DWF and manage these fishing activities. This paper aims to analyse the governance of distant-water fishing (DWF) in Madagascar. To do so, it looks at three aspects: the policy framework for managing DWF in the country, the challenges faced by stakeholders and some proposed pathways to make DWF more relevant to national needs within the blue economy narrative in Madagascar. This paper brings an important contribution to the literature on DWF, which often pictures developing coastal states as rather passive actors when it comes to managing DWF activities. It shows that Madagascar has made various policy efforts to manage DWF although such efforts have not always been coordinated or monitored for their impacts on management and local livelihoods.

2. Materials and Methods

This paper is based on two methods. To analyse the policy frameworks available for DWF, a review has been undertaken of national laws and policies relevant to distant-water fishing between 2012 and 2022 in Madagascar (Table 1) and of the content of the United Nations Convention on the Law of the Sea (UNCLOS) related to DWF. The review includes a presentation of the relevant texts as well as observations on their implementation, based on publicly available information and interviews. From the review, some challenges emerged that were also raised by stakeholders. Stakeholder insights were gathered from 40 interviews undertaken in Madagascar in 2017 and 2018. While interviews dated from 2017 and 2018, the author continued to have informal exchanges with various stakeholders interviewed either through expert advice to the NGOs and the Ministry of Fisheries (between 2019 and 2022) or fieldwork in fishing villages (in June 2022). There was no substantial change of views noticed regarding the insights shared in this paper. Actors included 22 local fishers based in four coastal towns (referenced in the text as ArtFisher for artisanal fishers operating offshore, SSFisher for small-scale fishers operating within coastal waters and AssoFisher for a fishing association representative), 13 government officials from local to national levels (referenced as GovRep in the text) and 5 NGO staff members working in fishery management (referenced as NgoRep in the text) (see Supplementary Material S1 for more details on the interviews). These stakeholders were chosen based on their long-term involvement in fishery management (more than 5 years for government representatives and NGOs) and for their involvement in offshore fishing activities or knowledge on these activities for the fishers. The stakeholders were asked three open questions: what they knew about DWF in the Malagasy EEZ (for fishers and NGOs), what were the actors and policies related to DWF (government officials) and what were their perspectives on the challenges and future regarding fishery management in Madagascar (all actors). The section on challenges is presented through the lens of institutional misfits [13] and transparency in fishery governance [14] as two complementary angles when looking at policy implementation in DWF. After the analysis of frameworks and challenges, a discussion section presents potential pathways towards improving DWF management. These pathways were inspired from solutions provided by the literature on institutional misfits and fishery governance. They were adapted to the case of Madagascar by mobilising insights from the interviews and perspectives from the author.

Table 1. List of national laws and policy reviewed.

Name of the Legal Document	Adoption Year	Area Covered by the Text
National strategy on the management of tuna fisheries in Madagascar [1]	2014	Main priorities for the management of DWF and improvement of governance
Law n° 2015-053 of December 2nd 2015 regarding the Fishery and Aquaculture Code [2]	2015	Modalities of DWF
Blue Policy paper [3]	2015	Main vision and aspirations for DWF in Madagascar
Law n° 2018-025 regarding maritime zones in the maritime space under the jurisdiction of the Republic of Madagascar [1]	2018	Contextualising fishing activities in the exclusive economic zone of Madagascar
Updated national strategy on the management of tuna fisheries in Madagascar [1]	2021	Updated priorities and actions for the management of DWF
Fisheries Transparency Initiative Standards [4]	2021	Standards regarding transparency of DWF agreements and modalities of operation
Malagasy blue economy strategy for the fishery and aquaculture sector	2022	Improvement of fishing access agreements and the fight against IUU fishing.

[1] Collected through email exchanges with the Ministry of Fisheries. [2] https://www.fao.org/faolex/results/details/fr/c/LEX-FAOC162704/ (accessed on 23 February 2022). [3] https://www.fao.org/faolex/results/details/fr/c/LEX-FAOC163970/#:~:text=Madagascar%20(Niveau%20national)-,Lettre%20de%20Politique%20Bleue.,principales%20orientations%20jusqu\T1\textquoterighten%202025 (accessed on 23 February 2022). [4] https://www.fiti.global/fiti-standard (accessed on 5 July 2022).

3. Results

3.1. Analysis of National and International Legal and Policy Frameworks Governing Distant-Water Fishing in Madagascar

3.1.1. The Fishery and Aquaculture Code (2015)

At the national level, this code provides the legal basis for allowing DWF in Madagascar's EEZ. While Article 25 states the conditions for national fleets to undertake fishing in the EEZ (including a national registration and being chartered by a Malagasy owner or entity), Articles 26 to 40 state the conditions under which foreign vessels can undertake fishing in Madagascar's EEZ. Table 2 presents the key articles that govern these fishing activities.

Table 2. Relevant clauses in the fishery law regarding DWF by foreign fleets.

	Content	Ref. in the Law	Comments on Implementation (as of September 2022)
Modalities of fishing	- Fishing zone must be beyond the territorial waters	Art. 26	- These modalities of fishing are present in the existing template of fishing access agreements and all EU public agreements - Art. 28, 30 and 31 relate to IUU fishing. White et al. (2021) [8] established that there is some DWF within marine protected areas while WWF (2023) [15] also established that more than 2500 metric tonnes of tuna catch might be under-reported in Madagascar. This evidence shows limited compliance with these legal clauses by DWF vessels.
	- Fishing must comply with existing management plans and national navigation laws	Art. 28 and 31	
	- Vessels must maintain a logbook of catch, fishing area, landings and any other information required by the ministry	Art. 30	
	- Ministry can reserve certain fishing areas to the national fleet	Art. 34.b	
	- Mandatory compliance of vessels with international and national laws	Art. 35.c	
Content of access agreements	- Parties who can enter into access agreements: country, organization of countries, fishing association, fishing company	Art. 35.a	The elements prescribed in Art. 35b are present in the existing template of fishing access agreements used by the ministry. These elements are also present in all EU public agreements. Agreements that allowed for a consultation also included a clause mentioning that agreements might not be renewed in the case of a lack of respect of the agreements' terms or lack of respect of fishing modalities as presented above.
	Mandatory key contents of agreements: - Number and types of vessels allowed - Payment and access fees - Vessel marking requirements - Mandatory communication of catch data - Flag state obligation to ensure compliance with agreement by vessels and assistance with monitoring and control - Penalties in case of non-compliance with agreement terms	Art. 35.b	
	- Madagascar must align with countries of the Indian Ocean region regarding terms and conditions of agreements	Art. 36	Implemented through the adoption of the Southwest Indian Ocean Fisheries Commission (SWIOFC) guideline for minimum terms and conditions on fishing access agreements in 2018
Fishing licenses	Content of licenses: - Types and quantity of fishing gear and equipment - Authorized periods and areas of fishing - Authorized quantity and minimum size and weight of species caught - Restrictions on bycatches and discards - Boarding of vessels by observers and scientists	Art. 38	- Only the EU public agreements include a reference to quantity of species caught through a reference tonnage that can potentially be caught - The rest of these items are present in the existing template of fishing access agreements and included in all EU public agreements

Table 2. *Cont.*

	Content	Ref. in the Law	Comments on Implementation (as of September 2022)
Fishing licenses	Conditions under which licenses will not be renewed: - Proven engagement of vessels in IUU activities - Vessels not complying with regional measures - Vessels without valid registration - Vessels without an authorization to fish beyond their flag state waters - Vessel owner or fishing company convicted of a fishery-related offence 5 years before license application - Non-compliance with national regulations and management plans	Art. 40	White et al. [8] reported that there is some illegal fishing within marine protected areas and territorial waters while authorities interviewed mentioned such occurrences have been limited over the years. WWF [15] also indicated that more than 2500 metric tonnes of tuna catch might be under-reported in Madagascar. If these fishing operators are identified, it could impact the renewal of their fishing licenses.

The Law n° 2015-053 of 2 December 2015 on the fishery and aquaculture code also specifies that the ministry has the right to adopt bylaws for further regulation, including modalities for granting, renewing, suspending and cancelling licenses to vessel operators. In 2016, the Malagasy government issued a decree to continue implementing the fishery law and provided more details on potential terms, including for access agreements such as encouragement for landings at ports, training of local fishers, infrastructure building, technology transfer and fishery development (Art. 18 of Decree 2016-1492 regarding the reorganization of marine fishing activities). As of 2022, despite the existence of access agreements (Table 4), some of these measures notably increase landings at ports and technology transfer did not materialise and remain in the realm of aspirations of the ministry in charge of fisheries. In the past 10 years, landings at ports have decreased rather than increased in the country [9]. Infrastructure building and training of local fishers have been ongoing under donor-led projects such as the World Bank South West Indian Ocean Fisheries Governance and Shared Growth Project (SWIOFISH2) allocating more than USD 20 million in building human and institutional capacity necessary to implement policies and management plans regarding fisheries [16]. The impact of these initiatives on the development of the sector is yet to be seen.

3.1.2. The Blue Policy Paper (2015)

In 2015, the Ministry of Fisheries adopted a Blue Policy paper (lettre de Politique bleue) outlining the objectives of the ministry regarding the "blue economy" in the fishery sector. It also included the core principles that should govern the fishery sector, which were ensuring the sustainable management of fisheries' resources, increasing the productivity and economic contribution of the sector, improving the food and nutritional security of small-scale fishers and fish farmers, fulfilling the national demand for seafood and promoting transparent and accountable governance [2]. Each of these principles were accompanied with proposed implementation measures. Under Section 10 of this policy, industrial fishing is considered one of the main engines to increase the productivity and economic contribution of the fishery sector. At the national level, measures to achieve this goal include promoting and supporting a national fleet focused on high commercial species such as tuna and demersal species. The ministry, therefore, wanted to improve its overall business framework and encourage new private investments. As of 2022, this was only partially achieved by the exploration of flagging 10 Chinese vessels to increase productivity (Table 4 below). In fact, the number of vessels within the national fleet has decreased from eight vessels in 2013 to five vessels in 2019 [17].

At the DWF level, the ministry aimed to improve economic benefits from fishing access agreements. The Blue Policy paper demonstrates the ministry's goal of gradually harmonizing agreements by standardizing their duration of validity, their access fees and their terms of implementation. As of 2022, this can be considered as achieved by the ministry through the adoption of a licence-fees grid by the government. The grid was, however, not applied to the latest EU agreement, which nonetheless included higher fees compared to past EU agreements. For standardization, the ministry already had a template for fishing access agreements that applies to all non-EU agreements.

3.1.3. The National Strategies for the Management of Tuna Fisheries (2014 and 2021)

In 2014, the ministry in charge of fisheries, with funding from the WWF, adopted a national strategy for tuna fisheries, which are the main industrial fisheries in Madagascar's EEZ. The ministry and various stakeholders adopted an updated strategy in 2021. The 2021 strategy set up five key areas for the management of tuna fisheries. These are mostly like the areas identified in the 2014 strategy and include the following:

Ensuring coherent, transparent and responsive governance of tuna fisheries (already present in the 2014 strategy).

Making all statistics and other necessary information available to the ministry for more informed management of tuna fisheries (already present in the 2014 strategy).

Improving the attractivity of Malagasy ports for foreign industrial fleets.

Developing the national tuna fishery.

Ensuring that the exploitation of tuna resources and the development of fisheries do not harm the environment and marine ecosystems (already present in the 2014 strategy).

Each area specifies measures to be undertaken such as building the capacity of the ministry and its statistical unit (the latter was, however, undermined by the dissolution of the unit later in the same year), clarifying legal texts, maintaining the presence of DWF fleets, developing a national fleet and reducing the environmental impacts of tuna fisheries.

In Table 3, the 2014 and 2021 tuna strategies were evaluated by comparing indicators of success against results as of 2022. As can be seen concerning DWF, most of the measures are either ongoing, partially implemented or not yet implemented. This shows that although Madagascar is aware of which measures are needed in its fishery sector, including for DWF, the country continues to struggle to fully implement its policies and laws.

3.1.4. Law Regarding Maritime Zones in the Maritime Space under the Jurisdiction of the Republic of Madagascar (2018)

In 2018, the government introduced legislation to define the maritime space under Madagascar's jurisdiction and permitted activities based on international law. Territorial waters were established at 12 miles from the baselines (low-water line or geodesic lines) and within which Madagascar has sovereign rights. The contiguous zone extends to 24 miles from the baselines and allows the state to prevent and pursue offences. The text sets the EEZ at 200 miles from the baselines, an area within which any occurring activity must comply with existing national laws such as the fishery code or the mining code. The legislation also specifies that Madagascar will engage in marine spatial planning following its national needs and the development of a "blue economy" (Art. 21). Furthermore, new conservation and exploitation measures within Madagascar's marine space are subject to environmental assessments (Art. 22). A strict interpretation of this legislation would mean that while existing DWF is accepted, new fleets arriving in Madagascar's EEZ could be subject to environmental assessments. Drafts of fishing access agreements shared with civil society in 2022 did not include such a clause (pers. obs.).

Table 3. Evaluation of the tuna strategies against indicators as of September 2022.

Key Elements of the Strategy	Indicators of Success	Year of Activity Launch	Achievement as of September 2022	Comments
Activities in the 2021 strategy				
R.1.2. Establish and operationalise a fishery consultation platform to align with the reform to improve fishery management in Madagascar	• Platform in place and active	2021	Ongoing [1]	• A platform of observers and CSOs is convened for access agreements and other similar processes; the platform still needs to be officialised.
R.2.1. Strengthening statistical and information systems	• Data collection mechanism • Capacity-building plan for statistic units	2022	Partially implemented	• This key activity has been stalled by the dissolution in 2021 of the statistic unit and the unit of socio-economic data on fisheries.
R.1.3. Systematic publication of licences, royalties and bilateral agreements	• Publications	2022	Partially implemented	• There have been publications on the Facebook page of the ministry, although they are not systematic with their content.
R.1.4. Reviewing the way fees are calculated	• New access-fee grid	2021	Implemented	• A new grid was adopted by the government in 2021. However, it was not used for the 2022 EU agreement since EU agreements in the Indian Ocean region were subject to a harmonised mode of access fees calculation, different to the grid.
R.5.1. Promote the implementation of the Environmental Impact Assessment (EIA) decree on fishing activities (fishing agreement)	• EIA conducted before DWF activity	-	Not implemented	• No mention of EIA in the three agreements concluded in 2022.
Activities in 2014 tuna strategy also present in 2021 strategy				
R.1.5. Improvement of agreements on fisheries by harmonizing these agreements at the national level	• Definition of a national model of agreements • Revision of agreements according to this model	2019	Ongoing	• The 2012 model needs to be updated and aligned with existing guidelines. • One of the blue economy projects aims to establish a new template.
R.1.3. Establishing a chart of dissuasive sanctions; classifying sanctions by fishery and type of offence	• Establishment and validation of a chart • Recording offenses	-	Not implemented	• Chart not established yet.
	• Using IOTC referral protocol to report IUU cases	2014	Ongoing	

Table 3. *Cont.*

Key Elements of the Strategy	Indicators of Success	Year of Activity Launch	Achievement as of September 2022	Comments
R.3.3. Strengthening MCS systems in Madagascar and in the region: VMS data, port inspections, offshore control, observer program, information exchange in the region	• Number of controls/inspections • Number of offences identified • Exchange of information	2014	Ongoing	• Madagascar takes part in regional missions and hosts the centre for data management of IOC countries. • The SADC Regional Fisheries Monitoring, Control and Surveillance Coordination Centre charter was signed in 2022.
R.3.1. Increasing landings, trans-shipments and stopovers of DWF fleets in Malagasy ports by consulting operators and service providers	• Landing at Malagasy port	-	Not implemented	• Since both strategies have been adopted, landings have not increased in Malagasy ports.

[1] As of the time of the revision of this paper (June 2023), this activity has now been implemented through the adoption of a bylaw (16026/2023) officialising the establishment of a consultative platform for the management of fisheries.

3.1.5. Commitment to the Fisheries Transparency Initiative in 2021

In September 2021, the Malagasy government via the ministry in charge of fisheries publicly declared its commitment to join the Fisheries Transparency Initiative (FiTI). The FiTI gathers 12 standards of transparency including on foreign fishing access agreements, large-scale fisheries (vessel registry, payments for fishing, recorded catch data) and beneficial ownership. The FiTI was launched in the country in May 2022. In addition to assessing the status of Madagascar on the standards, the FiTI also establishes a platform with all relevant stakeholders in the sector to exchange information including on DWF activities. While not a written legal or policy text, the FiTI still represents an important policy instrument for the country implementing its current vision of transparency in fisheries. Considering the recent adoption of this instrument, it is not possible, at the time of the submission of this paper, to assess the level of implementation of the standard by Madagascar. Joining the initiative involves committing to assessing adherence to the standard, which will be presented in transparency reports validated by the FiTI.

3.1.6. The Malagasy Blue Economy Strategy for the Fishery and Aquaculture Sector (2022)

In 2022, the ministry in charge of fisheries adopted its blue economy strategy, validated by national stakeholders. The strategy has two components that are highly relevant to DWF, which are the improvement of fishing access agreements and the fight against Illegal, Unreported and Unregulated (IUU) fishing. Activities linked to fishing access agreements included the establishment of a fishery agreement and operation analysis unit, increasing the added value of access agreements and improving ports and processing infrastructures at ports. Regarding IUU fishing, activities include the implementation of the national plan to combat IUU fishing, the establishment of mechanisms to combat IUU fishing and the strengthening of regional cooperation. The implementation of this strategy is yet to be seen. Considering the recent adoption of this policy, the level of implementation cannot be assessed. Although, the recent WWF report (2023) on IUU fishing in the Southwest Indian Ocean reflects that IUU remains an important challenge for Madagascar to address.

3.1.7. The UNCLOS as the International Instrument Framing Access by Distant-Water Fishing Nations (1982)

The most relevant international framework for distant-water fishing is the United Nations Convention on the Law of the Sea (UNCLOS). Through UNCLOS, countries were granted a bundle of rights regarding natural resources within their national waters: the right to access, use and manage resources within those limits, the right to determine who can have access and use rights and the right to lease or sell those rights. In addition to demarcating geographical boundaries, the UNCLOS established the right of distant-water fishing nations (DWFNs) to access resources in coastal countries' EEZs. The UNCLOS also prescribes the need for regional cooperation for the conservation and management of migratory species or shared stocks (Article 118). This led to the creation of regional fishery management organisations such as the Indian Ocean Tuna Commission (IOTC), to which Madagascar is a party since 1996.

Article 62 of the UNCLOS prescribes that if coastal countries are not able to fully exploit their marine resources, they can establish access agreements with DWFNs to exploit these resources. From UNCLOS, various types of fishing access arrangements developed between DWFNs and coastal states. They range from bilateral agreements with governments, industry associations or fishing companies to the allocation of access and/or reduced licensing costs in return for vessel flagging or investments in the country. Fishing access agreements determine the terms of access to tuna fishing grounds within the EEZ. They set the fees to be paid in exchange for access, and they specify the number of vessels that can be licensed to access the EEZ, the accessible fishing area, the species that can be fished and other conditions such as the obligation to install satellite devices onboard vessels.

Fishing access agreements are based on the assumption that countries have established their capacity to exploit their resources and can lease the surplus that they are not able to fish to DWF nations. Considering its challenging socio-economic context, which is similar to that of other developing countries, Madagascar was and is still not able to determine its capacity to harvest its marine resources. Similarly, its surplus has not been determined. In the case of offshore marine resources such as tuna or swordfish, which are targeted species in fishing access agreements, Madagascar relies on regional assessments at the Indian Ocean level by the Indian Ocean Tuna Commission (IOTC).

A range of stakeholders has legal access to Madagascar's EEZ through fishing access agreements. These include countries but also fishing associations and specific fishing companies. In 2019, operators from six countries were fishing in the waters of Madagascar: Spain, France, South Korea, Taiwan, Japan and China [5]. Most of the catch in the waters of Madagascar is caught by South Korean and Taiwanese longliners (6000 tons in 2016, mostly albacore, yellowfin and bigeye tuna), followed by European purse seiners (2600 tons in 2016, mostly skipjack, yellowfin and albacore). The remaining 1400 tons are caught by European, Japanese and Malaysian longliners [9]. The purse seiners operate in Madagascar's EEZ mainly between March and June, whereas longliners operate between October and March, especially in the eastern and southern parts of Madagascar's EEZ.

Madagascar has two types of fishing access agreements with DWF nations (Table 4). There are private agreements that are not accessible to the public (as of the time of the submission of this paper). Confidentiality of fishing access agreements has been practiced since the start of DWF within the EEZ and is commonly seen in other countries as well despite that such opacity has often been criticised [18]. There are also public agreements, of which the EU agreements are the only case and in line with the Common Fisheries Policy requiring transparency for all public EU agreements. Between 2016 and 2022, there have been between six and nine active agreements every year. Interestingly, Spain and France access the EEZ through a public access agreement between Madagascar and the EU with some of their operators (ANABAC and OPAGAC) also having private agreements with the Malagasy government through vessels flagged to Mauritius or Seychelles. In 2017 and 2018, only Spanish vessels were landing in Madagascar [9]; pers. obs.

Table 4. Types of access agreements with DWF fleets in Madagascar.

Party of the Agreement	Flag State	Type	Type of Vessel	Nbr of Vessels	Status (as of September 2022)	Comments
European Union	Spain and France	Public	Purse seiners	32	Signed [1]	4-year agreement
			Longliners	33		
Interatun	Seychelles/Mauritius	Private	Purse seiners	5	Signed	2-year agreement
Japan Tuna	Japan	Private	Longliners	10	Signed	2-year agreement
Dae Young Fisheries	South Korea/ Taiwan	Private	Purse seiners	3	Applied for renewal	
		Private	Longliners	72	Applied for renewal	
ANABAC	Seychelles/Mauritius	Private	Purse seiners	6	Applied for renewal	
OPAGAC	Seychelles/Mauritius	Private	Purse seiners	6	Applied for renewal	

Source: Data collected from public sources and governmental documents by September 2022 [5,19]. [1] At the time of the revision of this paper (June 2023), this agreement has not been enforced yet as it was in the validation process by the EU parliament.

Fishing access agreements represent a dilemma for the country. Madagascar's Ministry of Fisheries relies heavily on revenue from access agreements, representing approximately 80% of the department's revenue [6] and funding a majority of governmental activities. At the same time, Madagascar's access agreements have been highly criticized over the years. Le Manach et al. [20] reported that over the years, the EU has not paid a fair price for accessing the Malagasy EEZ as it did not consider the inflation nor currency devaluation or the landed value of tuna. Carver [21] reported on the opacity of the agreements and the potential risks to local fishers. Andriamahefazafy et al. [22] also showed that these agreements generate aid dependency and limit the geopolitical power of Madagascar in negotiating management measures regarding tuna fisheries at the Indian Ocean level. Gorez [23] argued that fishing access agreements could be a serious threat to the livelihoods of local fishers. Such critics are not unique to the case of Madagascar. More global research has raised similar issues. Access agreements contribute to the depletion of fish stocks, constitute a threat to national economic development and affect local fishing communities including by hindering food security [24–28].

Regarding conservation and management measures (CMMs) adopted at the IOTC, 59 CMMs were active as of September 2022, 56 being binding resolutions and 3 being recommendations [29]. These CMMs cover a vast array of topics from data reporting and management of certain gears to catch limits and harvest strategies. According to the compliance report of Madagascar, the country had 17 non-compliance issues, 10 of them being repeated non-compliance occurrences. These relate to a lack of data reporting regarding national catch and lack of data reporting regarding certain species of IOTC such as shark, marlin or sailfish [19]. Madagascar was, however, compliant for obligations linked to reporting of vessels in the EEZ, notably by complying with the obligation to provide the list of foreign vessels licensed in the EEZ and information on access agreements (the latter consisting of providing a list of signed agreements (ibid.)). Madagascar also participates actively in the negotiation of CMMs at the IOTC, although its stance has been variable depending on topics. Madagascar, for example, co-sponsored a ground-breaking resolution on the management on drifting fish aggregating devices [30] while it also objected to the resolution on the rebuilding plan for yellowfin tuna, which has been overfished since 2015 [31].

3.2. Governance Challenges Linked to Distant-Water Fishing in the Malagasy Waters

Madagascar faces various socio-economic and political challenges, such as high rates of poverty and corruption, which have greatly impacted its institutional efficiency [32]. In this context and adding other factors such as geopolitics, managing DWF remains a challenge for Madagascar. The country suffers from various inadequacies between policy making, implementation and local realities, also known as misfits in the socio-ecological systems literature [13,33] as well as implementation challenges found in the literature of fishery management mainly around transparency and the challenge of fighting IUU fishing [14,34].

3.2.1. Spatial and Social Misfits of DWF Policies

The first governance misfit that Madagascar faces is the spatial misfit, within which the institutional arrangements and policies adopted do not match the spatial scale of the socio-ecological system to be managed [13,35]. Due to Madagascar's large landmass and EEZ, DWF policy adopted at the national level can be challenging for coastal regions. When new laws are passed, the country has limited resources to ensure outreach at the regional and local levels. Regional representatives of the ministry are given the text of the law, while coastal communities are often informed through the work of NGOs in the country. Fishers interviewed presented anecdotes of encountering large vessels when going offshore without any knowledge about the legalities or modalities of DWF in the EEZ. Local and regional needs are often difficult to convey at the national level, where laws are developed. The state has limited resources to consult local communities; the ministry's regional offices are only located in the main cities. They are mainly in charge of the registration of all fishers and fishing vessels and implementing projects specifically targeted to their region (GovReps 4, 6 and 8). Over the past 10 years, various NGOs have reached out to isolated fishing communities and undertaken projects to organize fishers into fishing associations and cooperatives to amplify their voice at the national level (NgoReps 1, 3 and 5). The Ministry of Fisheries validated that approach in the 2015 fishery code by giving fishing associations the right to manage fishing resources at the local level through management transfers. However, the question of representation within associations remains an issue, as not all fishers want to be part of an association, leading to some associations having a short life span or being dominated by local elites in the fishing villages (ArtFisher 10, SSFishers 6 and 8). Therefore, to develop its policies and laws, the ministry relies on inputs from the NGOs working within fishing villages and from well-established or trusted fishing associations and cooperatives.

The spatial misfit also manifests with the paradox of priorities that the state faces between DWF and small-scale and artisanal fisheries. As seen in the analysis of the tuna strategies in the previous section, developing the domestic and DWF industrial sector remains a key priority for tuna fisheries in Madagascar. In this context, the impact of DWF activities on small-scale and artisanal fishers is not considered. On the other hand, the socio-economic contribution of small-scale and artisanal fishing to the local and national economy is less known and less visible in the national balance of benefits provided by different sectors of fisheries. Fishers have long demanded more support for their activities as resources continue to deplete [36]. Such a limited inclusion of local needs has often led to regulations and policies which, while they acknowledge the importance of local fisheries and fishing communities, are either difficult to implement at the local level or opposed by fishers. As a solution, the Ministry of Fisheries often passes bylaws and implements regulations. Currently, there are still some legal texts under development aimed at implementing the 2015 fishery code. Given the challenging context, these may add to the existing backlog of legal texts to establish and implement.

Madagascar also faces a social misfit within which views at different levels do not match the institutional arrangements in place [13]. This usually manifests with fishers' mistrust of the government including the ministry in charge of fisheries. Local fishers often see the government as a repressive rather than collaborative entity, one that often intervenes in fishing communities to prosecute offences and illegal activities (views shared by two to three small-scale fishers in each of the four coastal towns). Furthermore, fishers have very little knowledge of the contribution of fishing access agreements to their livelihoods even though some projects for local fisheries are funded by such agreements. Most fishers interviewed often talked of the need to improve their fishing equipment and vessels or the supply chain. The ministry has only been able to respond to these demands intermittently through projects that, once implemented, are not guaranteed to be monitored. The dependency on donor funding also limits the ability of the state to have a long-term vision for the various sectors of fisheries. As an illustration, 10 years ago (as of the submission of this paper), Madagascar received funding from the African Bank of Development to build landing sites for local fishers in various coastal villages. As of 2020, very few of these landing facilities were operational and they are currently managed by private-sector entities through a contract with the state government. This demonstrates the persistent challenge in Madagascar to ensure consistency in initiatives for fisheries. The impact of the recent efforts of the current ministry that has offered better support to local fishing communities is yet to be seen in the medium and long term.

The social misfit also manifests with the continuous adoption of legal text and policy fuelled by the various interests behind policy making. From the 1993 fishery code to the 2015 code and the blue economy strategy for the fisheries and aquaculture sector, improvements have been made to address issues regarding not only DWF. The adoption of new policies and laws is often influenced by various interests at the national level and aided by the willingness of the ministry in charge of fisheries to demonstrate its active involvement in policy change to various NGOs and funders. The tuna strategies, for example, were funded with support from the WWF as part of its own regional strategy to sustainably manage tuna fisheries in the Western Indian Ocean region. By passing recent laws on the maritime zone, the Malagasy government made a strong statement about national sovereignty. The adoption of a blue economy strategy for the sector and the creation of a blue economy department in 2021 was also a political signal to highlight the importance of fisheries for the economy. The ministry's core staff navigates these competing interests and initiatives to ensure the continuity of programs. The difficulties in implementing sustainable programs in Madagascar mean that new laws and policies keep being adopted with a limited evaluation of their effectiveness (NgoReps 2, 4 and 5).

3.2.2. Pertaining Issues of Transparency and Accountability

In fishery management, a lack of transparency and accountability in DWF can constitute a substantial barrier in advancing sustainable fisheries [14,37,38]. In both the signing and the implementation of fishing access agreements, significant issues remain. For the past 12 years or so, NGOs and CSOs have been asking the Ministry of Fisheries to open the negotiations of agreements to observers to ensure transparency and accountability and to eliminate the risk of corruption. Since 2018, the ministry has finally complied with these demands, but only slowly and only for EU negotiations (NgoRep 4). A group of experts comprising of CSOs, academics and advisors was also invited to prepare and observe the negotiation of the latest EU agreement, which started in 2019 [39]. While this was a substantial step, it was not replicated for the negotiations of other agreements, especially with Asian operators, which remain behind closed doors as of September 2022. Some members of civil society organisations were asked to provide comments once through email in 2021 regarding the renewal of the Japan Tuna agreement without any further follow up (pers. obs.).

In terms of accountability, the public, including local fishers, is not aware of how many fishing boats operate within the EEZ or who is operating them. The surveillance and monitoring centre ('Centre de Surveillance des Pêches'—CSP) and local fishers recount anecdotes of fishers complaining about big vessels entering territorial waters and not knowing if they are legal or illegal (interview insights from ArtFishers 8, SSFishers 5 and 6, AssoFisher 2, NgoRep 1, GovRep 5). The limited knowledge about which agreements are currently in effect and which vessels are present in the EEZ exacerbates the mistrust between local fishers and authorities. Since 2021, the ministry has made more efforts to publicise the conclusion of fishing access agreements and licenses through publications on its Facebook page. The commitment to the FiTI was also a step towards better publications of information linked to DWF. A database of such information is also part of the requirement of the FiTI standards. A substantial challenge also remains in terms of transparency of DWF in Madagascar especially regarding non-EU vessels. These vessels do not land at Malagasy ports and have an unsatisfactory rate of logbook submissions (interview insights from GovReps 1, 2, 3 and 10). There is little information on their activities in Madagascar's EEZ. One has to refer to the reporting of parties to the IOTC to obtain an idea of non-EU catch volumes, and these data are only based on fleet reports, not observer data. The lack of transparency in the activity of Asian fleets hinders any attempt at improving the management of DWF. As seen in other studies, dark fleet operations can have severe impacts on fish resources and missing revenues from under-reported catches [15,40].

3.2.3. A Still Limited Capacity to Fight IUU Fishing

Regarding its capacity to fight IUU fishing through Monitoring, Control and Surveillance (MCS), the CSP's surveillance fleet includes 15 vessels (2 vessels operating in Madagascar's EEZ and 13 in coastal waters). The rest of its fleet is composed of small boats and four-wheel drive vehicles. In addition to these, the CSP uses aerial surveillance, has inspectors and runs an observer program.

While the creation of the CSP in 1999 and investments in its infrastructure and resources have contributed to the reduction in illegal fishing, the CSP's current resources are too limited to ensure surveillance of the entire EEZ. This is due to its limited financial resources and its struggle to mobilise more of the national budget (GovRep 13). To monitor activities within the EEZ, the CSP mostly relies on vessel captains honouring their obligation to turn on their vessel monitoring system (VMS) while inside the EEZ. Furthermore, a report presented to the ministry in charge of fisheries in 2022 by civil society also highlighted that some fleets also had poor AIS transmissions, which prevents the monitoring of their activities in the EEZ even more. Only aerial surveillance can provide a full picture of fishing activities within the EEZ. However, aerial surveillance is very expensive for the ministry. In 2017 and 2018, no aerial surveillance was undertaken, while only 74 days of actual surveillance out of 120 days of planned surveillance took place in 2017 [40]. The observer program, an important component of the CSP's MCS capacity, has not performed as needed mainly due to the reduced number of observers, from 30 to 17 in 2017 (pers. comm.). Observer coverage is also low at around 24% of the domestic fleet and 10% of foreign fleets [41].

To overcome the limitations linked to the fight against IUU, Madagascar has relied on regional initiatives such as the regional MCS program, funded by the Indian Ocean Commission, which allows the use of logistical resources between country members of the IOC to conduct regional monitoring and surveillance. Thanks to this, illegal fishing in Madagascar's EEZ is considered to be less prolific by enforcing agents at CSP and other similar MCS units based in Mauritius and Seychelles [22]. In 2022, the government also signed the SADC MCS charter that aims to establish a regional SADC Fisheries Monitoring, Control and Surveillance Coordination Centre (MCSCC) [42]. This regional centre aims to mobilise regional intelligence in the SADC region, build capacity of MCS units and establish joint surveillance activities of EEZs between the SADC countries (ibid.).

4. Discussion on Pathways towards an Improved DWF Policy Framework

From the analysis of the policy framework and challenges for managing DWF in Madagascar, four pathways are proposed to improve the governance of distant-water fishing in Madagascar. These are based on pathways suggested in the literature on institutional misfits and fishery governance, adapted to the case of DWF in Madagascar.

4.1. Address Policy Coherence and Improve Monitoring

Addressing policy coherence, from policy design to implementation, can allow various stakeholders to advance fisheries and marine resources management that considers diverse views and interests [43,44]. In the case of DWF management in Madagascar, such coherence could be achieved not only by harmonising policies and laws but also by ensuring that policies are effectively implemented at different levels. While Madagascar can indeed be seen as progressive in its adoption of laws, policies and regional measures, implementation remains a challenge, especially for the Ministry of Fisheries. While NGOs have attempted to fill the gap in various coastal areas, the ministry is the main player when it comes to implementing its laws and policies. As new laws, policies and strategies continue to be adopted, harmonising them with existing laws could help determine which policies can be implemented and where policy gaps exist. Moreover, outreach activities toward regional offices, fishing communities and DWF operators could improve knowledge regarding existing legal texts and policies amongst local stakeholders. Fishers interviewed felt that more dialogue with local authorities would improve their feeling of integration and being part of national policy making and implementation. Since the ministry in charge of fisheries communicates new laws and policies directly to DWF operators, the latter should be fully aware of existing texts, which would allow for strong penalties in the case of offences.

Another area that requires considerable improvement is monitoring and evaluation. Madagascar, like many other developing countries, has been a living laboratory of all sorts of development projects and initiatives [45]. These have often been short-lived, with limited consolidation of lessons learnt (ibid.). In this context, there is limited consolidated knowledge of what works and especially what does not. Like for the management of terrestrial resources, for example, failed initiatives are often not recorded, and stakeholders' experiences are often not shared [46,47]. A continuous monitoring system for policy initiatives and projects relating to the development of fisheries could prevent repeating past mistakes. To this end, monitoring and evaluation of activities that are funded through revenues from DWF would help understand if they contributed to the improvement of the fishery sector. Stricter monitoring of DWF operations in the EEZ can also inform decision making on DWF management and contracts [48]. This could be achieved by improving observers' coverage along with ensuring their working conditions on vessels. All stakeholders working on data that were interviewed raised that better monitoring of catch data is needed, especially for non-EU fleets. This could be done through the implementation of more dissuasive sanctions in the case of late submissions of logbooks to the statistical unit. Finally, Madagascar is part of various initiatives that involve the sharing of data on fishing in the Indian Ocean; these data could be mobilised to inform the decision-making process during negotiations of access agreements. While the ministry's resources and funding might be limited, implementation, monitoring and evaluation activities could at least guide initiatives that can be effective or not for the future. Specific budget lines could be assigned to these activities while incoming revenues and funding could be directed towards such activities.

4.2. Realign National Policy and Actions with Local Needs

As DWF continues to face criticism at the national and global levels [26,27], its perpetuation in the waters of a developing country such as Madagascar remains paradoxical. Its contribution to the national economy, in particular fishing communities, is yet to be seen in tangible ways. To ensure that DWF has a limited negative impact on local livelihoods and communities, Madagascar's national policy and actions might need to be realigned with local needs. This could help address the spatial and social misfits identified in the challenges above. For example, the country could develop more long-term initiatives with the revenue from fishing access agreements and similar funding to promote activities that contribute to the well-being and livelihoods of fishing communities. The recent turn of the government towards flagging industrial vessels could be re-evaluated on the kinds and levels of benefits and trade-offs it brings to the economy and the coastal communities relying on fishing as a livelihood.

For artisanal and small-scale fishers, the combination of a steady decrease in marine resources and the need to focus on a day-to-day livelihood provides limited perspectives for long-term prosperity [49]. As national policies have begun to recognize the importance of artisanal and small-scale fishing, addressing broader socio-economic and political factors that prevent fishers from prospering could be beneficial. This includes rethinking the value of projects regarding fisheries based on intermittent donations of fishing gears and tackling the problem of sustaining livelihoods in the short and long term. This could also improve local fishers' trust in the state. Initiatives that improve learning exchanges in the management of marine resources among coastal communities exist [50]. These represent an avenue for the government to gain access to local fishers, although this also requires a strong willingness from the authorities to actively listen to the voices of small-scale fishers when they express their concerns.

4.3. Further Strengthen Transparency in DWF

The next suggestion aligns with several demands of NGOs at the national and international levels over the years to improve transparency in DWF, from negotiations to implementation and benefit sharing [51]. Transparency is also considered an important pillar of effective fishery governance [14]. Regarding negotiations of fishing access agreements, the efforts since 2018 to allow observers at EU negotiations represent a key step for Madagascar to open its negotiations to more transparency. This could be established as a practice to be applied to all types of fishing access agreements. Expanding transparency in negotiations would also help disassociate fishing access agreements from any development aid, which often influences negotiations and the development of aid flows in the country [52].

Another area where more transparency could be beneficial is regarding the use of revenue from DWF. An increased accountability in the allocation of funds could be achieved by publishing the multiyear plan for EU revenue dedicated to sectoral support and making public any report on activities undertaken with these funds. Consultations with national stakeholders on fishery priorities could help the ministry obtain a clear and consistent list of activities. A look at the multiyear plan of the 2014–2018 EU agreement reveals that some activities did not fit clearly into traditional fisheries or capacity-building categories. Providing a clearer direction on allowable activities and the allocation of funds could facilitate the monitoring of implementation. Transparency in this area would also help the country comply with the FiTI standard on DWF.

A last area where improved transparency could help shape policy regarding the management of DWF is on the impacts of DWF on the resources and the coastal communities of Madagascar. There is limited knowledge on the potential competition over resources between DWF and national fisheries and a need for evidence on the impact of DWF on the marine biomass and ecosystems within the EEZ. Such evidence could help the government decide on the need to maintain, improve or stop DWF in Madagascar's waters. Transparency on the value added and the potential negative impacts of DWF is key to decision making for the Ministry of Fisheries.

4.4. Mobilise Measures Adopted at the Regional Indian Ocean Level

While aligning regional and national fishery policies can be a challenge for coastal states, measures adopted at the regional level such as the IOTC also represent an opportunity to improve fishery management at the national level [53]. As Madagascar is involved in various management initiatives at the Indian Ocean level, there are opportunities to improve the management of DWF by implementing regional measures. Since Madagascar has been engaged in regional surveillance and control for several years, two other opportunities could be mobilised further: a better integration of measures adopted at the IOTC, and the implementation of the Southwest Indian Ocean Fisheries Commission (SWIOFC)'s guidelines on minimum terms and conditions for fishing access agreements (see Supplementary Material for a summary of the key clauses). Through adopted IOTC resolutions, Madagascar has attempted to improve its tuna fishery management. Continuing these efforts will require more resources to be sustained in the longer term. Implementing the IOTC's conservation measures is not always a straightforward process including for developing countries like Madagascar [54]. Therefore, integrating IOTC resolutions into national laws and policies implementing resolutions and conservation measures requires more investment. Ideally, a national budget could be assigned to such a process. However, there are also venues within fishing access agreements and donor-led funding that could also be redirected to such efforts.

5. Conclusions

While DWF is fully a part of Madagascar's blue economy, the current analysis shows that Madagascar faces challenges in managing this segment of the fishery sector. These challenges are not uncommon in developing coastal states. They include a limited harmonisation of approaches and policies through the years. While the importance of local fisheries is increasingly acknowledged, this is often contradicted by governmental actions that aim to industrialise the sector including by continuing to conclude fishing access agreements or through flagging foreign industrial vessels. Despite these, the drive of the country to develop and adopt frameworks to manage DWF needs to be recognised. Moreover, national laws and frameworks are open to improvement, as they all specify the possibility of adopting specific legislation or measures to implement better policies (such as the tuna strategy). The recent commitment to the Fisheries Transparency Initiative presents great potential to improve transparency in DWF operations and governance if implemented thoroughly. Madagascar has accumulated extensive experience in dealing with DWF. In addition to ongoing efforts (as of early 2023), there are opportunities for improvement and there is an institutional setting to enable changes in Madagascar. This includes Madagascar's openness to adopt new regulations and policies when needed, and the continued interest of DWF fleets in fishing in Madagascar's EEZ. In its current political and economic situation and its current drive for the blue economy, Madagascar would find it difficult to ban DWF in its waters. Therefore, DWF must be governed by a well-implemented coherent framework that can provide tangible benefits at the national level through revenue but also sustain coastal livelihoods and food security. A future reflection on the potential removal of DWF in the EEZ requires filling the knowledge and data gap on the contribution and impact of DWF. This could assist decision makers in weighing the value added by DWF compared to its impact on the future of the blue economy from the local to the national level.

Supplementary Materials: The following supporting information can be downloaded at: https://www.mdpi.com/article/10.3390/fishes8070361/s1, Key clauses of Southwest Indian Ocean Fisheries Commission (SWIOFC)'s guidelines on minimum terms and conditions for fishing access agreements.

Funding: The results of this review were part of larger research on the state of play regarding distant-water fishing in Madagascar. The report was funded by Bloom Association France: Contract 2021–2022.

Institutional Review Board Statement: Not applicable.

Informed Consent Statement: Informed consent was obtained from all the interviewees involved in the study.

Data Availability Statement: The data presented in this study are available within the paper and in the Supplementary Material S1.

Acknowledgments: I would like to thank Frederic Le Manach and Peter Bouckeart for their feedback on the results of the policy analysis that emerged from the Bloom Association report as these results framed the content of this paper.

Conflicts of Interest: The author declares no conflict of interest.

References

1. Breuil, C.; Grima, D. *Baseline Report Madagascar*; IOC: Ebene, Mauritius, 2014.
2. MRHP. *Blue Policy Letter—Lettre de Politique Bleue: Pour une Économie Bleue, Valorisant le Travail des Pêcheurs et Aquaculteurs, Durabilisant la Création de Ses Richesses, et Prenant en Compte le Bien Être Écologique des Ressources Halieutiques*; Ministère des Ressources Halieutiques et de la Pêche: Antananarivo, Madagascar, 2015. Available online: https://www.fao.org/faolex/results/details/fr/c/LEX-FAOC163970/#:~:text=Madagascar%20(Niveau%20national)-,Lettre%20de%20Politique%20Bleue.,principales%20orientations%20jusqu\T1\textquoterighten%202025 (accessed on 17 January 2022).
3. INSTAT. Enquête périodique auprès des ménages. In *Résultats Globaux du Recensement Général de la Population et de L'habitation de 2018 de Madagascar*; INSTAT: Antananarivo, Madagascar, 2020.
4. MRHP. *Enquete Cadre Nationale 2011–2012*; MRHP: Antananarivo, Madagascar, 2013.
5. MAEP 'Ministère de l'Agriculture, de l'Elevage et de la Pêche' (Ministry in charge of fisheries in 2021). Update of the National Strategy for the Management of Tuna Fisheries. 2021. Available at the tuna fisheries unit and the WWF. Collected via email exchanges with the consultant who wrote the strategy.
6. Rakotosoa, R. Impacts socioéconomiques de la Pêche thonière industrielle à Madagascar. In Proceedings of the Lancement de la Campagne Thonière 2017, Antsiranana, Madagascar, 24 May 2017.
7. Crosnier, A.F. La Pêche du Thon à Madagascar. In *Bulletin de Madagascar*; Haut Commissariat de la République Française à Madagascar et Dépendances, Service Général de L'information: Antananarivo, Madagascar, 1961; Volume 185. Available online: https://horizon.documentation.ird.fr/exl-doc/pleins_textes/pleins_textes_7/b_fdi_59-60/010026680.pdf (accessed on 14 January 2020).
8. White, E.R.; Baker-Médard, M.; Vakhitova, V.; Farquhar, S.; Ramaharitra, T.T. Distant water industrial fishing in developing countries: A case study of Madagascar. *Ocean Coast. Manag.* **2021**, *216*, 105925. [CrossRef]
9. USTA. *Bulletin Statistique Thonier 2017 de l'Unité Statistique Thonière d'Antsiranana*; USTA: Antsiranana, Madagascar, 2017.
10. IOTC. *Catch History between 1950 and 2016 for Albacore, Bigeye Tuna, Skipjack Tuna, Yellowfin Tuna and Swordfish*; IOTC: Victoria, Seychelles, 2018.
11. Campling, L. *The EU-Centered Commodity Chain in Canned Tuna and Upgrading in Seychelles*; Queen Mary University of London: London, UK, 2012.
12. SFA. *Fisheries Statistical Report, Semester 1, Year 2016*; SFA: Mahe, Seychelles, 2016.
13. Epstein, G.; Pittman, J.; Alexander, S.M.; Berdej, S.; Dyck, T.; Kreitmair, U.; Rathwell, K.J.; Villamayor-Tomas, S.; Vogt, J.; Armitage, D. Institutional fit and the sustainability of social–ecological systems. *Curr. Opin. Environ. Sustain.* **2015**, *14*, 34–40. [CrossRef]
14. Guggisberg, S.; Jaeckel, A.; Stephens, T. Transparency in fisheries governance: Achievements to date and challenges ahead. *Mar. Policy* **2022**, *136*, 104639. [CrossRef]

15. WWF. US $142.8 Million Potentially Lost Each Year to Illicit fishing in the South West Indian Ocean. 2023. Available online: https://www.wwf.eu/?10270441/US1428-million-potentially-lost-each-year-to-illicit-fishing-in-the-South-West-Indian-Ocean (accessed on 30 May 2023).
16. World Bank. Second South West Indian Ocean Fisheries Governance and Shared Growth Project—Region & Madagascar (P153370). 2017. Available online: https://www.swiofish2.mg/wp-content/uploads/2021/09/Project-Appraisal-Document-PAD-P153370-1.pdf (accessed on 30 May 2023).
17. IOTC. Madagascar 2020 National Report to the IOTC. 2020. Available online: https://www.iotc.org/documents/SC/23/NR11 (accessed on 23 March 2022).
18. Belhabib, D.; Sumaila, U.R.; Lam, V.W.Y.; Zeller, D.; Le Billon, P.; Kane, E.A.; Pauly, D. Euros vs. Yuan: Comparing European and Chinese Fishing Access in West Africa. *PLoS ONE* **2015**, *10*, e0118351. [CrossRef]
19. IOTC. Madagascar Compliance Report | IOTC. IOTC-2023-CoC20-CR14, 2023. Available online: https://iotc.org/documents/madagascar-24 (accessed on 20 June 2023).
20. Le Manach, F.; Andriamahefazafy, M.; Harper, S.; Harris, A.; Hosch, G.; Lange, G.-M.; Zeller, D.; Sumaila, U.R. Who gets what? Developing a more equitable framework for EU fishing agreements. *Mar. Policy* **2013**, *38*, 257–266. [CrossRef]
21. Carver, E. Madagascar: Opaque foreign fisheries deals leave empty nets at home. *Mongabay News*, 9 October 2019.
22. Andriamahefazafy, M.; Kull, C.A.; Campling, L. Connected by sea, disconnected by tuna? Challenges to regionalism in the Southwest Indian Ocean. *J. Indian Ocean Reg.* **2019**, *15*, 58–77. [CrossRef]
23. Gorez, B. Small Scale Fisheries at Risk: Madagascar Signs Destructive Fishing Agreements with Chinese Investors. Coalition for Fair Fisheries Arrangements. 17 November 2020. Available online: https://www.cffacape.org/publications-blog/small-scale-fisheries-at-risk-madagascar-signs-destructive-fishing-agreements-with-chinese-investors (accessed on 27 April 2023).
24. Gagern, A.; Bergh, J.v.D. A critical review of fishing agreements with tropical developing countries. *Mar. Policy* **2013**, *38*, 375–386. [CrossRef]
25. Gegout, C. Unethical power Europe? Something fishy about EU trade and development policies. *Third World Q.* **2016**, *37*, 2192–2210. [CrossRef]
26. Le Manach, F.; Andriamahefazafy, M.; Legroux, N.; Quentin, L. Questionning Fishing Access Agreements towards Social and Ecological Health in the Global South. AFD, Research Paper 203. 2021. Available online: https://www.afd.fr/en/ressources/questionning-fishing-access-agreements-towards-social-and-ecological-health-global-south (accessed on 15 February 2021).
27. Nash, K.L.; MacNeil, M.A.; Blanchard, J.L.; Cohen, P.J.; Farmery, A.K.; Graham, N.A.J.; Thorne-Lyman, A.L.; Watson, R.A.; Hicks, C.C. Trade and foreign fishing mediate global marine nutrient supply. *Proc. Natl. Acad. Sci. USA* **2022**, *119*, e2120817119. [CrossRef]
28. Standing, A. *Are The EU's Fisheries Agreements Helping to Develop African Fisheries?* CFFA-CAPE: Etterbeek, Belgium, 2016.
29. IOTC. Conservation and Management Measures (CMMs) | IOTC. 2023. Available online: https://iotc.org/cmms (accessed on 20 June 2023).
30. IOTC. On the Management of Drifting FADs -DFADS (Kenya et al.) | IOTC. IOTC-2022-S26-REF02. 2022. Available online: https://iotc.org/documents/management-drifting-fads-dfads-kenya-et-al (accessed on 20 June 2023).
31. IOTC. Objection from Madagascar to IOTC Resolution 21/01 | IOTC. Circular IOTC CIRCULAR 2021-50. 2021. Available online: https://www.iotc.org/documents/objection-madagascar-iotc-resolution-2101 (accessed on 17 February 2022).
32. Razafindrakoto, M.; Roubaud, F.; Wachsberger, J.-M. *L'énigme et le Paradoxe—Économie Politique de Madagascar*; Collection Synthèses; Marseilles; IRD Éditions: Marseilles, France, 2017.
33. Cox, M. Diagnosing Institutional Fit: A Formal Perspective. *Ecol. Soc.* **2012**, *17*, 54. [CrossRef]
34. Liddick, D. The dimensions of a transnational crime problem: The case of iuu fishing. *Trends Organ. Crime* **2014**, *17*, 290–312. [CrossRef]
35. Falco, F.L.; Preiss-Bloom, S.; Dayan, T. Recent Evidence of Scale Matches and Mismatches Between Ecological Systems and Management Actions. *Curr. Landsc. Ecol. Rep.* **2022**, *7*, 104–115. [CrossRef]
36. Andriamahefazafy, M. The Politics of Sustaining Tuna, Fisheries and Livelihoods in the Western Indian Ocean. A Marine Political Ecology Perspective, Université de Lausanne, Faculté des Géosciences et de L'environnement. 2020. Available online: https://serval.unil.ch/notice/serval:BIB_7E0D668DF275 (accessed on 26 January 2021).
37. Davis, R.A.; Hanich, Q. Transparency in fisheries conservation and management measures. *Mar. Policy* **2022**, *136*, 104088. [CrossRef]
38. Walton, G.W.; Keen, M.; Hanich, Q. Can Greater Transparency improve the Sustainability of Pacific Fisheries? *Mar. Policy* **2022**, *136*, 104251. [CrossRef]
39. Orange Actu. Accord de Pêche: Reprise des Négociations Avec l'Union Européenne. Orange actu Madagascar. 5 July 2022. Available online: https://actu.orange.mg/accord-de-peche-reprise-des-negociations-avec-lunion-europeenne/ (accessed on 30 May 2023).
40. Kroodsma, D.A.; Hochberg, T.; Davis, P.B.; Paolo, F.S.; Joo, R.; Wong, B.A. Revealing the global longline fleet with satellite radar. *Sci. Rep.* **2022**, *12*, 21004. [CrossRef] [PubMed]
41. IOTC. Madagascar—National Report 2019. National Report IOTC-2019-SC22-NR14. 2019. Available online: https://iotc.org/fr/documents/SC/22/NR14 (accessed on 2 June 2020).

42. SADC. Protecting Our Fisheries—Working Towards a Common Future. 2021. Available online: https://stopillegalfishing.com/publications/protecting-our-fisheries-working-towards-a-common-future/ (accessed on 10 March 2022).
43. Lindström, L.; De La Torre-Castro, M. Tuna or Tasi? Fishing for Policy Coherence in Zanzibar's Small-Scale Fisheries Sector. In *The Small-Scale Fisheries Guidelines*; Springer: Berlin/Heidelberg, Germany, 2017; pp. 79–94. [CrossRef]
44. Blasiak, R.; Wabnitz, C.C.; Daw, T.; Berger, M.; Blandon, A.; Carneiro, G.; Crona, B.; Davidson, M.F.; Guggisberg, S.; Hills, J.; et al. Towards greater transparency and coherence in funding for sustainable marine fisheries and healthy oceans. *Mar. Policy* **2019**, *107*, 103508. [CrossRef]
45. Waeber, P.O.; Wilmé, L.; Mercier, J.-R.; Camara, C.; Ii, P.P.L. How Effective Have Thirty Years of Internationally Driven Conservation and Development Efforts Been in Madagascar? *PLoS ONE* **2016**, *11*, e0161115. [CrossRef]
46. Eklund, J.; Cheek, J.Z.; Andriamaro, L.; Bakoliarimisa, T.M.; Galitsky, C.; Rabearivololona, O.; Rakotobe, D.J.; Ralison, H.O.; Randriamiharisoa, L.O.; Rasamoelinarivo, J.; et al. Insights from practitioners in Madagascar to inform more effective international conservation funding. *Madag. Conserv. Dev.* **2022**, *17*, 29–35. [CrossRef]
47. Scales, I. The future of conservation and development in Madagascar: Time for a new paradigm? *Madag. Conserv. Dev.* **2014**, *9*, 5. [CrossRef]
48. Kerry, C.R.; Exeter, O.M.; Witt, M.J. Monitoring global fishing activity in proximity to seamounts using automatic identification systems. *Fish Fish.* **2022**, *23*, 733–749. [CrossRef]
49. Fattebert, C. La Pêche Traditionnelle ou Petite Pêche Maritime à Madagascar: Un État des Lieux. CAPE-CFFA. 2020. Available online: https://static1.squarespace.com/static/5d402069d36563000151fa5b/t/5e996c4303457a60bf3b3a72/1587113109461/200415+Report+Madagascar.pdf (accessed on 1 June 2020).
50. Rocliffe, S.; Harris, A. Scaling success in octopus fisheries management in the Western Indian Ocean. In Proceedings of the Workshop, Stone Town, Zanzibar, 3–5 December 2014; p. 24.
51. CAPE-CFFA. 10 priorities for the future of Sustainable Fisheries Partnership Agreements. CAPE-CFFS. 2020. Available online: https://www.cffacape.org/publications-blog/ten-priorities-for-the-future-of-sustainable-fisheries-partnership-agreements (accessed on 2 June 2020).
52. Pittman, J.; Wabnitz, C.C.; Blasiak, R. A global assessment of structural change in development funding for fisheries. *Mar. Policy* **2019**, *109*, 103644. [CrossRef]
53. Wright, G.; Ardron, J.; Gjerde, K.; Currie, D.; Rochette, J. Advancing marine biodiversity protection through regional fisheries management: A review of bottom fisheries closures in areas beyond national jurisdiction. *Mar. Policy* **2015**, *61*, 134–148. [CrossRef]
54. Sinan, H.; Bailey, M.; Hanich, Q.; Azmi, K. Common but differentiated rights and responsibilities in tuna fisheries management. *Fish Fish.* **2022**, *23*, 202–212. [CrossRef]

fishes

MDPI

Article

Changes in Abundance and Distribution of the Sea Pen, *Funiculina quadrangularis*, in the Central Adriatic Sea (Mediterranean Basin) in Response to Variations in Trawling Intensity

Michela Martinelli [1], Lorenzo Zacchetti [1,*], Andrea Belardinelli [1], Filippo Domenichetti [1], Paolo Scarpini [1], Pierluigi Penna [1], Damir Medvešek [2], Igor Isajlović [2] and Nedo Vrgoč [2]

1 National Research Council, Institute of Marine Biological Resources and Biotechnologies, 60125 Ancona, Italy; michela.martinelli@cnr.it (M.M.); andrea.belardinelli@cnr.it (A.B.); filippo.domenichetti@cnr.it (F.D.); paolo.scarpini@cnr.it (P.S.); pierluigi.penna@cnr.it (P.P.)

2 Institute of Oceanography and Fisheries, 21000 Split, Croatia; medvesek@izor.hr (D.M.); igor@izor.hr (I.I.); vrgoc@izor.hr (N.V.)

* Correspondence: lorenzo.zacchetti@irbim.cnr.it

Abstract: Marine resources exploitation through bottom trawling affects marine ecosystems; thus, management should consider the presence of sensitive species as ecosystem health indicators. Epibenthic organisms such as sea pens are widely used to assess benthic conditions, as their populations are declining where trawling is intense. The Pomo/Jabuka Pits fishing ground in the Adriatic Sea, subject to various management measures over the years, is a nursery for European hake and hosts a small, but dense, population of Norway lobster and a remarkable abundance of pink shrimp. The sea pen *Funiculina quadrangularis* shares its habitat (sandy-muddy bottoms) with these crustaceans. Through UnderWater TeleVision surveys conducted from 2012 to 2019, *F. quadrangularis* abundance and distribution were quantified in relation to changes in the spatiotemporal distribution of fishing efforts. The average density (n/m^2) of colonies was calculated for three periods: BEFORE implementation of measures (before 1 July 2015), during an INTERMEDIATE period in which limitations changed (2 July 2015 to 31 August 2017), and AFTER the implementation of a Fishery Restricted Area (from 1 September 2017). *F. quadrangularis* revealed an increase in density where fisheries were closed, even after a short period. This showed how management measures can positively influence epibenthic communities and that sea pens can be indicators of impact and/or recovery of habitats.

Keywords: *Funiculina quadrangularis*; vulnerable marine ecosystem (VME) indicators; Jabuka/Pomo Pit; Fishery Restricted Area (FRA)

Key Contribution: Bottom trawling can impact marine ecosystems; benthic conditions can be evaluated by means of sensitive organisms such as sea pens. Through UnderWater TeleVision surveys conducted from 2012 to 2019, *F. quadrangularis*' abundance and distribution were quantified in relation to changes in the spatiotemporal distribution of fishing efforts in the Pomo/Jabuka Pits area (Adriatic Sea). This showed how management measures can positively influence epibenthic communities.

Citation: Martinelli, M.; Zacchetti, L.; Belardinelli, A.; Domenichetti, F.; Scarpini, P.; Penna, P.; Medvešek, D.; Isajlović, I.; Vrgoč, N. Changes in Abundance and Distribution of the Sea Pen, *Funiculina quadrangularis*, in the Central Adriatic Sea (Mediterranean Basin) in Response to Variations in Trawling Intensity. *Fishes* **2023**, *8*, 347. https://doi.org/10.3390/fishes8070347

Academic Editor: Ben Drakeford

Received: 29 May 2023
Revised: 23 June 2023
Accepted: 27 June 2023
Published: 1 July 2023

1. Introduction

Overexploitation of marine resources, especially by means of bottom trawling, can change or degrade habitats and have negative impacts on marine ecosystems [1–3]. Bottom trawling, in fact, may lead to long-term changes in benthic communities and affect the trophic condition of benthic ecosystems [4–7]. In addition to the effects directly linked to the removal of organisms, a secondary cause can be, for example, the resuspension of sediment, which can have a suffocating effect on non-target benthic fauna and flora [8–10].

In order to maintain the sustainability of fish resources over time and the conservation of ecosystems, careful management should be based on a solid understanding of the structure and functioning of its components, including the effects of human and natural disturbances [11,12]. A variety of management techniques can be adopted that also take into account the presence of sensitive species as ecosystem health indicators useful to evaluate the effectiveness of measures implemented [12,13]. Depending on the biological traits of some species, the anthropogenic impact might result in fluctuations in abundance, which might be reflected also in changes in the ecosystem components and biodiversity [14,15]. The selection of an organism as an indicator species is crucial, as this should represent a link between the objectives of management measures and the actions [16].

Physical disturbance to the seabed as a result of fishing activities, including effects on benthic communities, is addressed by the criteria under Descriptor 6 (Sea-floor integrity) of the Marine Strategy Framework Directive (MSFD) of the European Union [17]. In particular, criterion D6C3 requires Member States to investigate the adverse effects of physical disturbance on each habitat type and derived changes in its biotic and abiotic structure and functions, for example, through the analysis of changes in species composition and their relative abundance, absence of particularly sensitive or fragile species or species providing a key function, and the size structure of species. Epibenthic organisms, such as sponges and sea pens, are among the most used species to assess benthic conditions [13,18]. Sea pens are colonial cnidarians belonging to the subclass Octocorallia, order Pennatulacea [19]. Sea pen forests may provide important three-dimensional habitats for fish and invertebrate species, thus contributing to the preservation of ecosystem functions in marine benthic ecosystems [20–23]. They can host eggs and larvae, and serve as safe habitat for young fish [24].

The European Commission has classified sea pen forests as vulnerable marine ecosystems (VMEs) and essential fish habitats (EFHs) because of their significant ecological roles and the sensitivity of pennatulaceans to human activities [25–27]. The presence of sea pens and possible changes in their distribution provide information in support of the evaluation of management strategies aimed at safeguarding vulnerable species sensitive to human impacts [20]. The General Assembly of the United Nations adopted resolution 61/105 to safeguard VMEs in consideration of the significance of their marine ecological services [27]. To successfully ensure the long-term conservation and sustainable use of marine resources, Resolution 70/75 emphasized the urgent need to also safeguard VMEs and mitigate the effects of bottom trawling on them [21].

Funiculina quadrangularis (Pallas, 1766) is a tall sea pen having polyps that develop from a square-sectioned calcareous axial rod with a peduncle at the base [26]. Individuals can reach 200 cm in length and have an axis that is up to 25% immersed in the sediment [26]. They are frequently characterized by a feather-like appearance [22,26]. Between 20 and 2000 m of depth, *F. quadrangularis* is adapted to muddy environments and frequently develops dense meadows [22]. *F. quadrangularis* was found in marine lakes and open waters and is globally distributed (in the Atlantic Ocean, especially in the North Sea, Mediterranean Sea and throughout the Pacific Ocean [23,24]). *F. quadrangularis* is not a target for fishing activities but may be a by-catch due to the fact that it shares the same habitat (on sandy-muddy sediments) as *Nephrops norvegicus* (Linnaeus, 1758) and *Parapenaeus longirostris* (Lucas, 1846) [26], two of the most important commercial crustaceans, especially in the Mediterranean Sea [28]. *F. quadrangularis* is considered a critically endangered species [29]; in fact, its populations are declining in areas where trawling activities targeting the aforementioned species are intense [25].

Mapping *F. quadrangularis* density over time and space might, thus, be useful not only to determine their distribution in the Mediterranean, but also to assess the conditions of the VMEs to which they belong [30,31]. Sea pens are commonly sampled by using trawl nets, grab samplers, or scuba divers directly cutting/removing the organism [24,32,33]. Less invasive methodologies based on the analysis of photo/video recorded through Remotely Operated Vehicles (ROVs) or towed camera systems, such as the UnderWater

TeleVision (UWTV), are also used for the evaluation of the distribution of sea pens on the seabed [31,34–36]. In the Adriatic Sea, the UWTV methodology is used to derive *N. norvegicus* burrow densities in the Pomo/Jabuka Pits area (important for commercial fishing and subjected to various management strategies [37]) and was also trialled to assess the distribution in the area of other species of ecological interest, such as *F. quadrangularis*, and trawl marks [34]. The collected footage might indeed be helpful to conduct quantitative or semi-quantitative evaluation of species coexisting with *N. norvegicus* in its environment; OSPAR (the Oslo and Paris Convention for the Protection of the Marine Environment of the North-East Atlantic) underlined the possibility of using UWTV for the evaluation of sea pens [38].

The aim of this study is to assess changes over time in the abundance and distribution of *F. quadrangularis* in a particular region in which fishery restrictions changed in space and time, and therefore verify the actual potential of using this species as an indicator of human impact and/or the recovery of habitats after the implementation of management strategies.

2. Materials and Methods

2.1. Study Area

The Pomo/Jabuka Pits are three depressions delimited by the 200 m bathymetry located in the central Adriatic Sea ([39], Figure 1); this area is one of the main fishing grounds historically shared by the Italian and Croatian fleets [34,40]. The complex topography of the area and the Adriatic Sea oceanographic regimes make it a very particular environment in which the water exchange does not happen annually [41]. These conditions influence the nutrient cycle, with consequences on local biodiversity and on the trophic status of benthic communities [42]. Furthermore, alterations in species assemblages and possible consequences on trophic and ecosystem balances could probably be due to the synergistic action of fishing pressure and climate change [7,43–47].

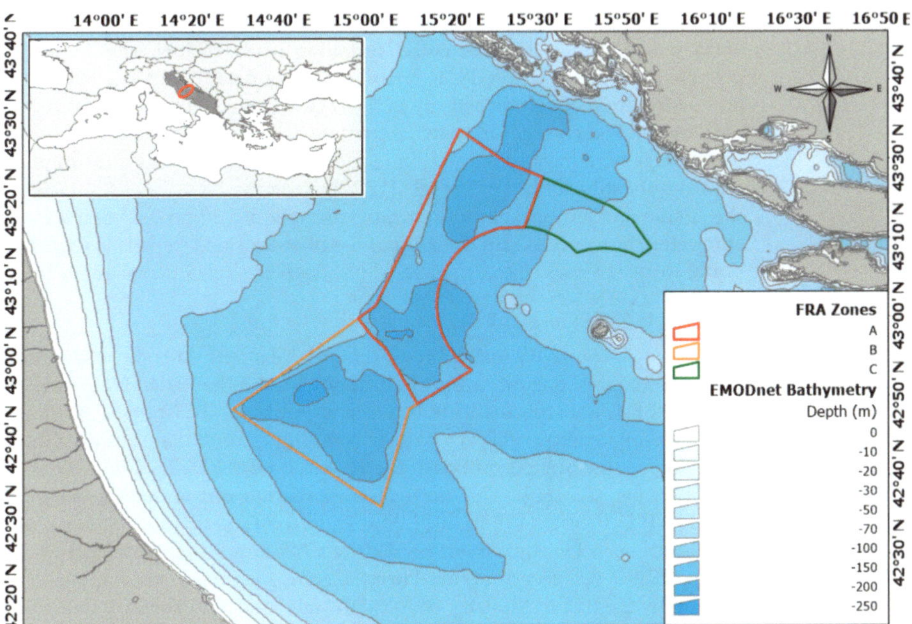

Figure 1. The location of the study area within the Mediterranean basin is indicated by a red circle in the top-left rectangle; the main map shows the bathymetry of the Central Adriatic Sea (source: [39]) and the Jabuka/Pomo Pits Fishery Restricted Area zones "A", "B", and "C", according to Recommendation GFCM/41/2017/3 [48].

This area is the main nursery for the European hake, *Merluccius merluccius (Linnaeus, 1758)*, in the Adriatic, and the presence of muddy bottoms and other exogenous factors make it the habitat of a population of small-but-dense Norway lobster, *N. norvegicus*, individuals [49–51]. Among the other crustacean species occurring in the area, a commercial and ecological relevance is attributable to the pink shrimp, *P. longirostris*, which showed periodic fluctuations in the area, probably linked to environmental parameter changes, and an abundance peak in 2017 [52,53]. Also possibly linked to climate change, a crustacean species shift also occurred in the area: the squat lobster *Munida intermedia* (Milne-Edwards and Bouvier, 1899), was replaced by *Iridonida speciosa* (von Martens, 1878) (previously known as *Munida speciosa*), first observed in 2003 [54]. Another gadoid species dwelling in the area, the blue whiting *Micromesistius poutassou* (Risso, 1827), was reported to experience fluctuations in abundance over time, probably as a result of environmental variations and fishing exploitation [55]. Therefore, the Pomo/Jabuka Pits represent a VME and an EFH worthy of implementation of appropriate management measures aimed at protecting the demersal fish stocks, enhancing the densities of organisms in terms of biomass and abundance, and protecting possible VMEs while ensuring the sustainable exploitation of the main resources [37,56]. For this reason, a series of management measures were implemented in the area over time [55,57], which led to the establishment, in 2017, of a Fishery Restricted Area (FRA) by the General Fisheries Commission for the Mediterranean (GFCM, [48]). The FRA is divided into three zones: "A," which is closed to all fishing activity, "B," where bottom trawling is regulated with licenses and a number of fishing days (two days allowed per licensed vessel per week, fishing vessels using twin bottom otter trawls can fish one day per week), and fishery is banned from 1 September to 31 October, and zone "C," where trawling is permitted through licenses on Saturdays and Sundays (from 5.00 am to 10.00 pm), whereas vessels fishing with bottom-set nets, set longlines and traps can fish from Monday 05.00 am till Thursday 22.00 pm, and fishing is also banned from 1 September to 31 October (fishing activity with purse seiners and pelagic trawlers targeting anchovy or sardine is prohibited in the FRA; Figure 1). The European Parliament and Council ratified the FRA with Regulation 2019/982 [58]. The Pomo/Jabuka Pits FRA became permanent with GFCM Recommendation 44/2021 [59]. Therefore, the intensity and distribution of fishing activity in the region changed over time as a consequence of the aforementioned changes in management measures. In fact, before the implementation of the management measures, the fishing pressure in the area was quite high and was exerted by a significant number of vessels [12,37]; since the FRA was established, in addition to the spatial and temporal limitations, the number of authorized vessels is regulated by the competent authorities and the annual list of authorized vessels is reported in the GFCM portal (https://www.fao.org/gfcm/data/fleet/fras/en/, accessed on 16 June 2023).

2.2. Sample Collection and Video Analysis

From 2009 to 2019 (except for 2011 and 2018), the Ancona section of the Institute of Marine Biological Resources and Biotechnologies of the Italian National Research Council (CNR-IRBIM) and the Institute of Oceanography and Fisheries (IOF) of Split (Croatia) partnered on an annual UWTV survey covering the entire area of the three meso-Adriatic depressions, conducted on board the research vessel Dallaporta (LOA 35.30 m, 258 GT, 1100 HP) [34,37,57,60]. The surveys were carried out under the auspices of the FAO–ADRIAMED project [60]; in 2013, thanks to the Italian National Flagship Program RITMARE, the UWTV equipment owned by CNR was completely renewed [61], and from 2015 to 2019 it was also supported (for the complementary experimental trawl fishery component) by the Italian Ministry of Agricultural, Food, and Forestry Policies (MIPAAF; [62]).

The 60 fixed stations firstly defined in the area according to a random stratified sampling design [34] were maintained over time and are still consistent with the stratification currently in use based on the FRA zones, as they mainly overlap the original strata [37]; not all stations were studied in all the surveys due to ship time and weather condition limitations [60]. During each survey, a Kongsberg Simrad colour camera mounted on a sledge was towed on the sea floor at about a 1-knot speed. The camera's field of view was set at a constant width of 80 cm. A custom data logger, synchronized with the video deck unit and with a CTD (Conductivity, Temperature, Depth) probe recording environmental parameters [63], acquired the vessel GPS (Global Positioning System) position (as a proxy of the sledge position) once per minute to allow the quantification of the surface viewed. After each survey, trained readers reviewed the collected footage following a specific protocol to derive estimates of the target features (e.g., burrows) and/or organisms [64,65]. Each station considered valid for *N. norvegicus* assessment purposes consisted of a minimum of 7 good minutes (normally out of about 10–20 min recorded per station) complying with the speed and visibility criteria set for the Adriatic in [60]. Recently, the surface viewed in all stations of the time series was recalculated to take into account the difference in the bottom surface covered by the sledge and the vessel route [37].

UWTV is used to evaluate *N. norvegicus* following specific standards in several European countries [66]. However, the fact that it uses a fixed field of view makes it suitable for collecting other ecological data potentially useful in the context of an ecosystem approach to fisheries management [34]. Figure 2 is composed of frames extrapolated from UWTV footage collected in the Adriatic Sea showing the environment and some recorded organisms. Recently, the UWTV footage collected in the Pomo Pits area was in fact further analysed with the aim to derive information on the epibenthic communities subjected to physical perturbations and the objects of interest in the context of Descriptor 6 of the MSFD, among which in particular is *F. quadrangularis* [67].

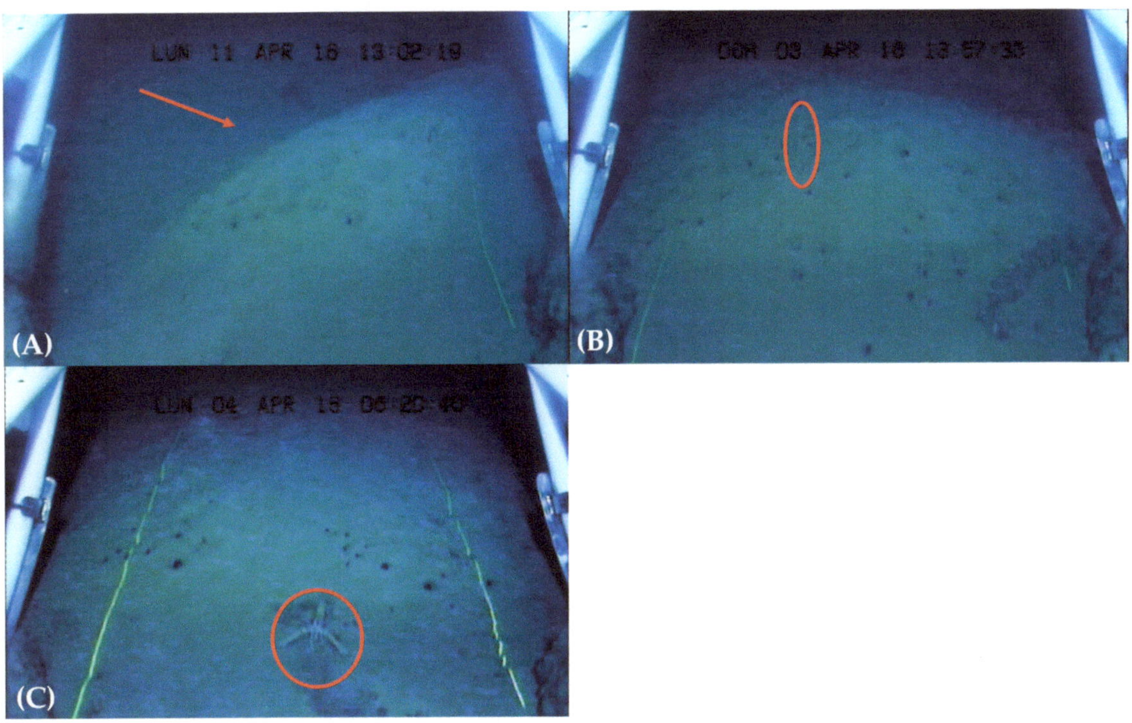

Figure 2. Images obtained by means of UWTV: (**A**) trawl track created by the passage of the otter door of a bottom trawler; (**B**) inside the red circle is a specimen of *F. quadrangularis*; (**C**) the red circle shows a specimen of *N. norvegicus*, one of the species with whom sea pens share the habitat; in all frames are visible burrows made by various organisms.

2.3. Data Analysis

All data acquired during the UWTV surveys are stored in a database created using the Geographic Information System (GIS) Manifold® System Release 8 (http://www.georeference.org/doc/manifold.htm, accessed on 16 April 2023), which allows for the validation and processing of GPS data to determine the swept area for each minute, mapping, and various types of analyses [37,60,62]. The analyses carried out within this study took into account the densities of *F. quadrangularis* colonies (n/m^2) calculated for each of the recorded minutes considered valid during the footage readings carried out to determine the densities of *N. norvegicus* burrows. Because *F. quadrangularis* is distributed in patches and the density values per station obtained could be very low, in order to obtain a finer spatial distribution, in this work, estimates were calculated at the level of one minute. The density (n/m^2) of colonies per minute was calculated using the number of individuals of *F. quadrangularis* counted divided by the swept area. In order to assess the *F. quadrangularis* density variation as a potential effect of the management measures implemented over time and space, an approach was adopted that is very similar to that used in Chiarini et al. [55]; in the latter study, in fact, a short-term evaluation of changes in fisheries management measures occurring in the Pomo/Jabuka Pits Area was carried out through a Before–Intermediate–After Multiple Sites (BIAMS) approach (a variant of the classic "before-after-control-impact" model design also adopted in an attempt to compare contiguous strata with different characteristics overcoming the unavailability of an adequate independent control site). To facilitate a possible comparison with the results obtained in the area for other species, within this study, only UWTV data collected from 2012 onward were considered; therefore, 8 UWTV surveys conducted in the period 2012–2019 (except for 2018) were taken into account. Overall, 3244 min of video were analysed, for a total of about 85,541 m^2 of the seabed. The same three time steps defined in Chiarini et al. [55] were also adopted to allow possible comparisons of the species response: a period "BEFORE" the implementation of the first management measures (from 1 January 2012 to 1 July 2015), the "INTERMEDIATE" stage (from 2 July 2015 to 31 August 2017), in which management measures have changed over time following national regulations [68,69], and "AFTER" the adoption of Italian and Croatian regulations analogous to the subsequent FRA application regulation [70] (from 1 September 2017 to 1 January 2020). The average density of colonies over the considered periods (i.e., "BEFORE", "INTERMEDIATE", and "AFTER") in each of the three zones ("A", "B", and "C") was analysed to assess any variations in the *F. quadrangularis* population. Average densities are not theoretically influenced by differences in surface; furthermore, by taking into consideration periods rather than single survey years, the limitation caused by the difference in the number (and location) of minutes recorded in each zone every year should be reduced. The standard deviation was also calculated to account for variability. In addition, the homogeneity of variance was assessed by a Levene test and, to address homoscedasticity, a one-way ANOVA (type II) was performed on the dataset and the Tukey test was used as a post hoc analysis. Regarding heteroscedasticity, a non-parametric ANOVA (Kruskal–Wallis test) and Games–Howell post hoc test were adopted. Each of the aforementioned periods were considered as the level of a temporal factor accounting for the fisheries management measures adopted in the study area from 2012 to 2019. To determine significance, the reference *p*-value was set at 0.05. These statistical analyses, and the production of related graphics, were performed using R software and the rstatix, car, dplyr and ggplot2 packages [71–74].

Furthermore, a spatial grid of 2 × 2 nautical mile cells (surface corresponding to 13.72 km^2 each) was built for the Pomo/Jabuka Pits area by means of the GIS. Figure 3 highlights the cells in the entire study area which were effectively monitored throughout the time frame under examination. Mean density (n/m^2) of colonies per cell was calculated using all data from 2012 to 2019 and also for the three considered periods (i.e., "BEFORE", "INTERMDIATE", and "AFTER") in order to obtain density maps.

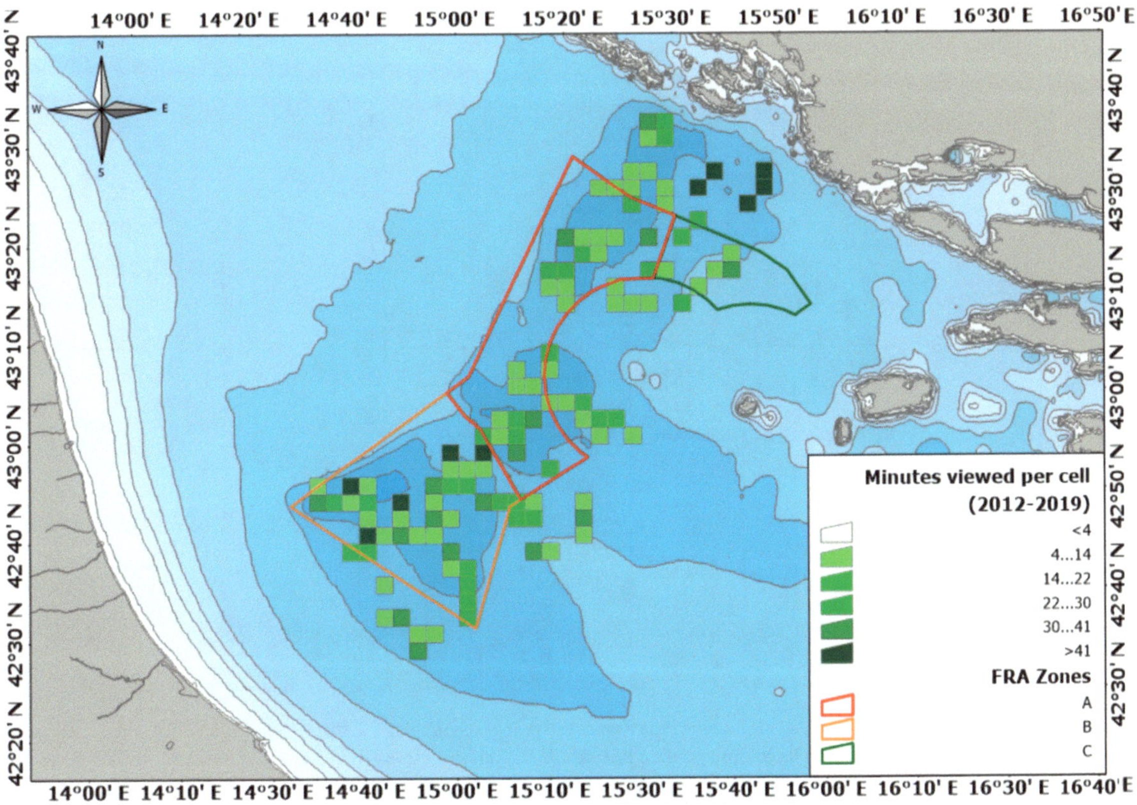

Figure 3. Map of the study area: the polygons indicate the three zones of the Jabuka/Pomo Pits FRA, cells of 2 × 2 nautical miles show the area covered by the UWTV surveys, while their green palette indicates the number of minutes recorded within each.

Persistence per cell from 2012 to 2019 was also calculated using an adaptation of the Getis G statistic adopted to identify spatial hotspots [75,76]. The persistence estimates were obtained by dividing the number of surveys in which the species was found in a particular cell by the total number of surveys in which that cell was visited. Only cells visited more than 1 time (i.e., in at least 2 surveys) and in which *F. quadrangularis* was recorded at least once were considered to calculate this index. Persistence was also calculated over the single period "BEFORE" in order to create a baseline showing the initial situation before the implementation of the management measures, to be used for comparison in future analyses of medium- to long-term effects.

3. Results

3.1. Spatial Distribution of Colonies and Density by Period for Each Zone ("A","B", and "C")

In the entire study area, in total, 138 colonies of *F. quadrangularis* were found, and the resulting average density per minute was 0.004 n/m^2 (±0.002). The spatial distribution of the average density of colonies per cell over the three considered periods and over the entire time frame of this study is shown in Figure 4. Figure 5, instead, shows the average densities of *F. quadrangularis* colonies calculated for each considered zone and period.

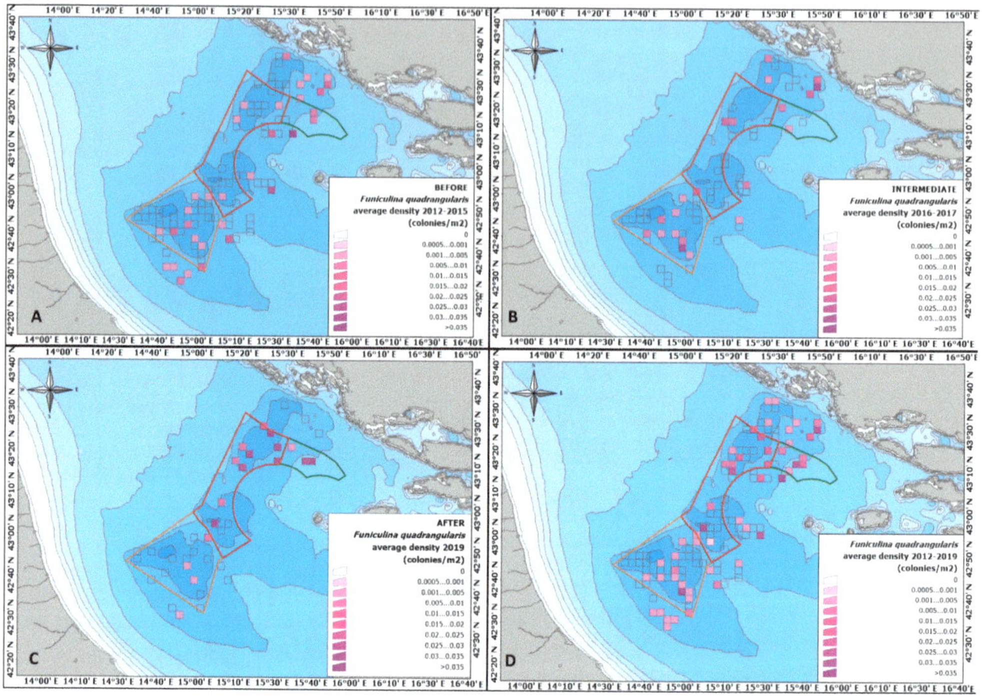

Figure 4. Map of the study area: the polygons indicate the three zones of the FRA, the 2 × 2 nautical mile cells coloured with a purple palette indicate the average density of *F. quadrangularis* colonies (no colonies were ever recorded in the empty cells); panels (**A–C**) refer respectively to periods "BEFORE", "INTERMEDIATE", and "AFTER", while panel (**D**) shows values calculated per each cell over the entire time frame of the study (2012–2019).

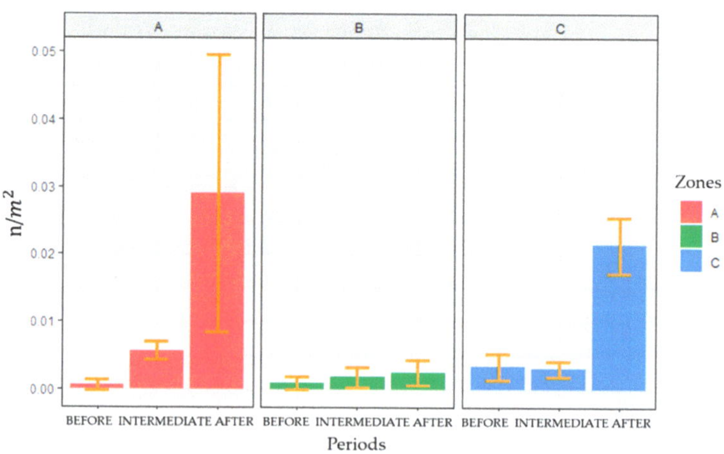

Figure 5. Average densities (n/m^2) of *F. quadrangularis* colonies recorded in the three FRA zones ("A", "B", "C") for the three considered periods ("BEFORE", "INTERMEDIATE", "AFTER"); yellow bars indicate the standard deviation.

3.1.1. Area "A" across the Periods

In total, 75 colonies of *F. quadrangularis* were counted in zone "A", and the average density value calculated over the entire time frame of the study was 0.004 n/m^2 (±0.01). The highest value of density was found in the period "AFTER" (maximum density = 0.176 n/m^2), whereas the lowest was in the period "BEFORE" (minimum density other than 0 values = 0.031 n/m^2). Given the heterogeneity of the variance, the Kruskal–Wallis test was performed (p-value < 0.001). The Games–Howell post hoc test proved that the period "AFTER" is significantly different from the other periods, showing an increase in density over time (p-value = 0.006 obtained for the comparison of "BEFORE" and "AFTER" and p-value = 0.014 for that between "INTERMEDIATE" and "AFTER"; Figure 5).

3.1.2. Area "B" across the Periods

In zone "B", 37 colonies were found, and the average density value calculated over the entire time frame of the study was 0.003 n/m^2 (±0.003). Both the highest and the lowest density values were found in the period "BEFORE" (maximum density = 0.095 n/m^2, minimum density other than 0 values = 0.032 n/m^2). Given the homogeneity of the variance, a one-way ANOVA was performed, but no significant differences among the periods were found, even if Figure 5 shows a slight increase in density over time.

3.1.3. Area "C" across the Periods

A total of 26 colonies were found in zone "C", and the average density value calculated over the entire time frame of the study was 0.003 n/m^2 (±0.001). The highest density was found in the period "AFTER" (maximum density = 0.091 n/m^2), whereas the lowest was found in the period "BEFORE" (minimum density other than 0 values = 0.029 n/m^2). Given the heterogeneity of the variance, the Kruskal–Wallis test was performed (p-value = 0.002). The Games–Howell post hoc test proved that the period "AFTER" is significantly different from the others, showing an increase in density over time (p-value = 0.004 obtained for the comparison of "BEFORE" and "AFTER" and p-value = 0.005 for that between "INTERMEDIATE" and "AFTER"; Figure 5).

3.2. Persistence per Cell

The persistence index, calculated for each cell within the entire study area across the entire time frame of the study (Figure 6A), showed the highest values (>0.9) in zones "A" and "C". The minimum value of persistence was recorded in Croatian territorial waters (=0.166), whereas the maximum value was found in zone "A" (=1). The persistence index calculated in each cell for the "BEFORE" period is shown in Figure 6B. The lowest recorded value (=0.333) in this case was in zone "B"; the highest was in Croatian territorial waters (=1).

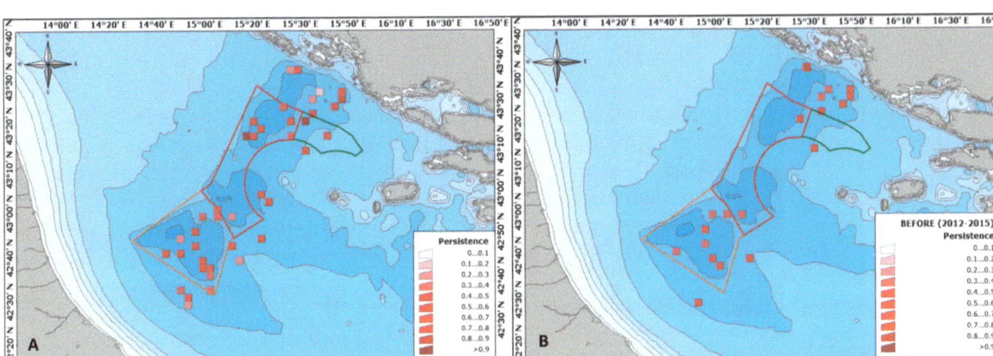

Figure 6. Map of *F. quadrangularis* colonies' persistence in the Pomo/Jabuka Pits area calculated across the whole considered UWTV time series (2012–2019; panel (**A**)) and only for the "BEFORE" period (panel (**B**)); the 2 × 2 nautical mile cells indicate portions of seabed visually inspected at least one time and in which colonies were seen at least once; the red palette indicates values calculated for each cell (from 0 to 1); the polygons indicate the three zones of the FRA.

4. Discussion

This study advances the understanding of *F. quadrangularis'* geographic range in the Adriatic Sea. Twelve species of sea pens are actually present in the Mediterranean Sea, of which six are distributed in the Adriatic Sea, with a southward decline in abundance [30]. The presence of *F. quadrangularis* in the Off Ancona area, in the Pomo/Jabuka Pits, and off the Montenegrin coast is worthy of note [30,31]. Sessile organisms are damaged by severe bottom trawling, endangering the benthic environment and its related fish resources [26,77]. The abundance and location of sea pens in the Adriatic basin may have been influenced by the high level of fishing activity [30,78]. Historical information on the abundance and distribution of Pennatulacea in the Adriatic Sea is lacking, and it is challenging to identify an original, undisturbed state. The quantity and geographic distribution of sea pens may have been reduced in the Mediterranean basin, especially in the Adriatic Sea, due to intense trawling targeting commercial species [30]. The use of Sentinels of Seabed (SoS) indicators could help to map VMEs in the Mediterranean and evaluate anthropogenic impacts, as described by Serrano et al. [13]. The historical analysis of the distribution of *F. quadrangularis* in the Pomo/Jabuka Pits area is therefore noteworthy; being nowadays a FRA, it can be used to test the recovery times of species of high ecological importance.

4.1. *Effects of Different Protection Levels on Funiculina quadrangularis*

The development of *F. quadrangularis* distribution maps could provide indications on the state of the benthic component impacted by bottom trawling [13,23,31,34]. The slow growth rate, late sexual maturity and lack of withdrawl capacity make this species sensitive to trawl impact [24,26,79]. In fact, Downie et al. [33] reported *F. quadrangularis* to be less resilient to trawling in the Greater North Sea and Celtic Seas than other sea pens such as *Pennatula phosphorea* (Linnaeus, 1758) and *Virgularia mirabilis* (Müller, 1776). However, according to Pierdomenico et al. [80], the relationship between trawling intensity and the abundance of this species is not straightforward in the southern Tyrrhenian Sea (i.e., the lowest abundances were recorded in areas with strong fishing pressure, but relatively high values were observed in areas subject to intermediate effort), as it is for another VME indicator such as the bamboo coral *Isidella elongata* (Esper, 1788). While performing ROV video survey transects in a no-take reserve and some control areas in the northwestern Mediterranean Sea, Vigo et al. [36] found a higher abundance of Norway lobster and other fish species within the no-take reserve than in the control area, but only a slightly higher, but not significant, abundance of *F. quadrangularis* in the no-take area after 2.5 years since the implementation. The study area considered in the present work was instead subjected to various management measures, and consequent levels of fishing pressure, which varied over time and space. This condition allowed to evaluate the effects of different levels of trawl impact on sea pen colonies. The average density of *F. quadrangularis* colonies showed variations over the entire study period (2012–2019), particularly in zones "A" and "C", which appeared to be positively influenced by the management measures. In fact, the average *F. quadrangularis* density in these two zones consistently increased, especially in the "AFTER" period following the establishment of the FRA. Although some environmental differences between the two zones, such as the difference in bathymetry and circulation of the Adriatic Sea, do not allow to directly compare them, it is evident that despite a theoretically similar level of protection in the latter period, zones "B" and "C" showed different results. The modest increase in zone "B" is likely the result of a fishing pressure still strong enough to condition the colonies' presence and growth compared to the other zones. Therefore, the results of this study suggest that under effective trawling management and, likely, favourable environmental conditions, *F. quadrangularis* can increase in density within a few years. These findings are quite in agreement with those resulting for other benthic species (e.g., *N. norvegicus*) in Chiarini et al. [55,57].

4.2. Ecological Aspects and Colonies' Persistence

Although it turns out to be a species sensitive to trawling, the unbranched form and elasticity of *F. quadrangularis* allow the persistence of this species even in heavily trawled regions, but at a lower density than in well-preserved populations [80,81]. In this study, an index was tested to determine which regions within the Pomo/Jabuka Pits had more persistent *F. quadrangularis* patches. A possible limitation of this index is that it does not account for the time sequence of the records within each cell over the different years, so it probably should not be applied in its current form to long time series when two records could paradoxically be at the beginning and end. However, in shorter (and more homogeneous) time series, as in the case of the "BEFORE" period, it can provide useful information to compare with subsequent time steps. In fact, when considering the entire time frame of the study, zones "A" and "C" recorded the highest persistence values. This might be correlated with management measures, but a further comparison with persistence calculated for the "AFTER" period, considering more than one survey/year, would be interesting to confirm this.

It would be interesting to also include in future medium- to long-term evaluations the distribution and displacement of fishing efforts directly estimated by means of Vessel Monitoring Systems (VMSs) and/or Automatic Identification Systems (AISs) [82–85]. To allow a better understanding of the response of the abundance and distribution of sea pen colonies to fishing pressure, more complex analyses could also be performed on the acquired datasets, including the main physical characteristics of the study area (e.g., bathymetry) and possible local variations over time of environmental parameters influencing benthic communities (e.g., temperature, salinity, oxygen, etc. [57,63]).

F. quadrangularis is a significant habitat-forming species that helps create three-dimensional bathyal environments, improves ecological functionality, and provides essential habitat for fish and invertebrates [30,77,86,87]. Therefore, the assessment of its persistence can offer important information for the correct management of the study area; in fact, the possibility of geographically locating the most persistent patches, associated with the measurement of density, can indicate which are the zones most in need of protection.

The use of non-invasive methodologies such as ROVs or UWTV for the collection of information on benthic indicator species is well established and, for the evaluation of the long-term effects of the management measures in place in the study area, it would be desirable to maintain the UWTV time series in the future, or at least repeat video surveys from time to time [26,36,56].

5. Conclusions

The Adriatic Sea geographic range of *F. quadrangularis* has been improved by this work, which also supports the mapping of VMEs and EFHs. This study demonstrates that effective management measures can have a positive influence on epibenthic communities and highlights the potential of sea pens as indicators of impact on and/or recovery of exploited habitats. These findings might be used in the planning and monitoring of sensitive marine areas. Furthermore, these results could also be used in the context of the Blue Economy, allowing for the exploitation of the benefits offered to commercial fishing due to the maintenance of intact benthic habitats. However, more studies, using video surveys, are required to assess the long-term effects of the management strategies applied in the Pomo/Jabuka Pits area on the sea pen community. It would be relevant to include, in future, analyses of the abundance and distribution of commercially and ecologically important species, and more variables, such as environmental parameters or VMS and AIS data for authorized fleets operating in zones "B", "C", and surrounding areas, which are also useful to monitor the displacement of fishing efforts.

Author Contributions: Conceptualization, M.M.; Data curation, M.M. and L.Z.; Formal analysis, M.M. and L.Z.; Funding acquisition, M.M.; Investigation, F.D., A.B., L.Z., P.S., P.P., D.M., I.I. and M.M.; Methodology, M.M. and L.Z.; Project administration, M.M. and N.V.; Supervision, M.M., I.I. and N.V.; Writing—original draft, M.M. and L.Z. All authors have read and agreed to the published version of the manuscript.

Funding: Funding was provided for the collection and processing of the data used in this study by the FAO AdriaMed regional project, the RITMARE Flagship Project of the Italian Ministry of University and Research, the Direzione Generale della Pesca Marittima e dell'Acquacoltura of the Italian Ministry of Agricultural, Food, and Forestry Policies (through a series of agreements with CNR-IRBIM; CUP J52I15003990001, J53C17000540001), and by the Italian Institute for Environmental Protection and Research (through an agreement with CNR-IRBIM for the study of epimegabenthic communities subjected to physical perturbations carried out in the context of Descriptor 6 of the European Marine Strategy Framework Directive).

Institutional Review Board Statement: This study did not involve the removal of animals from the environment; the institutional review board statement is not applicable.

Data Availability Statement: All data are stored at CNR IRBIM Ancona and can be shared upon the reasonable request of qualified researchers. A portion of the dataset was also collected with permits of the Croatian Institutions. The authors received formal authorization to use the data for scientific purposes by the competent authorities, but no special access privileges were granted.

Acknowledgments: The authors want to thank for their precious contributions Giordano Giuliani, Giovanni Canduci, Camilla Croci, and the crew of RV Dallaporta during the scientific surveys.

Conflicts of Interest: The authors declare no conflict of interest. The funders had no role in the design of the study, the collection, analyses, or interpretation of data, the writing of the manuscript, or in the decision to publish the results.

References

1. Clark, M.R.; Althaus, F.; Schlacher, T.A.; Williams, A.; Bowden, D.A.; Rowden, A.A. The Impacts of Deep-Sea Fisheries on Benthic Communities: A Review. *ICES J. Mar. Sci.* **2016**, *73*, i51–i69. [CrossRef]
2. Farella, G.; Tassetti, A.N.; Menegon, S.; Bocci, M.; Ferrà, C.; Grati, F.; Fadini, A.; Giovanardi, O.; Fabi, G.; Raicevich, S.; et al. Ecosystem-Based MSP for Enhanced Fisheries Sustainability: An Example from the Northern Adriatic (Chioggia-Venice and Rovigo, Italy). *Sustainability* **2021**, *13*, 1211. [CrossRef]
3. Van Denderen, P.D.; Hintzen, N.T.; Van Kooten, T.; Rijnsdorp, A.D. Temporal Aggregation of Bottom Trawling and Its Implication for the Impact on the Benthic Ecosystem. *ICES J. Mar. Sci.* **2015**, *72*, 952–961. [CrossRef]
4. Pusceddu, A.; Bianchelli, S.; Martín, J.; Puig, P.; Palanques, A.; Masqué, P.; Danovaro, R. Chronic and Intensive Bottom Trawling Impairs Deep-Sea Biodiversity and Ecosystem Functioning. *Proc. Natl. Acad. Sci. USA* **2014**, *111*, 8861–8866. [CrossRef] [PubMed]
5. Pusceddu, A.; Bianchelli, S.; Danovaro, R. Quantity and Biochemical Composition of Particulate Organic Matter in a Highly Trawled Area (Thermaikos Gulf, Eastern Mediterranean Sea). *Adv. Oceanogr. Limnol.* **2015**, *6*, 21–32. [CrossRef]
6. Pommer, C.D.; Olesen, M.; Jhansen, Ø.L.S. Impact and Distribution of Bottom Trawl Fishing on Mud-Bottom Communities in the Kattegat. *Mar. Ecol. Prog. Ser.* **2016**, *548*, 47–60. [CrossRef]
7. Jennings, S.; Pinnegar, J.K.; Polunin, N.V.C.; Warr, K.J. Impacts of Trawling Disturbance on the Trophic Structure of Benthic Invertebrate Communities. *Mar. Ecol. Prog. Ser.* **2001**, *213*, 127–142. [CrossRef]
8. Hutchings, P. Review of the Effects of Trawling on Macrobenthic Epifaunal Communities. *Mar. Freshw. Res.* **1990**, *41*, 111–120. [CrossRef]
9. Kaiser, M.J.; Collie, J.S.; Hall, S.J.; Jennings, S.; Poiner, I.R. Modification of Marine Habitats by Trawling Activities: Prognosis and Solutions. *Fish Fish.* **2002**, *3*, 114–136. [CrossRef]
10. Lauria, V.; Garofalo, G.; Fiorentino, F.; Massi, D.; Milisenda, G.; Piraino, S.; Russo, T.; Gristina, M. Species Distribution Models of Two Critically Endangered Deep-Sea Octocorals Reveal Fishing Impacts on Vulnerable Marine Ecosystems in Central Mediterranean Sea. *Sci. Rep.* **2017**, *7*, 8049. [CrossRef]
11. Rodríguez-Rodríguez, D.; Rodríguez, J.; Abdul Malak, D. Development and Testing of a New Framework for Rapidly Assessing Legal and Managerial Protection Afforded by Marine Protected Areas: Mediterranean Sea Case Study. *J. Environ. Manage.* **2016**, *167*, 29–37. [CrossRef]
12. Bastardie, F.; Angelini, S.; Bolognini, L.; Fuga, F.; Manfredi, C.; Martinelli, M.; Nielsen, J.R.; Santojanni, A.; Scarcella, G.; Grati, F. Spatial Planning for Fisheries in the Northern Adriatic: Working toward Viable and Sustainable Fishing. *Ecosphere* **2017**, *8*, e01696. [CrossRef]

13. Serrano, A.; de la Torriente, A.; Punzón, A.; Blanco, M.; Bellas, J.; Durán-Muñoz, P.; Murillo, F.J.; Sacau, M.; García-Alegre, A.; Antolínez, A.; et al. Sentinels of Seabed (SoS) Indicator: Assessing Benthic Habitats Condition Using Typical and Sensitive Species. *Ecol. Indic.* **2022**, *140*, 108979. [CrossRef]
14. Coll, M.; Piroddi, C.; Steenbeek, J.; Kaschner, K.; Lasram, F.B.R.; Aguzzi, J.; Ballesteros, E.; Bianchi, C.N.; Corbera, J.; Dailianis, T.; et al. The Biodiversity of the Mediterranean Sea: Estimates, Patterns, and Threats. *PLoS ONE* **2010**, *5*, e11842. [CrossRef] [PubMed]
15. Smith, C.; Papadopoulou, N.; Sevastou, K.; Franco, A.; Teixeira, H.; Piroddi, C.; Katsanevakis, S.; Fürhaupter, K.; Beauchard, O.; Cochrane, S.; et al. Report on Identification of Keystone Species and Processes across Regional Seas. 2014, pp. 1–103. Available online: https://hal.science/hal-01790558 (accessed on 27 April 2023).
16. Degnbol, P. Indicators as a Means of Communicating Knowledge. *ICES J. Mar. Sci.* **2005**, *62*, 606–611. [CrossRef]
17. European Commission. *Commission Decision (EU) 2017/848 of 17 May 2017-Laying down Criteria and Methodological Standards on Good Environmental Status of Marine Waters and Specifications and Standardized Methods for Monitoring and Assessment*; European Commission: Brussels, Belgium, 2017; Official Journal L 125; pp. 43–74. Available online: https://eur-lex.europa.eu/legal-content/EN/TXT/?uri=CELEX%3A32017D0848 (accessed on 27 April 2023).
18. Burgos, J.M.; Buhl-Mortensen, L.; Buhl-Mortensen, P.; Ólafsdóttir, S.H.; Steingrund, P.; Ragnarsson, S.; Skagseth, Ø. Predicting the Distribution of Indicator Taxa of Vulnerable Marine Ecosystems in the Arctic and Sub-Arctic Waters of the Nordic Seas. *Front. Mar. Sci.* **2020**, *7*, 131. [CrossRef]
19. Williams, G.C. The Global Diversity of Sea Pens (Cnidaria: Octocorallia: Pennatulacea). *PLoS ONE* **2011**, *6*, e22747. [CrossRef]
20. Cogswell, A.; Kenchington, E.; Lirette, C.; Murillo, F.J.; Campanis, G.; Campbell, N.; Ollerhead, N. Layers Utilized by an ArcGIS Model to Approximate Commercial Coral and Sponge By-Catch in the NAFO Regulatory Area. 2011. Available online: https://archive.nafo.int/open/sc/2011/scr11-072.pdf (accessed on 26 April 2023).
21. United Nations. *Resolution 70/75: Sustainable Fisheries, Including through the 1995 Agreement for the Implementation of the Provisions of the United Nations Convention on the Law of the Sea of 10 December 1982 Relating to the Conservation and Management of Straddling Fiss*; United Nations: New York, NY, USA, 2016; p. 35. Available online: https://sdgs.un.org/documents/ares7075-sustainable-fisheries-including-thr-23183 (accessed on 28 April 2023).
22. Hughes, D.J. Sea Pens and Burrowing Megafauna: An Overview of Dynamics and Sensitivity Characteristics for Conservation Management of Marine SACs. *Science* **1998**, *3*, 1–105.
23. Felder, D.L.; Camp, D.K. Gulf of Mexico: Origin, Waters, and Biota. 2009, Volume 1. Available online: https://decapoda.nhm.org/pdfs/31409/31409.pdf (accessed on 25 April 2023).
24. Wright, E.P.; Kemp, K.; Rogers, A.D.; Yesson, C. Genetic Structure of the Tall Sea Pen *Funiculina quadrangularis* in NW Scottish Sea Lochs. *Mar. Ecol.* **2015**, *36*, 659–667. [CrossRef]
25. Ardizzone, G.D. *An Introduction to Sensitive and Essential Fish Habitats Identification and Protection in the Mediterranean Sea*; STECF: Rome, Italy, 2006; p. 63.
26. Greathead, C.F.; Donnan, D.W.; Mair, J.M.; Saunders, G.R. The Sea Pens *Virgularia mirabilis*, *Pennatula phosphorea* and *Funiculina quadrangularis*: Distribution and Conservation Issues in Scottish Waters. *J. Mar. Biol. Assoc. UK* **2007**, *87*, 1095–1103. [CrossRef]
27. Rogers, A.D.; Gianni, M. The Implementation of UNGA Resolutions 61/105 and 64/72 in the Management of Deep-Sea Fisheries on the High Seas. 2010. Available online: http://www.stateoftheocean.org/pdfs/61105-Implemention-finalreport.pdf (accessed on 18 April 2023).
28. FAO; GFCM. *Fishery and Aquaculture Statistics. GFCM Capture Production 1970-2020 (FishStatJ)*; FAO Fisheries and Aquaculture Division: Rome, Italy, 2022; Available online: https://www.fao.org/fishery/en/topic/166235?lang=en (accessed on 10 April 2023).
29. International Union for Conservation of Nature. *The IUCN Red List of Threatened Species*; IUCN Version 2022-2; IUCN: Gland, Switzerland, 2022; Available online: https://nc.iucnredlist.org/redlist/content/attachment_files/2022-2_RL_Stats_Table_1a.pdf (accessed on 25 April 2023).
30. Bastari, A.; Pica, D.; Ferretti, F.; Micheli, F.; Cerrano, C. Sea Pens in the Mediterranean Sea: Habitat Suitability and Opportunities for Ecosystem Recovery. *ICES J. Mar. Sci.* **2018**, *75*, 1722–1732. [CrossRef]
31. Mačić, V.; Đorđević, N.; Đurović, M.; Petović, S.; Russo, T. Improving Knowledge of *Funiculina quadrangularis* and Vulnerable Marine Ecosystems in the South Adriatic. *Mediterr. Mar. Sci.* **2022**, *23*, 805–816. [CrossRef]
32. Edwards, D.C.B.; Moore, C.G. Reproduction in the Sea Pen *Funiculina quadrangularis* (Anthozoa: Pennatulacea) from the West Coast of Scotland. *Estuar. Coast. Shelf Sci.* **2009**, *82*, 161–168. [CrossRef]
33. Downie, A.L.; Noble-James, T.; Chaverra, A.; Howell, K.L. Predicting Sea Pen (Pennatulacea) Distribution on the UK Continental Shelf: Evidence of Range Modification by Benthic Trawling. *Mar. Ecol. Prog. Ser.* **2021**, *670*, 75–91. [CrossRef]
34. Martinelli, M.; Morello, E.B.; Isajlović, I.; Belardinelli, A.; Lucchetti, A.; Santojanni, A.; Atkinson, R.J.A.; Vrgoč, N.; Arneri, E. Towed Underwater Television towards the Quantification of Norway Lobster, Squat Lobsters and Sea Pens in the Adriatic Sea. *Acta Adriat.* **2013**, *54*, 3–12.
35. Miatta, M.; Snelgrove, P.V.R. Sea Pens as Indicators of Macrofaunal Communities in Deep-Sea Sediments: Evidence from the Laurentian Channel Marine Protected Area. *Deep. Res. Part I Oceanogr. Res. Pap.* **2022**, *182*, 103702. [CrossRef]
36. Vigo, M.; Navarro, J.; Aguzzi, J.; Bahamon, N.; García, J.A.; Rotllant, G.; Recasens, L.; Company, J.B. ROV-Based Monitoring of Passive Ecological Recovery in a Deep-Sea No-Take Fishery Reserve. *Sci. Total Environ.* **2023**, *883*, 163339. [CrossRef]
37. ICES. Working Group on *Nephrops* Surveys (WGNEPS; Outputs from 2021). *ICES Sci. Rep.* **2022**, *4*, 183. [CrossRef]

38. OSPAR. Meeting of the Intersessional Correspondence Group on the Implementation Follow up of Measures for the Protection and Conservation of Species and Habitats (ICG-POSH), Paris, France, 9–11 October 2018. Roadmap for the Implementation of Collective Actions W. 2018, pp. 1–8. Available online: https://www.ospar.org/site/assets/files/38906/0205a5_actionsheet13_mpashabitats.pdf (accessed on 2 May 2023).
39. EMODnet; EMODnet Bathymetry Consortium. *EMODnet Digital Bathymetry (DTM), Ifremer*; EMODnet: Oostende, Belgium, 2016. [CrossRef]
40. Russo, T.; Morello, E.B.; Parisi, A.; Scarcella, G.; Angelini, S.; Labanchi, L.; Martinelli, M.; D'Andrea, L.; Santojanni, A.; Arneri, E.; et al. A Model Combining Landings and VMS Data to Estimate Landings by Fishing Ground and Harbor. *Fish. Res.* **2018**, *199*, 218–230. [CrossRef]
41. Marini, M.; Maselli, V.; Campanelli, A.; Foglini, F.; Grilli, F. Role of the Mid-Adriatic Deep in Dense Water Interception and Modification. *Mar. Geol.* **2016**, *375*, 5–14. [CrossRef]
42. Taviani, M.; Angeletti, L.; Beuck, L.; Campiani, E.; Canese, S.; Foglini, F.; Freiwald, A.; Montagna, P.; Trincardi, F. Reprint of "On and off the Beaten Track: Megafaunal Sessile Life and Adriatic Cascading Processes". *Mar. Geol.* **2016**, *375*, 146–160. [CrossRef]
43. Jennings, S.; Kaiser, M.J. The Effects of Fishing on Marine Ecosystems. *Adv. Mar. Biol.* **1998**, *34*, 201–212. [CrossRef]
44. Lucey, S.M.; Nye, J.A. Shifting Species Assemblages in the Northeast US Continental Shelf Large Marine Ecosystem. *Mar. Ecol. Prog. Ser.* **2010**, *415*, 23–33. [CrossRef]
45. Albouy, C.; Guilhaumon, F.; Araújo, M.B.; Mouillot, D.; Leprieur, F. Combining Projected Changes in Species Richness and Composition Reveals Climate Change Impacts on Coastal Mediterranean Fish Assemblages. *Glob. Chang. Biol.* **2012**, *18*, 2995–3003. [CrossRef] [PubMed]
46. Hillebrand, H.; Soininen, J.; Snoeijs, P. Warming Leads to Higher Species Turnover in a Coastal Ecosystem. *Glob. Chang. Biol.* **2010**, *16*, 1181–1193. [CrossRef]
47. Lavergne, S.; Mouquet, N.; Thuiller, W.; Ronce, O. Biodiversity and Climate Change: Integrating Evolutionary and Ecological Responses of Species and Communities. *Annu. Rev. Ecol. Evol. Syst.* **2010**, *41*, 321–350. [CrossRef]
48. GFCM. *Recommendation GFCM/41/2017/3 on the Establishment of a Fisheries Restricted Area in the Jabuka/Pomo Pit in the Adriatic Sea*; GFCM: Rome, Italy, 2017.
49. Zupanovic, S.; Jardas, I. A Contribution to the Study of Biology and Population Dynamics of the Adriatic Hake, *Merluccius merluccius* (L.). *Acta Adriat.* **1986**, *27*, 97–146.
50. Angelini, S. Towards an Ecosystem Approach to Fisheries for Nephrops norvegicus and Merluccius merluccius Inhabiting the Central Adriatic Sea. 2016. Available online: http://amsdottorato.unibo.it/7380/1/Angelini_Silvia_tesi.pdf (accessed on 3 May 2023).
51. Angelini, S.; Martinelli, M.; Santojanni, A.; Colella, S. Biological Evidence of the Presence of Different Subpopulations of Norway Lobster (*Nephrops norvegicus*) in the Adriatic Sea (Central Mediterranean Sea). *Fish. Res.* **2020**, *221*, 105365. [CrossRef]
52. Colloca, F.; Mastrantonio, G.; Lasinio, G.J.; Ligas, A.; Sartor, P. *Parapenaeus longirostris* (Lucas, 1846) an Early Warning Indicator Species of Global Warming in the Central Mediterranean Sea. *J. Mar. Syst.* **2014**, *138*, 29–39. [CrossRef]
53. Martinelli, M.; Angelini, S.; Belardinelli, A.; Caccamo, G.; Cacciamani, R.; Calì, F.; Canduci, G.; Chiarini, M.; Croci, C.; Domenichetti, F.; et al. *Accordo Tra MIPAAF e CNR-IRBIM Ancona in Merito Alla Proposta Progettuale Relativa Alle Attività Di Monitoraggio Periodico Delle Fosse Di Pomo e All'attuazione Di Misure Che, Nel Rispetto Dei Piani Di Gestione, Comportino Il Mantenimento Delle Condizioni Ambientali Idonee Alla vita e All'accrescimento dei Molluschi Bivalvi, Ponendo in Essere Misure Supplementari Tese a Proteggere le Diverse Fasi del Ciclo Biologico Delle Specie Interessate (CUP J41F19000080001)-Parte Monitoraggio Fosse di Pomo periodo 2019-2020*; Secondo interim report-Luglio 2020; CNR-IRBIM: Ancona, Italy, 2020.
54. Froglia, C.; Gramitto, M.; Martinelli, M.; Betulla, M. Long Term Changes in the Decapod Crustaceans Assemblage in the Western Meso-Adriatic Depression (Pomo Pit). *Crustac. Soc. Mid-Year* **2017**, *1*, 19–22.
55. Chiarini, M.; Guicciardi, S.; Zacchetti, L.; Domenichetti, F.; Canduci, G.; Angelini, S.; Belardinelli, A.; Croci, C.; Giuliani, G.; Scarpini, P.; et al. Looking for a Simple Assessment Tool for a Complex Task: Short-Term Evaluation of Changes in Fisheries Management Measures in the Pomo/Jabuka Pits Area (Central Adriatic Sea). *Sustainability* **2022**, *14*, 7742. [CrossRef]
56. GFCM. *GFCM-SAC Report of the Working Group on Vulnerable Marine Ecosystems and Essential Fish Habitats (WGVME-EFH)*; FAO: Rome, Italy, 2023; p. 41.
57. Chiarini, M.; Guicciardi, S.; Angelini, S.; Tuck, I.D.; Grilli, F.; Penna, P.; Domenichetti, F.; Canduci, G.; Belardinelli, A.; Santojanni, A.; et al. Accounting for Environmental and Fishery Management Factors When Standardizing CPUE Data from a Scientific Survey: A Case Study for *Nephrops norvegicus* in the Pomo Pits Area (Central Adriatic Sea). *PLoS ONE* **2022**, *17*, e0270703. [CrossRef] [PubMed]
58. EU. Regulation (EU) 2019/982 of the European Parliament and of the Council of 5 June 2019 Amending Regulation (EU) No 1343/2011 on Certain Provisions for Fishing in the GFCM (General Fisheries Commission for the Mediterranean) Agreement Area; 2019, Official Journal L 164, pp. 1–22. Available online: http://data.europa.eu/eli/reg/2019/982/oj (accessed on 5 May 2023).
59. GFCM. *Recommendation GFCM/44/2021/2 on the Establishment of a Fisheries Restricted Area in the Jabuka/Pomo Pit in the Adriatic Sea (Geographical Subarea 17), Amending Recommendation*; GFCM: Rome, Italy, 2021.
60. Martinelli, M.; Belardinelli, A.; Guicciardi, S.; Penna, P.; Domenichetti, F.; Croci, C.; Angelini, S.; Medvesek, D.; Froglia, C.; Scarpini, P. *Support the Monitoring of Fisheries and Fisheries Resources in the Adriatic Sea" Report of Task 2 "To Perform the Appraisal of Nephrops Norvegicus in the Central Adriatic Sea (GFCM GSA 17) through Underwater Television Surveys*; ISMAR: Ancona, Italy, 2017.

61. Martinelli, M.; Belardinelli, A.; Guicciardi, S.; Penna, P.; Domenichetti, F.; Croci, C.; Angelini, S.; Medvesek, D.; Scarpini, P.; Micucci, D.; et al. *SP2_LI1_WP1_UO05_D01-Rapporto Della Campagna 2015 (Ex SP2_WP1_AZ3_UO05_D03-Report 3° UWTV Survey–RITMARE)-RITMARE La Ricerca ITaliana per Il MARE*; ISMAR: Ancona, Italy, 2016.
62. Martinelli, M.; Angelini, S.; Belardinelli, A.; Chiarini, M.; Croci, C.; Domenichetti, F.; Guicciardi, S.; Scarpini, P.; Santojanni, A.; Lorenzo, Z. *Final report Modulo 6. Monitoraggio Fosse Di Pomo Periodo 2017-2018 (Esteso Primavera 2019) Convenzione Tra MIPAAFT e CNR-ISMAR Ancona per Uno Studio Propedeutico Al Rinnovo Dell'affidamento Della Gestione Della Pesca Dei Molluschi Bivalvi Ai Consorzi Di Gestione–CUP J53C17000540001*; IRBIM-CNR: Ancona, Italy, 2019.
63. Penna, P.; Grilli, F.; Belardinelli, A.; Domenichetti, F.; Scarpini, P.; Martinelli, M. Pomo Pits Pressure/Temperature/Salinity/Oxygen Profiles Spring Dataset 2012–2021. *Seanoe* **2022**. [CrossRef]
64. ICES. Working Group on *Nephrops* Surveys (WGNEPS; Outputs from 2019). *ICES Sci. Rep.* **2020**, *2*, 85.
65. Aguzzi, J.; Bahamon, N.; Doyle, J.; Lordan, C.; Tuck, I.D.; Chiarini, M.; Martinelli, M.; Company, J.B. Burrow Emergence Rhythms of *Nephrops norvegicus* by UWTV and Surveying Biases. *Sci. Rep.* **2021**, *11*, 5797. [CrossRef]
66. Aguzzi, J.; Chatzievangelou, D.; Robinson, N.J.; Bahamon, N.; Berry, A.; Carreras, M.; Company, J.B.; Costa, C.; del Rio Fernandez, J.; Falahzadeh, A.; et al. Advancing Fishery-Independent Stock Assessments for the Norway Lobster (*Nephrops norvegicus*) with New Monitoring Technologies. *Front. Mar. Sci.* **2022**, *9*, 969071. [CrossRef]
67. Martinelli, M.; Scarcella, G.; Sabatini, L.; Zacchetti, L.; Luzi, F.; Domenichetti, F. *Convenzione tra ISPRA e CNR-IRBIM per la realizzazione di attività condivise, finalizzate a dare attuazione alle previsioni del d. lgs 13 ottobre 2010 n. 190, nell'ambito della Strategia Marina nel triennio 2021–2023. Modulo comunità epimegabentoniche–Mar Adriatico (GSA 17), Final report*; IRBIM-CNR: Ancona, Italy, 2023.
68. MIPAAF. *Modalità Attuative per La Pesca Nella Fossa Di Pomo; Official Gazette of the Italian Republic No 2 of 03/01/2017*; Ministero Delle Politiche Agricole Alimentari e Forestali: Rome, Italy, 2017.
69. MIPAAF. *Arresto Temporaneo Obbligatorio Delle Unità Autorizzate All'esercizio Della Pesca Con Il Sistema Strascico-Annualità 2015; Official Gazette of the Italian Republic No 162 of 15/07/2015*; Ministero Delle Politiche Agricole Alimentari e Forestali: Rome, Italy, 2015.
70. MIPAAF. *Misure per La Pesca Nella Fossa Di Pomo*; Ministero Delle Politiche Agricole Alimentari e Forestali: Rome, Italy, 2017; Available online: https://www.politicheagricole.it/flex/cm/pages/ServeAttachment.php/L/IT/D/9%252F0%252F4%252FD.583cb74efe7278af0dc3/P/BLOB%3AID%3D11345/E/pdf?mode=download (accessed on 12 May 2023).
71. Wickham, H. *ggplot2 Elegant Graphics for Data Analysis*, 3rd. ed; Springer: New York, NY, USA, 2016; pp. 3–10.
72. Wickham, H.; François, R.; Henry, L.; Müller, K. A Grammar of Data Manipulation [R Package Dplyr Version 1.0.0]. *Media.* 2020. Available online: https://cran.r-project.org/web/packages/dplyr/index.html (accessed on 10 April 2023).
73. Fox, J.; Weisberg, S. *An {R} Companion to Applied Regression*, 3rd ed.; Sage: Thousand Oaks, CA, USA, 2019.
74. Kassambara, A. Rstatix:Pipe-Friendly Framework for Basic Statistical Tests. R Package Version 0.7.1. 2022. Available online: https://CRAN.R-project.org/package=rstatix (accessed on 12 April 2023).
75. Colloca, F.; Garofalo, G.; Bitetto, I.; Facchini, M.T.; Grati, F.; Martiradonna, A.; Mastrantonio, G.; Nikolioudakis, N.; Ordinas, F.; Scarcella, G.; et al. The Seascape of Demersal Fish Nursery Areas in the North Mediterranean Sea, a First Step towards the Implementation of Spatial Planning for Trawl Fisheries. *PLoS ONE* **2015**, *10*, e0119590. [CrossRef] [PubMed]
76. Getis, A.; Ord, J.K. The Analysis of Spatial Association by Use of Distance Statistics. *Geogr. Anal.* **1992**, *24*, 189–206. [CrossRef]
77. Mastrototaro, F.; Chimienti, G.; Acosta, J.; Blanco, J.; Garcia, S.; Rivera, J.; Aguilar, R. *Isidella elongata* (Cnidaria: Alcyonacea) Facies in the Western Mediterranean Sea: Visual Surveys and Descriptions of Its Ecological Role. *Eur. Zool. J.* **2017**, *84*, 209–225. [CrossRef]
78. Lotze, H.K.; Coll, M.; Dunne, J.A. Historical Changes in Marine Resources, Food-Web Structure and Ecosystem Functioning in the Adriatic Sea, Mediterranean. *Ecosystems* **2011**, *14*, 198–222. [CrossRef]
79. Eno, N.C.; MacDonald, D.S.; Kinnear, J.A.M.; Amos, S.C.; Chapman, C.J.; Clark, R.A.; Bunker, F.S.P.D.; Munro, C. Effects of Crustacean Traps on Benthic Fauna. *ICES J. Mar. Sci.* **2001**, *58*, 11–20. [CrossRef]
80. Pierdomenico, M.; Russo, T.; Ambroso, S.; Gori, A.; Martorelli, E.; D'Andrea, L.; Gili, J.M.; Chiocci, F.L. Effects of Trawling Activity on the Bamboo-Coral *Isidella elongata* and the Sea Pen *Funiculina quadrangularis* along the Gioia Canyon (Western Mediterranean, Southern Tyrrhenian Sea). *Prog. Oceanogr.* **2018**, *169*, 214–226. [CrossRef]
81. Bo, M.; Canese, S.; Spaggiari, C.; Pusceddu, A.; Bertolino, M.; Angiolillo, M.; Giusti, M.; Loreto, M.F.; Salvati, E.; Greco, S.; et al. Deep Coral Oases in the South Tyrrhenian Sea. *PLoS ONE* **2012**, *7*, e49870. [CrossRef] [PubMed]
82. Russo, T.; D'Andrea, L.; Parisi, A.; Cataudella, S. VMSbase: An R-Package for VMS and Logbook Data Management and Analysis in Fisheries Ecology. *PLoS ONE* **2014**, *9*, e100195. [CrossRef]
83. Russo, T.; D'Andrea, L.; Parisi, A.; Martinelli, M.; Belardinelli, A.; Boccoli, F.; Cignini, I.; Tordoni, M.; Cataudella, S. Assessing the Fishing Footprint Using Data Integrated from Different Tracking Devices: Issues and Opportunities. *Ecol. Indic.* **2016**, *69*, 818–827. [CrossRef]
84. Ferrà Vega, C.; Tassetti, A.N.; Grati, F.; Scarcella, S.; Fabi, G. AIS as a Useful System to Support the Identification of Fisheries Restricted Areas. Licence: CC-NC-SA 3.0 IGO. FAO Fish Forum, B. Abstr.. 2018, p. 338. Available online: https://www.fao.org/3/cb7622en/cb7622en.pdf (accessed on 10 May 2023).
85. Tassetti, A.N.; Ferrà, C.; Fabi, G. Rating the Effectiveness of Fishery-Regulated Areas with AIS Data. *Ocean Coast. Manag.* **2019**, *175*, 90–97. [CrossRef]

86. De Clippele, L.H.; Buhl-Mortensen, P.; Buhl-Mortensen, L. Fauna Associated with Cold Water Gorgonians and Sea Pens. *Cont. Shelf Res.* **2015**, *105*, 67–78. [CrossRef]
87. Grinyó, J.; Francescangeli, M.; Santín, A.; Ercilla, G.; Estrada, F.; Mecho, A.; Fanelli, E.; Costa, C.; Danovaro, R.; Company, J.B.; et al. Megafaunal assemblages in deep-sea ecosystems of the Gulf of Cadiz, northeast Atlantic ocean. *Deep Sea Res. Part I Oceanogr. Res. Pap.* **2022**, *183*, 103738. [CrossRef]

fishes

MDPI

Article

(Un)wanted Fish: Potential Consumers' Acceptability of Landings in the Portuguese Case

Jorge Ramos [1,*], Pedro G. Lino [2], Jaime Aníbal [3,4] and Eduardo Esteves [3,5]

[1] CinTurs—Research Centre for Tourism, Sustainability and Well-Being, Universidade do Algarve, 8005-139 Faro, Portugal
[2] IPMA—Instituto Português do Mar e da Atmosfera, Av. 5 Outubro, s/n, 8700-305 Olhão, Portugal; plino@ipma.pt
[3] Departamento de Engenharia Alimentar, Instituto Superior de Engenharia, Universidade do Algarve, 8005-139 Faro, Portugal; janibal@ualg.pt (J.A.); eesteves@ualg.pt (E.E.)
[4] CIMA—Centro de Investigação Marinha e Ambiental & ARNET, Universidade do Algarve, 8005-139 Faro, Portugal
[5] CCMAR—Centro de Ciências do Mar do Algarve, Universidade do Algarve, 8005-139 Faro, Portugal
* Correspondence: jhramos@ualg.pt; Tel.: +351-289-800-900

Abstract: The Landing Obligation (LO), introduced in 2015 by the Common Fisheries Policy of the European Union (EU-CFP), has been subject to a transitional period until recently. The rationale behind the measure is that all fish species subject to a total allowable catch (TAC) must be landed to increase the sustainability of fishing activities. Through the analysis of official statistical data, it is possible to find out which species of fish were landed and their relative importance, including their monetary value, and verify the potential for consumer acceptance. Some insights are drawn from the interconnection between these three factors (i.e., social acceptability, landings of main fish species, and their market value) with empirical results and the scientific literature using data from Portugal.

Keywords: circular economy; discards; marine fish species; non-edible; waste

Key Contribution: The purpose of this study is to analyze the literature in search of potential processing options for fish (or their parts) lacking a clear market, providing them a justifiable use while reducing waste of resources that are already under some pressure to run out. The approach's primary objective is to make a relationship with the species caught and relate to the potential outlet according to the landings market.

Citation: Ramos, J.; Lino, P.G.; Aníbal, J.; Esteves, E. (Un)wanted Fish: Potential Consumers' Acceptability of Landings in the Portuguese Case. *Fishes* **2023**, *8*, 324. https://doi.org/10.3390/fishes8060324

Academic Editor: Ben Drakeford

Received: 29 May 2023
Revised: 12 June 2023
Accepted: 16 June 2023
Published: 18 June 2023

1. Introduction

Although progress has been made, there is a huge difficulty in managing fishing resources in a sustainable way. There are many species that have suffered immense pressure from overfishing and, as a result, became overexploited [1,2]. It is somehow consensual among the scientific community that, in general, the world's fisheries have already reached their exploitable limit regarding commercial species and are currently in a phase of stagnation [3–5]. Many such fishable resources—i.e., several commercial species—are subject to various pressures [6,7]. Likewise, with the sustainability issues of renewable resources, it has been important to consider and incorporate not only mitigating measures of certain less correct fishing practices, but also to introduce the concept of circularity within a blue economy scope. To this end, sustainability approaches are posing particular emphasis on the aspect of avoiding discards, as well as taking advantage of species and parts of the fish/specimens that are underused or even unused [8–10].

In the European Union, landings are mandatory for species subject to quotas [11–13]. This legislation—i.e., Regulation (EU) No 1380/2013, where all species subject to catch limits must be landed—has been gradually put into practice since 1 January 2019 [14,15].

Nevertheless, this type of public policy positioning is not only difficult to implement, but has also generated some controversy [16–18]. Upstream of the problem, there are issues of various order, such as landing through unofficial channels of less desirable or undersized species. Downstream, there is the issue of post-landing storage, especially when there are no immediate buyers or other outlets for the landed species [19].

The blue economy is an expression that relates to the exploration, preservation, and regeneration of the marine environment. In the particular case of marine fish landed in Portugal, it is important to know the quantities caught in order to be able to assess the status of fish stocks and their potential sustainability or vulnerability over time [20].

Illegal, unreported, and unregulated (IUU) fishing has been identified all over the world. IUU is thought to occur in many fisheries and may amount to as much as 30% of total catches in certain fisheries [21]. In Portugal, there are some estimates of unreported fishing—i.e., fishing that has not been declared or that has been incorrectly reported—to national authorities. It is a recurrent practice and difficult to enforce the applicable law and regulations [22].

In the reported landings of the Portuguese fleet that occur along the coast, there are about 40 main species that have commercial value and consequently good marketability, while over 100 species are captured as by-catch [23–26]. Some of the by-catch species—i.e., with no apparent market—do not have any destination, because it is not known how to value them [27]. This finding is easily corroborated by studies that are conducted in the field of fisheries research [28,29]. Recent time series [30] show that the marine fish species that have the best commercial value in the wholesale market are about a dozen (sold for more than 10 EUR/kg), while, in the tail, there are some species whose average value of market is normally low (less than 1 EUR/kg).

The general aim of this work is to point out solutions to greater sustainability in responsible production and consumption of marine fish. In order to address the above, this work is divided into three additional parts. First, we frame what is perceived as social acceptability of seafood (with a focus on marine fish). Second, based on what is effectively and officially reported in terms of landings, we draw up a list of marine fish species that are landed by the commercial fleet in Portugal and categorize if they are subjected to quota. Third, we estimate the wholesale market value per unit of weight for each of those species.

2. Materials and Methods

2.1. General Background and Hypothesis

In a very general way, markets are the result of the balance that exists between demand—for certain goods and services—and their supply [31]. In a free market, but regulated and well established in time, the transactions of goods and services have a certain value depending on the needs of consumers and the possibility of satisfying them, being subject to rules of competition and scarcity [32]. Herein, it is assumed that the wholesale fishery market behaves in this way.

In the present case study, it is hypothesized that the potential acceptability that consumers of marine fish have are dependent on what is landed in Portuguese wholesalers and the average value at auction that each of the landed species has. For this, we developed a conceptual framework to be able to integrate the main factors and their interaction.

2.2. Conceptual Framework

The literature search suggests that, in general, the concept of social acceptability is usually more focused towards paradigmatic changes. In that scope, social acceptability has been related to forest conditions [33], renewable energy [34–36], marine protected areas [37–39], and others. In terms of marine-related themes, there are studies on social acceptability of aquaculture developments and their food products [40,41].

The present conceptual framework was developed in an original way, given the data available and what was hypothesized to be analyzed. Herein, one approach was adopted that focuses on the importance of establishing a conceptual framework to integrate three factors (Figure 1).

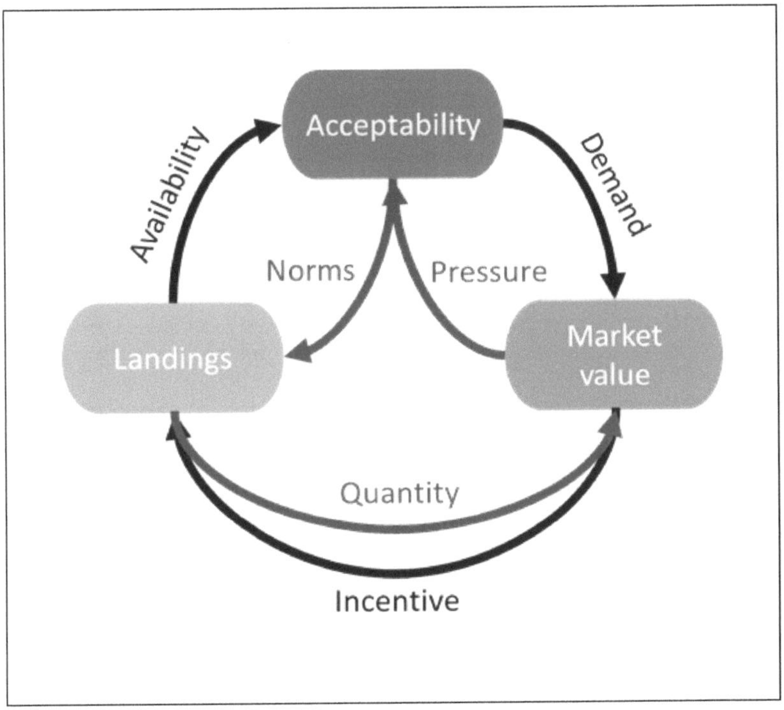

Figure 1. Conceptual framework for the analysis approach carried out in this work.

The framework involves three factors. The starting factor in the present study consists of the social acceptability linked to the consumers of fish/seafood. In turn, this factor should be integrated into a system linked to another factor consisting of the abundance of marine fish species (estimated) from official landings. The final factor consists of the average market value of landed species, according to the quantity supplied and the acceptability, that is, admitting that the value of the distinct species is a function of several factors (e.g., scarcity, freshness, and perceived taste). We assumed that there is an interconnection between the three main factors mentioned: social acceptability, landings of marine fish species, and the value of the species in the wholesale market.

From the social acceptability factor as a starting point, it can be assumed that consumers are aware of their own consumption habits [42]. The latter may depend on well-established norms, whether cultural, religious, health, and/or nutritional value of fish species [43]. The productive sector—which, in this case, is the fishing industry—supplies the catches according to what the social acceptability of all potential consumers suggests, depending on the marine resources available, and the legal, technical, or other constraints [44].

These attitudes and preferences of consumers regarding fish are determined by tastes, nutritional aspects, and the quality and freshness of the product [45–47]. The demand for fish products is also very dependent on regulations, as households will avoid consuming endangered species or those which have limitations in terms of size or weight or which come from protected areas or polluted sites or that may eventually cause any health hazard in any way [48,49].

To some extent, consumers' demand determines the landing prices (i.e., at 1st auction, in the fish wholesale market) that are adjusted over time. Market prices reciprocally exert some effect on consumers' choices [50]. Demanding consumers or those with greater purchasing power can pay more for species that reach a higher market value. Species with lower market value are purchased by less demanding consumers or those less willing to pay for fish [51,52].

In general, the value that species reach in the wholesale market depends on the historical record of that species and the quantity available [53]. If a species usually sold on the market occasionally appears in great abundance and has little demand, the market value will inevitably drop considerably. In contrast, if a species has a substantial market demand but there is a shortage to supply it, there will be an incentive for the fishing fleet to catch that species to satisfy the demand, but with the inevitable increase in costs that will be reflected in the final product [54].

2.3. *Acceptability of Marine Fish*

When people make choices about fish species in general, several acceptability determinants can be considered (Table 1). Olsen [55] established three main determinants: attitudes and preferences, norms, and control or barriers. In the present work, we will use Olsen's terminology but with some adaptations to the Portuguese case study.

Table 1. Conditions affecting marine fish acceptability in the Portuguese case. Source: adapted from Olsen [55].

	Acceptability	
Attitudes and Preferences	**Norms**	**Control/Barriers**
- Taste - Negative effect - Nutrition - Quality/freshness	- Social expectations - Moral obligations - Health involvement	- Price/cost - Convenience/availability - Knowledge
	Motivation to consume and/or buy	
	Propensity to consume (behavior)	

The first determinant highlights the possibility that people's decisions to eat fish can be made solely on how the species tastes. In contrast, people also react negatively to attributes of a certain species that are not appealing for consumption. There is also the nutritional aspect, where factors such as the quantity and quality of the fats (namely fatty acids) and proteins consumed should be considered, for instance, for health reasons. The quality of the fish that is available for consumption, including its freshness and general appearance, is another factor.

The second determinant refers to the norms that can be found in societies. As well as social expectations about what people expect from the appearance of the fish, the texture, the edible parts, the source of production (wild caught fish or farmed fish from aquaculture), there are also moral obligations regarding aspects of fishing or production, such as ethically responsible sourcing of fish, sustainable capture, avoiding suffering of organisms. Consumers also want to know whether the waters and surrounding environment where the fish was caught are places that have a Good Environmental Status of conservation (GES) or simply free of pollution.

Finally, the third determinant is related to the control or barrier conditions. When there is a high demand for a certain species (of marine fish) and a finite supply, the market price is solely determined by the industry's capacity to produce and supply that species. Demand can be a motivator to produce or catch more of that species even when there are technological or legal restrictions or when production costs are high. As a result, the availability of species is influenced by their abundance in natural environments and by how easily the business can meet consumer demand. Imports are used when local production is unable to meet demand. Scientific or empirical knowledge can also be important in influencing the acceptability of consumption of some species at a given time.

To codify acceptability, we used a simple traffic light code, where green is usually positive, orange is neutral, and red is commonly negative.

2.4. Reported Official Landings of Marine Fish

Based on official data from the Portuguese National Institute of Statistics (INE) [30], a list of the species landed at fish auctions in Portugal (including mainland and archipelagos of Azores and Madeira) for the year 2021 was compiled. The classification of the species listed was made in descending order of the quantity landed.

It is also important to point out which species are subject to quotas. These are distributed from agreements made in Brussels with other countries within the EU [56,57].

2.5. Market Value

As not all species landed have the same commercial interest, it was necessary to classify them according to their average value at the 1st auction market [58–60]. To this end, a simple method was used, which consisted of dividing the total value in EUR of each species by its total quantity landed, thus obtaining an average value of EUR per kilogram. This listing was subsequently classified into five wholesale market value categories (i.e., up to EUR 0.99, EUR 1 to EUR 1.99, EUR 2 to EUR 4.99, EUR 5 to EUR 9.99, and over EUR 10) empirically from the most valuable to the lowest set of fish species.

3. Results

3.1. Marine Fish Acceptability

Table 2 presents the results of applying a method derived from Olsen [55], which adopted a simple traffic light categorization, to the 42 landed species in terms of reasons affecting acceptability.

Table 2. Empirical classification of fish species by level of commercial acceptability. For acceptability, we used a traffic light code, where green is usually positive, orange is neutral, and red is commonly negative. Sources: nutrition (% protein: >15 g/100), social acceptability (% edible part: >60%), and moral obligations (gear used: line/hook) are based on [61], whereas health involvement (mercury: <0.5 mg/kg) is based on [62]; all other scores are based on authors' own scoring and adapted from acceptability defined by Olsen [55].

Marine Fish Species	**Acceptability**									
	Attitudes/Preferences				**Norms**			**Control/Barriers**		
	Taste	**Negative Effect**	**Nutrition**	**Freshness**	**Social Expectations**	**Moral Obligations**	**Health Involvement**	**Price/Cost**	**Availability**	**Knowledge**
Wreckfish (*Polyprion americanus*)			[61]		[61]	[61]				
Alfonsino (*Beryx decadactylus*)			[61]		[61]	[61]				
Red mullets (*Mullus spp.*)										
Turbot (*Psetta maxima*)										
Red seabream (*Pagrus major*)			[61]		[61]	[61]				
John Dory (*Zeus faber*)										
Brill (*Scophthalmus rhombus*)										
Snappers (*Pagrus pagrus*)			[61]		[61]	[61]				
Gilt-head seabream (*Sparus aurata*)			[61]		[61]	[61]	[62]			
Sea bass (*Dicentrarchus labrax*)			[61]		[61]	[61]				
Flounders (*Microchirus spp.*)			[61]		[61]	[61]				
Meagres (*Argyrosomus spp.*)			[61]		[61]	[61]				
Grouper (*Epinephelus marginatus*)										
Monkfish (*Lophius piscatorius*)			[61]		[61]	[61]	[62]			

Table 2. *Cont.*

Marine Fish Species	Acceptability									
	Attitudes/Preferences				Norms			Control/Barriers		
	Taste	Negative Effect	Nutrition	Freshness	Social Expectations	Moral Obligations	Health Involvement	Price/Cost	Availability	Knowledge
Common pandora (*Pagellus erythrinus*)										
Redfish (*Sebastes spp.*)										
Axillary seabream (*Pagellus acarne*)			[61]		[61]	[61]				
Whiting (*Merlangius merlangus*)										
Flounders (*Hippoglossus spp.*)			[61]		[61]	[61]				
Whiteseabream (*Diplodus spp.*)										
Forkbeard (*Phycis phycis*)			[61]		[61]	[61]				
Scabbardfish (*Lepidopus caudatus*)			[61]		[61]	[61]	[62]			
Dogfish (Squaliformes)			[61]		[61]	[61]	[62]			
Hake (*Merluccius merluccius*)			[61]		[61]	[61]	[62]			
Blacksword fish (*Aphanopus carbo*)			[61]		[61]	[61]	[62]			
Scaldfishes (*Arnoglossus imperialis*)										
Conger (*Conger spp.*)			[61]		[61]	[61]				
Atlantic pomfret (*Brama brama*)										
Skates (*Raja spp.*)			[61]		[61]	[61]	[62]			
Tuna and similar (*Thunnus spp.* and other)										
Gurnards (Triglidae)										
Anchovy (*Engraulis encrasicolus*)										
Pout (*Trisopterus luscus*)										
Horse mackerel (*Trachurus trachurus*)			[61]		[61]	[61]	[62]			
Atlantic mackerel (*Scomber scombrus*)			[61]		[61]	[61]	[62]			
Mullets (*Liza spp.* and *Mugil spp.*)										
Sardine (*Sardina pilchardus*)			[61]		[61]	[61]				
Blue whiting (*Micromesistius poutassou*)										
Black horse mackerel (*Trachurus picturatus*)										
Toadfish (*Sarpa salpa*)										
Mackerel (*Scomber japonicus*)			[61]		[61]	[61]	[62]			
Bogue (*Boops boops*)										
Other										

3.2. Marine Fish Landings

More than 100 commercial species are landed in Portugal [24,26]. However, just over 40 are reported in the official statistics. Figure 2 shows the official statistics of marine fish landed at auction for the year 2021 [30]. The species landed in greater quantity are essentially (small) pelagic fishes. In Figure 2, the species that are subject to quotas are also identified in a distinct color. Most of the larger and some of the smaller pelagic species landed are subject to a quota, namely sardine, horse mackerel, anchovy, or tuna and similar.

3.3. Marine Fish Landings Value in the Wholesale Market

From the data available in landing statistics for the year 2021 [30], it was possible to know the value (in EUR) by each species according to the quantity landed. Thus, we defined the average value (EUR/kg) per species and five categories were empirically defined depending on the average price obtained in the wholesale market (Figure 3).

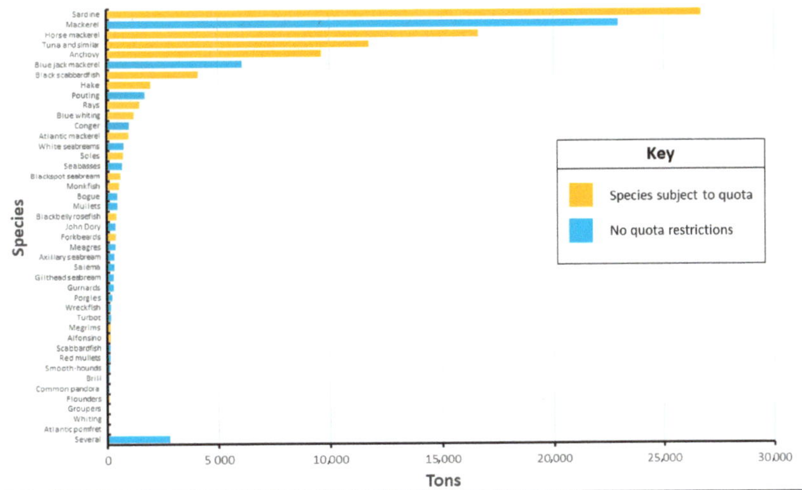

Figure 2. Listing of fish species landed in mainland Portugal. Note: major tuna species have quotas but "Tunas and similar" includes species without quota.

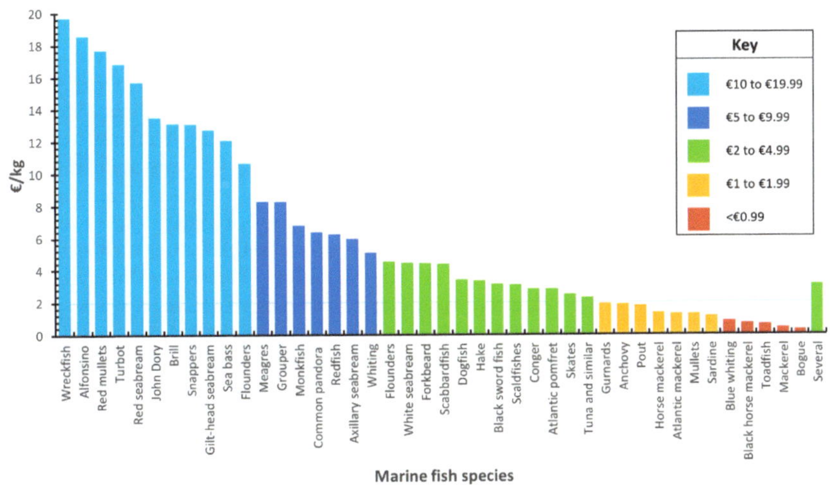

Figure 3. Marine fish wholesale market value (EUR/kg).

4. Discussion

Regarding the acceptability of fish landed in Portugal, there may be much variability. Portugal, despite being a relatively small country, has a large level of consumption of fish, being over 50 kg/per capita average for 2017–2019 [63], one of the countries that consume the most fish in the world, only being surpassed by Japan, Iceland, and some island states in the Pacific Ocean [64]. Nevertheless, there is a wide variation in fish consumption per region; understandably, in coastal areas, there are generally numerous supply sources both for residents and for tourists [65].

Moreover, in Portugal, the high consumption of fish is related to sociocultural reasons [66]. Inland, fish consumption is much scarcer, only having some notoriety at festive occasions, such as Christmas (cod and octopus) and summer (sardines) and popular or saints' festivities (*Santos Populares*) (particularly sardines). However, this study does not include the statistics for cod or octopus. Cod is almost entirely imported from Northern European countries. Likewise, all marine species other than fish (the octopus is a cephalopod mollusk) were also excluded. Sardines are an extremely popular fish only when they are fat, that is, in the late spring/summer, which also defines the beginning of the fishing season. Coincidentally, sardines are also the fish with the largest volume landed in Portugal. Sardines, as well as some other small-sized pelagic species, are also consumed in canned form [67]. Hence, perhaps the fact that they have lower market prices, as they are purchased in massive quantities by the processing industry. Anchovy stands out in this field and, for about two decades, mackerel has also been on the list of these species, with a significant effort on the part of various entities to promote the species because it is very nutritionally rich and because it has proven benefits to health [68].

There are also species, such as tuna, which are also highly demanded by consumers but for different, sometimes opposing reasons. On one hand, due to the nutritional and exquisite characteristics of tuna but generous size of these fish, they are consumed in terms of fresh or frozen tuna steaks. On the other hand, tuna is processed in canning factories [66].

The fish that command the greatest prices on the market are typically intended for the hospitality and tourism industry, where freshness is a crucial factor, particularly for the tourism sector [69]. According to the data analysis, 11 species that were landed in Portugal in 2021 have an average wholesale value of more than EUR 10 per kilogram (in a descending order: wreckfish, alfonsino, red mullets, turbot, blackspot seabream, John dory, brill, porgies, gilthead seabream, seabasses, and soles). Except for sea bass and sea bream, which may be produced in aquaculture, these species are scarcer on the market, which contributes to their high market value. They are also in high demand in the tourism and hospitality sectors [70].

The species from landings that have a lower value, on the other hand, include species that are substantially less acceptable to consumers [71] because they are sometimes associated as living in polluted waters, e.g., mullets [72]; they are herbivores and consume plants that accumulate substances that can be harmful and eventually transmit problems to human health, e.g., salema or bogue, [73]; and some herbivorous fish also have the problem of not having a pleasant taste if they are not eviscerated immediately after being caught [74]. Fish with lower market value are not always associated with their great abundance. There are other conditions that influence this value, such as low demand from households, either due to lack of knowledge or for other reasons of acceptability in relation to one aspect or another that is associated as being negative (e.g., less pleasant taste and being captured in places whose waters are of poor quality) [58].

Associated with all these catches there is also the question of the parts of the organisms that are not consumed as human food. Usually these parts include scales, bones, spines, eyes, viscera, and gonads. All the above-mentioned species are consumed fresh and whole, but some of them are also processed to create other products, such as canned (e.g., all small pelagic species, such as sardine, mackerel, and anchovy, and some large ones, such as tuna). Some parts are also sometimes used in canned products (e.g., gonads of sardine, mackerel, and hake). The current societal pressure in terms of sustainability, rejects products and practices that generate damages to the environment, the stocks, the animals and produce excessive waste, and supports the enhancement of inedible parts [75,76]. Therefore, it is particularly important to consider in the future for better management of living marine resources not only fish of lower value that is not consumed as food, but also to find ways to value fish as much as possible in the sense of circularity of the economy [9,77].

5. Conclusions

The fish that is landed and traded at first auction or the wholesale market derives from several factors related to the efficiency of the fishing fleet and its different gears for catching fish (supply). Fishing activities are subject to a series of technical, legal, and environmental constraints, which are reciprocally determined by demand on the part of consumer acceptability. The value of fish is realized in the prices practiced in the wholesale market derived from a historical record that is continuously adjusted, considering the balance between demand and supply of fresh fish.

It can be pointed out that there are some species in which there is interest in continuing to study possibilities to value them (e.g., pouting, blue whiting, and mullets). This valuation should be based on their relative abundance, that is, that they are not subject to excessive pressure and that, biologically, they recover easily. In the same way, for the consumer/household, awareness must be made in the sense of social acceptability for consumption, showing, for example, the benefits for health and the Good Environmental Status (GES) of the species. Examples in this sense are mackerel and Atlantic mackerel, whose appreciation is recent in Portugal.

In order to circularize the blue economy, in the future, it is important to focus attention on the following specific objectives:

- Of the least valued species, scrutinize those with the most potential for household consumption.
- Find ways of preservation where less energy is spent.
- Enhance the parts of the fish that are not edible.

Author Contributions: Conceptualization, J.R., P.G.L., J.A. and E.E.; methodology, P.G.L., J.A. and E.E.; validation, J.R., P.G.L., J.A. and E.E.; writing—original draft preparation, J.R.; writing—review and editing, P.G.L., J.A. and E.E. All authors have read and agreed to the published version of the manuscript.

Funding: This work is financed by Portuguese National Funds via FCT—Foundation for Science and Technology through projects: UIDB/04020/2020 CinTurs, UIDB/04326/2020, UIDP/04326/2020, and LA/P/0101/2020 CCMAR, LA/P/0069/2020 granted to the Associate Laboratory ARNET and UID/00350/2020 CIMA.

Institutional Review Board Statement: Not applicable.

Informed Consent Statement: Not applicable.

Data Availability Statement: Publicly available datasets were analyzed in this study (National Institute of Statistics portal: https://www.ine.pt/), accessed on 25 May 2023.

Acknowledgments: The authors would like to thank two anonymous reviewers of this journal for helpful comments on an early version of this manuscript.

Conflicts of Interest: The authors declare that no conflict of interest were found.

References

1. Baum, J.K.; Myers, R.A.; Kehler, D.G.; Worm, B.; Harley, S.J.; Doherty, P.A. Collapse and conservation of shark populations in the Northwest Atlantic. *Science* **2003**, *299*, 389–392. [CrossRef] [PubMed]
2. Neubauer, P.; Jensen, O.P.; Hutchings, J.A.; Baum, J.K. Resilience and recovery of overexploited marine populations. *Science* **2013**, *340*, 347–349. [CrossRef] [PubMed]
3. Pitcher, T.J.; Cheung, W.W. Fisheries: Hope or despair? *Mar. Pollut. Bull.* **2013**, *74*, 506–516. [CrossRef]
4. Deutsch, L.; Troell, M.; Limburg, K. Global trade of fisheries products: Implications for marine ecosystems and their services. In *Ecosystem Services and Global Trade of Natural Resources*; Routledge: Abingdon, UK, 2013; pp. 136–163.
5. Srinivasan, U.T.; Watson, R.; Sumaila, U.R. Global fisheries losses at the exclusive economic zone level, 1950 to present. *Mar. Policy* **2012**, *36*, 544–549. [CrossRef]
6. Probst, W.N.; Kloppmann, M.; Kraus, G. Indicator-based status assessment of commercial fish species in the North Sea according to the EU Marine Strategy Framework Directive (MSFD). *ICES J. Mar. Sci.* **2013**, *70*, 694–706. [CrossRef]
7. Pentz, B.; Klenk, N. When is a commercial fish species recovered? *J. Environ. Manag.* **2022**, *301*, 113918. [CrossRef]
8. Agh, N.; Jasour, M.S.; Noori, F. Potential Development of Value-Added Fishery Products in Underutilized and Commercial Fish Species: Comparative Study of Lipid Quality Indicators. *J. Am. Oil Chem. Soc.* **2014**, *91*, 1171–1177. [CrossRef]
9. Coppola, D.; Lauritano, C.; Palma Esposito, F.; Riccio, G.; Rizzo, C.; de Pascale, D. Fish waste: From problem to valuable resource. *Mar. Drugs* **2021**, *19*, 116. [CrossRef]
10. Abecasis, R.C.; Schmidt, L.; Longnecker, N.; Clifton, J. Implications of community and stakeholder perceptions of the marine environment and its conservation for MPA management in a small Azorean Island. *Ocean Coast. Manag.* **2013**, *84*, 208–219. [CrossRef]
11. Hatcher, A. Implications of a discard ban in multispecies quota fisheries. *Environ. Resour. Econ.* **2014**, *58*, 463–472. [CrossRef]

12. Batsleer, J.; Hamon, K.G.; van Overzee, H.M.; Rijnsdorp, A.D.; Poos, J.J. High-grading and over-quota discarding in mixed fisheries. *Rev. Fish Biol. Fish.* **2015**, *25*, 715–736. [CrossRef]
13. Bueno-Pardo, J.; Pierce, G.J.; Cabecinha, E.; Grilo, C.; Assis, J.; Valavanis, V.; Queiroga, H. Trends and factors of marine fish landings in Portugal since its entrance in the European Union. *ICES J. Mar. Sci.* **2020**, *77*, 988–1001. [CrossRef]
14. Harte, M.; Tiller, R.; Kailis, G.; Burden, M. Countering a climate of instability: The future of relative stability under the Common Fisheries Policy. *ICES J. Mar. Sci.* **2019**, *76*, 1951–1958. [CrossRef]
15. Veiga, P.; Pita, C.; Rangel, M.; Gonçalves, J.M.; Campos, A.; Fernandes, P.G.; Erzini, K. The EU landing obligation and European small-scale fisheries: What are the odds for success? *Mar. Policy* **2016**, *64*, 64–71. [CrossRef]
16. Van Hoof, L.; Kraan, M.; Visser, N.M.; Avoyan, E.; Batsleer, J.; Trapman, B. Muddying the waters of the Landing Obligation: How multi-level governance structures can obscure policy implementation. In *The European Landing Obligation*; Springer: Berlin/Heidelberg, Germany, 2019; p. 179.
17. Leitão, F.; Baptista, V. The discard ban policy, economic trends and opportunities for the Portuguese fisheries sector. *Mar. Policy* **2017**, *75*, 75–83. [CrossRef]
18. Campos, A.; Henriques, V.; Erzini, K.; Castro, M. Deep-sea trawling off the Portuguese continental coast—-Spatial patterns, target species and impact of a prospective EU-level ban. *Mar. Policy* **2021**, *128*, 104466. [CrossRef]
19. Fitzpatrick, M.; Frangoudes, K.; Fauconnet, L.; Quetglas, A. Fishing industry perspectives on the EU Landing Obligation. In *The European Landing Obligation*; Springer: Berlin/Heidelberg, Germany, 2019; p. 71.
20. Barroso, S.; Pinto, F.R.; Silva, A.; Silva, F.G.; Duarte, A.M.; Gil, M.M. The circular economy solution to ocean sustainability: Innovative approaches for the blue economy. In *Research Anthology on Ecosystem Conservation and Preserving Biodiversity*; IGI Global: Hershey, PA, USA, 2022; pp. 875–901.
21. Temple, A.J.; Skerritt, D.J.; Howarth, P.E.; Pearce, J.; Mangi, S.C. Illegal, unregulated and unreported fishing impacts: A systematic review of evidence and proposed future agenda. *Mar. Policy* **2022**, *139*, 105033. [CrossRef]
22. Leitão, F.; Baptista, V.; Zeller, D.; Erzini, K. Reconstructed catches and trends for mainland Portugal fisheries between 1938 and 2009: Implications for sustainability, domestic fish supply and imports. *Fish. Res.* **2014**, *155*, 33–50. [CrossRef]
23. Gonçalves JM, S.; Bentes, L.; Lino, P.G.; Ribeiro, J.; Canario, A.V.; Erzini, K. Weight-length relationships for selected fish species of the small-scale demersal fisheries of the south and south-west coast of Portugal. *Fish. Res.* **1997**, *30*, 253–256. [CrossRef]
24. Borges, T.C.; Erzini, K.; Bentes, L.; Costa, M.E.; Goncalves, J.M.; Lino, P.G.; Ribeiro, J. By-catch and discarding practices in five Algarve (southern Portugal) métiers. *J. Appl. Ichthyol.* **2001**, *17*, 104–114. [CrossRef]
25. Monteiro, P.; Araújo, A.; Erzini, K.; Castro, M. Discards of the Algarve (southern Portugal) crustacean trawl fishery. *Hydrobiologia* **2001**, *449*, 267–277. [CrossRef]
26. Fernandes, A.C.; Pérez, N.; Prista, N.; Santos, J.; Azevedo, M. Discards composition from Iberian trawl fleets. *Mar. Policy* **2015**, *53*, 33–44. [CrossRef]
27. Bjørndal, T.; Lappo, A.; Ramos, J. An economic analysis of the Portuguese fisheries sector 1960–2011. *Mar. Policy* **2015**, *51*, 21–30. [CrossRef]
28. Machado, I.; Moura, T.; Figueiredo, I.; Chaves, C.; Costa, J.L.; Cabral, H.N. Effects of scale on the assessment of fish biodiversity in the marine strategy framework directive context. *Ecol. Indic.* **2020**, *117*, 106546. [CrossRef]
29. Murthy, A.; Galli, A.; Madeira, C.; Moreno Pires, S. Consumer Attitudes towards Fish and Seafood in Portugal: Opportunities for Footprint Reduction. *Sustainability* **2023**, *15*, 1363. [CrossRef]
30. Instituto Nacional de Estatística (INE). Estatísticas da Pesca 2021. Statistics Portugal. 2023. Available online: https://www.ine.pt/xportal/ (accessed on 25 May 2023). (In Portuguese).
31. Kawarazuka, N.; Béné, C. Linking small-scale fisheries and aquaculture to household nutritional security: An overview. *Food Secur.* **2010**, *2*, 343–357. [CrossRef]
32. Trondsen, T. Value chains, business conventions, and market adaptation: A comparative analysis of Norwegian and Icelandic fish exports. *Can. Geogr.. Can.* **2012**, *56*, 459–473. [CrossRef]
33. Shindler, B.A. *Social Acceptability of Forest Conditions and Management Practices: A Problem Analysis*; US Department of Agriculture, Forest Service, Pacific Northwest Research Station: Juneau, AL, USA, 2002; Volume 537.
34. Lundheim, S.H.; Pellegrini-Masini, G.; Klöckner, C.A.; Geiss, S. Developing a theoretical framework to explain the social acceptability of wind energy. *Energies* **2022**, *15*, 4934. [CrossRef]
35. Fournis, Y.; Fortin, M.J. From social 'acceptance'to social 'acceptability'of wind energy projects: Towards a territorial perspective. *J. Environ. Plan. Manag.* **2017**, *60*, 1–21. [CrossRef]
36. Moula MM, E.; Maula, J.; Hamdy, M.; Fang, T.; Jung, N.; Lahdelma, R. Researching social acceptability of renewable energy technologies in Finland. *Int. J. Sustain. Built Environ.* **2013**, *2*, 89–98. [CrossRef]
37. Voyer, M.; Gollan, N.; Barclay, K.; Gladstone, W. 'It's part of me'; understanding the values, images and principles of coastal users and their influence on the social acceptability of MPAs. *Mar. Policy* **2015**, *52*, 93–102. [CrossRef]
38. Gall, S.C.; Rodwell, L.D. Evaluating the social acceptability of Marine Protected Areas. *Mar. Policy* **2016**, *65*, 30–38. [CrossRef]
39. Thomassin, A.; White, C.S.; Stead, S.S.; David, G. Social acceptability of a marine protected area: The case of Reunion Island. *Ocean Coast. Manag.* **2010**, *53*, 169–179. [CrossRef]
40. Whitmarsh, D.; Palmieri, M.G. Social acceptability of marine aquaculture: The use of survey-based methods for eliciting public and stakeholder preferences. *Mar. Policy* **2009**, *33*, 452–457. [CrossRef]

41. Katranidis, S.; Nitsi, E.; Vakrou, A. Social acceptability of aquaculture development in coastal areas: The case of two Greek Islands. *Coast. Manag.* **2003**, *31*, 37–53. [CrossRef]
42. Moreno-Sánchez RD, P.; Maldonado, J.H. Adaptive capacity of fishing communities at marine protected areas: A case study from the Colombian Pacific. *Ambio* **2013**, *42*, 985–996. [CrossRef]
43. Stelzenmüller, V.; Fernández, T.V.; Cronin, K.; Röckmann, C.; Pantazi, M.; Vanaverbeke, J.; Van Hoof, L. Assessing uncertainty associated with the monitoring and evaluation of spatially managed areas. *Mar. Policy* **2015**, *51*, 151–162. [CrossRef]
44. Bjørndal, T.; Herrero, I.; Newman, A.; Romero, C.; Weintraub, A. Operations research in the natural resource industry. *Int. Trans. Oper. Res.* **2012**, *19*, 39–62. [CrossRef]
45. Christenson, J.K.; O'Kane, G.M.; Farmery, A.K.; McManus, A. The barriers and drivers of seafood consumption in Australia: A narrative literature review. *Int. J. Consum. Stud.* **2017**, *41*, 299–311. [CrossRef]
46. Tomić, M.; Matulić, D.; Jelić, M. What determines fresh fish consumption in Croatia? *Appetite* **2016**, *106*, 13–22. [CrossRef]
47. Rickertsen, K.; Alfnes, F.; Combris, P.; Enderli, G.; Issanchou, S.; Shogren, J.F. French consumers' attitudes and preferences toward wild and farmed fish. *Mar. Resour. Econ.* **2017**, *32*, 59–81. [CrossRef]
48. Kemmerer, L. *Eating Earth: Environmental Ethics and Dietary Choice*; Oxford University Press: Oxford, UK, 2014.
49. Barcia, L.G.; Argiro, J.; Babcock, E.A.; Cai, Y.; Shea, S.K.; Chapman, D.D. Mercury and arsenic in processed fins from nine of the most traded shark species in the Hong Kong and China dried seafood markets: The potential health risks of shark fin soup. *Mar. Pollut. Bull.* **2020**, *157*, 111281. [CrossRef] [PubMed]
50. Swartz, W.; Sumaila, U.R.; Watson, R.; Pauly, D. Sourcing seafood for the three major markets: The EU, Japan and the USA. *Mar. Policy* **2010**, *34*, 1366–1373. [CrossRef]
51. Asche, F.; Bellemare, M.F.; Roheim, C.; Smith, M.D.; Tveteras, S. Fair enough? Food security and the international trade of seafood. *World Dev.* **2015**, *67*, 151–160. [CrossRef]
52. Witkin, T.; Dissanayake, S.T.; McClenachan, L. Opportunities and barriers for fisheries diversification: Consumer choice in New England. *Fish. Res.* **2015**, *168*, 56–62. [CrossRef]
53. Bjørndal, T.; Child, A.; Lem, A.; Dey, M.M. Value chain dynamics and the small-scale sector: A summary of findings and policy recommendations for fisheries and aquaculture trade. *Aquac. Econ. Manag.* **2015**, *19*, 148–173. [CrossRef]
54. Delgado, C.L. *Fish to 2020: Supply and Demand in Changing Global Markets*; WorldFish: Penang, Malaysia, 2003; Volume 62.
55. Olsen, S.O. Antecedents of seafood consumption behavior: An overview. *J. Aquat. Food Prod. Technol.* **2004**, *13*, 79–91. [CrossRef]
56. Hoefnagel, E.; de Vos, B.; Buisman, E. Quota swapping, relative stability, and transparency. *Mar. Policy* **2015**, *57*, 111–119. [CrossRef]
57. Bellido, J.M.; García-Rodriguez, M.; García-Jiménez, T.; González-Aguilar, M.; Carbonell-Quetglas, A. Could the obligation to land undersized individuals increase the black market for juveniles: Evidence from the Mediterranean? *Fish Fish.* **2017**, *18*, 185–194. [CrossRef]
58. Karadzic, V.; Antunes, P.; Grin, J. Adapting to environmental and market change: Insights from Fish Producer Organizations in Portugal. *Ocean Coast. Manag.* **2014**, *102*, 364–374. [CrossRef]
59. Wise, L.; Murta, A.G.; Carvalho, J.P.; Mesquita, M. Qualitative modelling of fishermen's behaviour in a pelagic fishery. *Ecol. Model.* **2012**, *228*, 112–122. [CrossRef]
60. Sumaila, U.R. *Infinity Fish: Economics and the Future of Fish and Fisheries*; Academic Press: Cambridge, MA, USA, 2021.
61. Bandarra, N.; Calhau, M.A.; Oliveira, L.; Ramos, M.; Dias MD, G.; Bártolo, H.; Irineu, B. Composição e valor nutricional dos produtos da pesca mais consumidos em Portugal. In *Publicações Avulsas do IPIMAR*; IPIMAR: Lisboa, Portugal, 2004. (In Portuguese)
62. Nunes, M.L.; Irineu, B.; Bandarra, N.; Morais MD, G.; Rodrigues, P.O. Produtos da pesca: Valor nutricional e importância para a saúde e bem-estar dos consumidores. In *Publicações Avulsas do IPIMAR*; IPIMAR: Lisboa, Portugal, 2008. (In Portuguese)
63. FAO. *The State of World Fisheries and Aquaculture 2022. Towards Blue Transformation*; FAO: Rome, Italy, 2022. [CrossRef]
64. Bjørndal, T.; Brasão, A.; Ramos, J.; Tusvik, A. Fish processing in Portugal: An industry in expansion. *Mar. Policy* **2016**, *72*, 94–106. [CrossRef]
65. Cardoso, C.; Lourenço, H.; Costa, S.; Gonçalves, S.; Nunes, M.L. Survey into the seafood consumption preferences and patterns in the Portuguese population. Gender and regional variability. *Appetite* **2013**, *64*, 20–31. [CrossRef]
66. Almeida, C.; Karadzic, V.; Vaz, S. The seafood market in Portugal: Driving forces and consequences. *Mar. Policy* **2015**, *61*, 87–94. [CrossRef]
67. Almeida, C.; Vaz, S.; Ziegler, F. Environmental life cycle assessment of a canned sardine product from Portugal. *J. Ind. Ecol.* **2015**, *19*, 607–617. [CrossRef]
68. Bimbo, A.P. Sources of omega-3 fatty acids. In *Food Enrichment with Omega-3 Fatty Acids*; Woodhead Publishing: Sawston, UK, 2013; pp. 27–107.
69. Hall, C.M.; Sharples, L. The consumption of experiences or the experience of consumption? An introduction to the tourism of taste. In *Food Tourism around the World*; Routledge: Abingdon, UK, 2004; pp. 1–24.
70. Regnier, E.; Bayramoglu, B. Competition between farmed and wild fish: The French sea bass and sea bream markets. *Aquac. Econ. Manag.* **2017**, *21*, 355–375. [CrossRef]
71. Villasante, S.; Pita, P.; Antelo, M.; Neira, J.A. Socio-economic impacts of the landing obligation of the European Union Common Fisheries Policy on Galician (NW Spain) small-scale fisheries. *Ocean Coast. Manag.* **2019**, *170*, 60–71. [CrossRef]

72. Ferreira, M.; Moradas-Ferreira, P.; Reis-Henriques, M.A. The effect of long-term depuration on phase I and phase II biotransformation in mullets (*Mugil cephalus*) chronically exposed to pollutants in River Douro Estuary, Portugal. *Mar. Environ. Res.* **2006**, *61*, 326–338. [CrossRef]
73. Mayer-Pinto, M.; Johnston, E.L.; Hutchings, P.A.; Marzinelli, E.M.; Ahyong, S.T.; Birch, G.; Hedge, L.H. Sydney Harbour: A review of anthropogenic impacts on the biodiversity and ecosystem function of one of the world's largest natural harbours. *Mar. Freshw. Res.* **2015**, *66*, 1088–1105. [CrossRef]
74. D'Arcy, P. *The People of the Sea: Environment, Identity, and History in Oceania*; University of Hawaii Press: Honolulu, HI, USA, 2006.
75. Helander, H.; Bruckner, M.; Leipold, S.; Petit-Boix, A.; Bringezu, S. Eating healthy or wasting less? Reducing resource footprints of food consumption. *Environ. Res. Lett.* **2021**, *16*, 054033. [CrossRef]
76. Kasavan, S.; Siron, R.; Yusoff, S.; Fakri, M.F.R. Drivers of food waste generation and best practice towards sustainable food waste management in the hotel sector: A systematic review. *Environ. Sci. Pollut. Res.* **2022**, *29*, 48152–48167. [CrossRef] [PubMed]
77. Farmery, A.K.; Alexander, K.; Anderson, K.; Blanchard, J.L.; Carter, C.G.; Evans, K.; Nowak, B. Food for all: Designing sustainable and secure future seafood systems. *Rev. Fish Biol. Fish.* **2022**, *32*, 101–121. [CrossRef] [PubMed]

fishes

Article

Investigating the Relationship between Aquaculture Investments, Training, and Environmental Factors in Guangdong: An Alternative Perspective

Peiwen Wang [1,*], Isabel Mendes [1] and Ramona Franić [2]

[1] SOCIUS/CSG, ISEG (Lisbon School of Economics & Management), Universidade de Lisboa, 1249-078 Lisboa, Portugal; midm@iseg.ulisboa.pt
[2] Department of Agricultural Economics and Rural Development, Faculty of Agriculture, University of Zagreb, 10000 Zagreb, Croatia; ramonaf@agr.hr
* Correspondence: peiwen.wang@phd.iseg.ulisboa.pt

Abstract: This study investigates the interplay between investment, training, and environmental factors in the aquaculture industry in the Guangdong region of China. Using NIPALS regression to address multicollinearity, we identify the factors that significantly impact losses of aquaculture products due to environmental factors. Our findings highlight the importance of targeted training and education for fisherfolks and extension staff to enhance environmental management practices and reduce losses. We also emphasize the need to consider regional variability and challenges in developing universal models. Based on our results, we propose using innovative technology, fostering public–private partnerships, and adapting to regional variability to address environmental challenges. Finally, we suggest establishing a comprehensive monitoring and evaluation system to assess the effectiveness of interventions and promote evidence-based decision-making for sustainable development in the region's aquaculture sector.

Keywords: aquaculture economy; education for fisherfolk; nonlinear iterative partial least squares

Key Contribution: This article contributes to a better understanding of the complex relationships between investment, training, and environmental factors in the aquaculture industry in the Guangdong region of China. The study identifies the key factors that influence losses of aquaculture products due to environmental factors and proposes targeted interventions to promote sustainable development in the region's aquaculture sector.

Citation: Wang, P.; Mendes, I.; Franić, R. Investigating the Relationship between Aquaculture Investments, Training, and Environmental Factors in Guangdong: An Alternative Perspective. *Fishes* **2023**, *8*, 237. https://doi.org/10.3390/fishes8050237

Academic Editor: Ben Drakeford

Received: 28 March 2023
Revised: 22 April 2023
Accepted: 28 April 2023
Published: 1 May 2023

1. Introduction

China is a major player in the global aquaculture industry and its seafood production has grown significantly in recent years [1]. China's sustained process of accelerated economic growth and urbanization has brought tremendous opportunities as well as new challenges to agriculture and rural society [2]. While China's economic growth has brought many benefits to the country, it has also led to a number of challenges that need to be addressed in order to ensure the sustainable development of the agriculture and fishing sectors. New concerns have emerged in various areas such as food safety and agricultural surface pollution [3]. However, the country's rapid economic growth and urbanization have also brought challenges. Overfishing is a critical issue that must be addressed to avoid negative impacts on marine ecosystems and biodiversity. In addition, there have been concerns about the use of illegal, unregulated, and unreported (IUU) fishing practices, which can lead to overfishing and undermine the sustainability of marine resources [4].

Science and technology (S&T) have become increasingly relevant for the management and economic development of China's marine resources, particularly in the aquaculture industry [5]. The development of new technologies has allowed for a greater understanding of the oceans and their ecosystems, as well as more efficient and sustainable ways to harvest marine resources. In recent decades, the development of technology has been proceeding at a rapid pace, with science opening new possibilities and technology enabling a higher level of human activity in the oceans [6]. For example, satellite technology to track and monitor fishing vessels is now being used to prevent overfishing and ensure that fishing is carried out in a sustainable manner [7]. Advances in aquaculture techniques have been a key area of focus in S&T, as they have the potential to improve the efficiency and sustainability of seafood production [8]. Recirculating Aquaculture Systems (RAS) is an example of such a technique, which is a closed-loop system that recycles water and uses advanced filtration methods to maintain optimal water quality for the fish [9]. This reduces the need for large amounts of water and can be performed in controlled environments, such as indoors or on land, making it less dependent on natural conditions [10]. Another example is Integrated Multi-Trophic Aquaculture (IMTA), a unique method of aquaculture that involves cultivating different species of fish, shellfish, and seaweed together in a symbiotic relationship. Unlike aquaponics, which typically only involves the cultivation of fish and plants, IMTA also incorporates the cultivation of shellfish and seaweed, and the waste produced by each species is used to sustain the others in a closed-loop system [11]. This helps to reduce the environmental impact of aquaculture and can increase production efficiency. Selective breeding, fish vaccination, and bio floc technology are other examples of advances in aquaculture techniques that have been developed in recent years and have the potential to improve the efficiency and sustainability of seafood production [12].

In December 2021, the Ministry of Agriculture released the 14th Five-Year Plan for Fishery Science and Technology Development, which proposes to further improve the level of aquaculture science and technology by 2035, with the contribution rate of scientific and technological progress in aquaculture reaching up to 67%. The national aquaculture germplasm resources protection and utilization system will be initially established, a number of new aquatic species will be cultivated, and the self-sufficiency rate of core seed sources would reach 80%. The mechanization rate of aquaculture has reached more than 50%. This has included support for technology development and innovation, as well as the establishment of research centres and institutes to study various aspects of these industries. The government has also implemented a number of policies and measures to promote the development of the fishery and aquaculture sectors, including tax incentives, financial subsidies, and other forms of support [13]. These efforts have helped to drive the growth of these industries in recent years and have contributed to the modernization and upgrading of the industry as a whole. Overall, the incorporation of the fishery and aquaculture industries into China's national development strategy reflects the importance of these sectors to the country's economy and food security; it also underscores the government's commitment to supporting their growth and development. The status of S&T investments in aquaculture directly affects the development of China's aquaculture industry, aquatic environmental management, and extreme environmental prevention. Meanwhile education and training improve the skills and productivity of farmers and fisherfolk; this is a point that cannot be overlooked, as the education and training of fisherfolk also plays a very important part in S&T [14].

Agricultural S&T investments is an important contributor to productivity growth and is a fundamental factor for sustained economic growth. It is well-established that investments in science and technology (S&T) can have a significant impact on the agricultural sector and the overall economy [15]. Therefore, it is extremely important to analyse whether the S&T of Chinese aquaculture in the past 10 years has contributed to the improvement of the aquaculture products production environment. S&T investments can help to increase agricultural productivity and efficiency, leading to higher crop yields and livestock production [16]. In turn, S&T investments can help to improve food security, reduce poverty, and contribute to economic growth [15]. Through the direct effect of agricultural S&T investments on the agricultural economy or socioeconomic growth, it has been discovered that agricultural S&T investments have a strong impact on food production, with a stronger positive interaction response and a more significant and stable response in the long run [16–20]. Deng's research [17] found that although China's agricultural S&T investments have become a decisive influencing factor of agricultural economic growth, there are still problems such as insufficient scale and intensity, unbalanced regional inputs, unreasonable resource allocation, and inefficient utilization remain [17]. Additionally, from the analysis of the relationship between China's agricultural S&T investments and effects, Song's study showed that the value added of agricultural products and government inputs are inextricably connected [18]. These studies are sufficient to prove that S&T investments are important for the sustained growth of China's economy. Nordhaus' study also found interactions between marine economic systems, marine environmental systems, and carbon cycle systems; furthermore, S&T investments have been detected to mitigate the impact of economic activities on the marine environment [19]. Holdren's study demonstrates that technological progress has also been demonstrated to increase productivity and reduce marine resource consumption and environmental impacts [20]. An important point in promoting aquaculture S&T investments is to transmit effective information and knowledge about scientific farming to the fisherfolk. There are many scholars who have demonstrated that increased education and training of farmers can be effective in increasing the productivity of agricultural products [21–23]. Kirtti analysed the impact of education on agricultural productivity in India and found that education and agricultural productivity have a direct impact [21]. Sharada analysed the effects of education on farmer productivity in rural Ethiopia, showing that there may be considerable opportunities to exploit the externalities of schooling to increase agricultural productivity if school enrolment increases in rural areas [22]. Abdulai's study's analysis of data from a survey of 342 rice farmers in northern Ghana showed that farmers' education can significantly increase rice yields and net returns [23].

China is a vast country with diverse regional characteristics, and studying a particular region can provide unique insights into the relationship between aquaculture investments, training, and environmental factors [3]. In this study, we focus on the Guangdong region, which is a significant producer of both freshwater and marine aquaculture products [24]. Guangdong's aquaculture industry is known for its advanced and diverse techniques, making it an interesting case study for exploring the relationship between investments, training, education, and environmental factors [25]. Moreover, Guangdong's coastal location and its proximity to major urban centres, such as Guangzhou and Shenzhen, make it an ideal case study for examining the impact of S&T investments on the aquaculture sector [26]. In the past decade, China has made significant S&T investments in the agricultural sector, and it is important to analyse whether these investments have contributed to improving aquaculture production and environmental management in the Guangdong region [5]. Previous studies have shown that S&T investments can increase agricultural productivity [27–29], improve food security, reduce poverty, and contribute to economic growth. However, challenges such as insufficient scale and intensity, unbalanced regional inputs, unreasonable resource allocation, and inefficient utilization remain in the agricultural sector [30]. Therefore, it is critical to examine how S&T investments have affected the aquaculture industry in the Guangdong region and explore the factors that may influence the effectiveness of such investments. In addition to S&T investments, education and training of fisherfolk are also important factors that can enhance aquaculture productivity and reduce losses caused by environmental factors. Previous studies have shown that increased education and training of farmers can be effective in increasing the productivity of agricultural products [21–23]. Therefore, in this study, we will investigate how education and training programs for fisherfolk, and extension staff may contribute to improved aquaculture practices and reduced losses caused by environmental factors in the Guangdong region. Overall, this study aims to provide an alternative perspective on the relationship between aquaculture investments, training, and environmental factors in the Guangdong region, and to offer insights into the factors that may influence the effectiveness of S&T investments and education and training programs in the aquaculture sector.

The hypotheses presented in this study were developed based on a review of the existing literature on aquaculture investments, training, education, and environmental factors. In particular, we drew on studies that have investigated the relationship between these variables in various contexts.

To adapt these findings to the Guangdong context, we conducted a thorough review of the available data on aquaculture production, environmental conditions, and education and training programs in Guangdong. Through this process, we identified two key hypotheses that we believe are relevant to understanding the relationship between aquaculture investments, training, education, and environmental factors in Guangdong. These hypotheses suggest that increased investments in staff funds and operating funds, enhanced training and education programs, and regional factors such as government policies and resource allocation are all important determinants of environmental management and losses due to environmental factors in Guangdong's aquaculture industry.

Hypothesis 1 (H1). *Increased investments in staff funds (SF) and operating funds (OF) in the Guangdong region's aquaculture sector will lead to better environmental management practices, resulting in a reduction of losses caused by environmental factors. This hypothesis assumes that increased financial support for the aquaculture industry will result in better environmental management practices, leading to reduced losses due to environmental factors. The focus will be on understanding how investments in staff funds and operating funds can be used to improve environmental management practices and the potential benefits that these investments can bring to industry in Guangdong.*

Hypothesis 2 (H2). *Enhanced training and education of fisherfolks and extension staff in the Guangdong region will lead to improved aquaculture practices, resulting in a decrease in losses caused by environmental factors. This hypothesis assumes that improved training and education of fisherfolks and extension staff will result in better aquaculture practices, leading to a reduction of losses due to environmental factors. The focus will be on identifying the most effective training and education methods for fisherfolks and extension staff in Guangdong and examining how these methods can be used to improve aquaculture practices.*

Overall, these hypotheses aim to explore the relationship between investments, training, and environmental factors in the aquaculture industry of the Guangdong region. By examining these relationships, the study aims to provide insights into how these factors can be leveraged to improve environmental management practices and reduce losses caused by environmental factors.

2. Materials and Methods

2.1. Introduction of the NIPALS Algorithm

The NIPALS (Nonlinear Iterative Partial Least Squares) algorithm is a multivariate statistical technique used to build predictive models. It is similar to principal component regression (PCR) and multiple linear regression (MLR) [31], but it can handle multicollinearity, nonlinear relationships, and high dimensional data better than these other methods. The NIPALS (Nonlinear Iterative Partial Least Squares) algorithm commences with a leave-one-out validation method to select the factor with the least mean predicted sum of squares of residuals (PRESS). This method is based on the van der Voet test (T^2), a randomization test that compares the residuals of the predicted series with different models and selects the number of factors with the lowest residuals that are not significantly larger than the residuals of the model with the minimum PRESS [32]. Compared to other decompositions of the covariance technique, this multivariate analysis method was used because of its numerical accuracy in terms of results and predictions. It is often used in a variety of fields, including chemistry, engineering, and economics, to build predictive models and understand the underlying relationships between variables [33,34]. During model estimation, both the original data variables [X (predictor variable) and Y (response)] are preprocessed with zero mean and standard deviation of 1 (hence, scale and cantered). This transformation addresses potential problems with unit roots among the variables.

2.2. Data Set and Variables

The study employs time series variables from China Fisheries Statistical Yearbook 2012–2022 [35], which covers the period 2011–2021. This paper covers only the Guangdong region. The red area in Figure 1 shows the geographical location of the Guangzhou region in China, similar to the study area of this study. The variables include loss of aquaculture products by environmental factors, funds for staff engaged in aquaculture, operating funds related to aquaculture, technical training for fisherfolk, operational training for aquaculture science and technology extension staff, practitioners in aquaculture research institutions, and aquaculture financial allocation. Table 1 contains information on how the variables are measured.

Figure 1. Location of the Guangdong region in China. (The map of China was generated by the standard map online service; URL link: http://bzdt.ch.mnr.gov.cn (accessed on 21 April 2023)).

Table 1. Variable definition.

Indicator Name	Indicator Code	Unit
Loss of aquaculture products by environmental factors in Guangdong region	ALGD	Tons
Funds for staff engaged in aquaculture in Guangdong region	SFGD	Million yuan
Operating funds related to aquaculture in Guangdong region	OFGD	Million yuan
Technical training for fisherfolk in Guangdong region	TFGD	People per training session
Operational training for aquatic science and technology extension staff in Guangdong region	TPGD	People per training session
Practitioners in aquaculture research institutions in Guangdong region	FOEGD	Population
Aquaculture financial allocation in Guangdong region	FAGD	Million yuan

Source: China Fisheries Statistical Yearbook 2012 to China Fisheries Statistical Yearbook 2022.

In ALGD (loss of aquaculture products by environmental factors in the Guangdong region), environmental factors include loss of aquaculture products affected by typhoons and flooding, loss of aquaculture products affected by diseases, loss of aquaculture products affected by droughts, loss of aquaculture products affected by pollution, and loss of aquaculture products affected by environmental factors other than typhoons, floods, disease, drought, and pollution. Due to limitations in data acquisition, the study only covers aquaculture and fisherfolk engaged in aquaculture production in the Guangdong region. All the statistical analysis in this study was performed using SAS JMP Pro16 software (SAS Institute Inc., Cary, NC, USA). This software was used for data exploration, visualization, and statistical analysis.

2.3. Analysis of Data

This study models the loss of aquaculture products by environmental factors as a function of fund for staff engaged in aquaculture, operating funds related to aquaculture, technical training for fisherfolk, operational training for aquaculture science and technology extension staff, and practitioners in aquaculture research institutions aquaculture financial allocation. Mathematically, this is depicted in Equation (1).

$$\mathrm{ALGD}_t = \beta_0 + \beta_1\mathrm{SFGD}_t + \beta_2\mathrm{OFGD}_t + \beta_3\mathrm{TFGD}_t + \beta_4\mathrm{TPGD}_t + \beta_5\mathrm{FOEGD}_t + \beta_6\mathrm{FAGD}_t + \varepsilon_t \quad (1)$$

where β_0 represents the constant, β_1, ..., β_6 denote the coefficients of the independent variables in year t, and ε designates the error term.

The model in this study is based on Song's model [18]. In Song's model, the number of agricultural research staff, operating funds related to agriculture, and funds for staff engaged in agriculture were included. We made improvements based on this model, and in this study, funds for staff engaged in aquaculture and operating funds related to aquaculture were retained. We also added variables for educational orientation, such as technical training for fisherfolk and operational training for aquaculture science and technology extension staff and practitioners in aquaculture research institutions. SFGD (funds for staff engaged in aquaculture in the Guangdong region), OFGD (operating funds related to aquaculture in the Guangdong region), and FAGD (aquaculture financial allocation in the Guangdong region) are the investment components of aquaculture S&T, because the S&T investment is one of the important indicators reflecting the aquaculture S&T section, and sufficient funding is also an important source to bring into play in the dynamics of research. The role of aquaculture researchers is not simply to conduct scientific research, as scientific research necessarily confronts the further deepening of economic reform, an increasing number of researchers are simultaneously engaged in the promotion of scientific and technological achievements and technologies, which require a large amount of funds to support, through the three types of investments chosen, the verification of Hypothesis (H1) of the study. Subsequently, TFGD (Technical training for fisherfolk in the Guangdong region), TPGD (Operational training for aquaculture science and technology extension staff in the Guangdong region), and FOEGD (Practitioners in aquaculture in the Guangdong region) were selected as the educational component of the study of aquaculture S&T to validate Hypothesis (H2). In contrast to the current empirical literature, this study performed an initial statistical test about the relevance of each regressor before proceeding to the final model estimation. Through this approach, the present study provides an explicit way to handle the problem of multicollinearity.

Table 2 shows the OLS estimates of the loss of aquaculture products by environmental factors. It shows a multiple regression model to explore the relationship between six predictor variables (SFGD, OFGD, TFGD, TPGD, ALGD, and FOEGD) and a single dependent variable (ALGD). Although the F-test for the model was statistically significant (Prob > F = 0.0054 *), further analysis of the t-values and associated p-values revealed that SFGD, OFGD, TFGD, and FOEGD did not have statistically significant effects on the dependent variable. However, we found that TPGD and ALGD had statistically significant effects on the dependent variable, with a one-unit increase in TPGD associated with an increase of 11.144931 and a one-unit increase in ALGD associated with an increase of 0.1373072 in the dependent variable, holding all other predictors constant.

Table 2. Results of the linear regression model.

Term	Estimate	Std Error	t Ratio	Prob > |t|	VIF
Intercept	−106,003.9	196,623.7	−0.54	0.6272	.
SFGD	2.0828588	3.678749	0.57	0.6109	111.56817
OFGD	0.4803825	7.807627	0.06	0.9548	109.70116
TFGD	0.307097	0.402901	0.76	0.5014	9.447883
TPGD	11.144931	9.022349	1.24	0.3047	2.8802129
ALGD	0.1373072	0.015592	8.81	0.0031 *	4.7373421
FOEGD	−5.512824	82.71853	−0.07	0.9511	31.341263
Source	**DF**	**Sum of Squares**	**Mean Square**	**F Ratio**	
Model	6	1.9014×10^{14}	3.1689×10^{13}	424,229	
Error	3	224,096,307	74,698,769	Prob > F	
C. Total	9	1.9238×10^{14}		0.0054 *	

Source: Authors' own calculations based on data from China Fisheries Statistical Yearbook 2012 to China Fisheries Statistical Yearbook 2022.c. Note: * indicates that the results are statistically significant at a significance level of 0.05.

The presence of multicollinearity, as indicated by high VIF values (rule of thumb: VIF < 10), can impact the interpretation of individual coefficients in the model. Therefore, we addressed the multicollinearity issue by using NIPALS regression, which is designed to handle issues related to variables with strong covariance and non-1st-order integration. Our aim was to reduce the impact of multicollinearity and identify the most important predictor variables driving the relationship with the dependent variable.

While we believe that NIPALS regression is a suitable alternative to OLS in this context, it is important to note that it may not always be the best approach for addressing multicollinearity. We carefully considered the nature of our data and the goals of our analysis before selecting PLS regression as our method for addressing multicollinearity. Our findings provide valuable insights into the effects of TPGD and ALGD on the dependent variable and highlight the importance of addressing multicollinearity in regression analysis. Further research in this area can lead to a better understanding of the underlying relationships between the predictor variables and the dependent variable.

Figure 2 shows that the minimum root mean PRESS is 0.53051 and the minimizing number of factors is 1. After confirming the number of optimal factors, the study proceeded to examine the variable importance of projection (VIP) to select the variables that were important (VIP > 0.8).

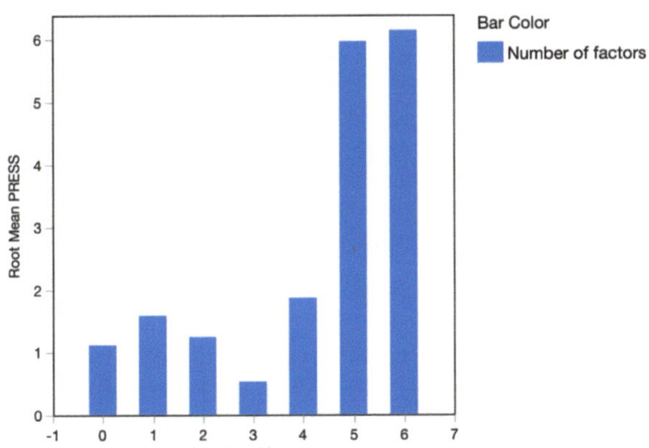

Figure 2. Factors of the NIPALS model. Source: Author's computation.

Figure 3 shows that most of the variables were important in explaining the loss of aquaculture products by environmental factors in China. According to the classification of Eriksson, Johansson [36], this study classified the important variables as "Highly influential" (VIP > 1), "Moderately influential" (0.8 < VIP < 1) and "Slightly influential" (VIP < 0.8). Table 3 shows that SFGD (funds for staff engaged in aquaculture in the Guangdong region), OFGD (operating funds related to aquaculture in the Guangdong region), TPGD (technical training for fisherfolk in the Guangdong region), and TFGD (operational training for aquatic science and technology extension staff in the Guangdong region) were moderately influential, while FOEGD (practitioners in aquaculture research institutions in the Guangdong region) was slightly influential in explaining the loss of aquaculture products by environment factors. Finally, only FAGD (aquaculture financial allocation in the Guangdong region) was highly influential.

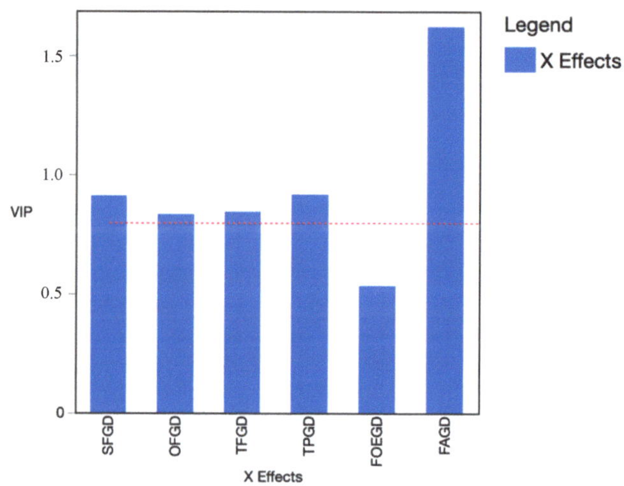

Figure 3. Variable Importance Plot. Source: Author's computation.

Table 3. Classification of variable importance.

X	VIP	Plot	Classifications
SFGD	0.9119		Moderately influential
OFGD	0.8344		Moderately influential
TFGD	0.8450		Moderately influential
TPGD	0.9172		Moderately influential
FOEGD	0.5347		Slightly influential
FAGD	1.6220		Highly influential

Source: Author's computation.

Subsequently, we investigated the drivers of the loss of aquaculture products due to environmental factors. Table 4 shows the sensitivity analysis of the loss of aquaculture products by environmental factors and their corresponding estimated coefficients. Based on the coefficients of the PLS regression model, we could determine the relative impact of each predictor variable on the dependent variable (ALGD). The coefficients show the direction and strength of the relationship between each predictor variable and the dependent variable.

Table 4. NIPALS estimation results.

Coefficient	ALGD
Intercept	0.0000
SFGD	−0.1793
OFGD	−0.0195
TFGD	0.0872
TPGD	−0.1866
FOEGD	−0.0478
FAGD	0.7650

Source: Author's computation.

SFGD: This variable has a negative coefficient, which suggests that an increase in funds for staff engaged in aquaculture in the Guangdong region was associated with a decrease in the loss of aquaculture products by environmental factors.

OFGD: This variable has a negative coefficient, which suggests that an increase in operating funds related to aquaculture in the Guangdong region was associated with a decrease in the loss of aquaculture products by environmental factors.

TFGD: This variable has a positive coefficient, which suggests that an increase in operational training for aquatic science and technology extension staff in the Guangdong region was associated with an increase in the loss of aquaculture products by environmental factors.

TPGD: This variable has a negative coefficient, which suggests that an increase in technical training for fisherfolk in the Guangdong region was associated with a decrease in the loss of aquaculture products by environmental factors.

FOEGD: This variable has a negative coefficient, which suggests that an increase in the number of practitioners in aquaculture research institutions in the Guangdong region was associated with a decrease in the loss of aquaculture products by environmental factors.

FAGD: This variable has a positive coefficient, which suggests that an increase in the aquaculture financial allocation in the Guangdong region was associated with an increase in the loss of aquaculture products by environmental factors.

It is important to keep in mind that the coefficients represent the relationship between each predictor variable and the dependent variable after accounting for the effects of the other variables in the model. So, for example, even though an increase in TFGD was associated with an increase in the loss of aquaculture products, this relationship may be different if we were to hold the other variables constant. Overall, the coefficients can provide insight into the relative importance of each predictor variable in predicting the loss of aquaculture products by environmental factors in the Guangdong region.

2.4. *Justification*

Applying the NIPALS algorithm is the best solution when there are more explanatory variables than observations, highly correlated explanatory variables and responses, and a large number of explanatory variables [37]. High-dimensional and nonlinearities are very common in different variables of agricultural production [38,39]. Strong nonlinear relationships may exist between different data sets, but when the nonlinearities are severe, they often behave unacceptably, and a feature of PLS is that the relationships between sets of observed variables are modelled by latent variables that are not usually directly observed and measured [40]. The NPLS model provides relatively stable modelling performance. This is mainly because it provides a nonlinear regression between each pair of latent variables while retaining the simple and linear external PLS framework [41].

NIPALS regression has also been widely used in agricultural practices. Samuel et al. [42] applied NIPALS regression to examine the assessment of the impact of energy, agricultural, and socioeconomic indicators on CO2 emissions in Ghana. Samuel [43] also used a similar approach to study the effects of energy, agriculture, macroeconomic, and anthropogenic indicators on environmental pollution from 1971 to 2011, and the regression demonstrated that increased economic growth in Ghana may lead to a decrease in environmental pollution.

3. Results

From the coefficients of the aquaculture product loss model presented in Table 4, we can conclude that an increase in funding (SFGD and OFGD) and training (TPGD and TFGD) is associated with a decrease in the loss of aquaculture products (ALGD). This means that allocating more funds for staff engaged in aquaculture, operating funds related to aquaculture, and providing technical and operational training for fisherfolk and aquatic science and technology extension staff can lead to a reduction in the loss of aquaculture products in the Guangdong region. It is important to note that the coefficients only provide an association, and other factors not included in the model may also impact the loss of aquaculture products.

The present study investigated the relationship between investments, training, and environmental factors in the aquaculture industry of the Guangdong region. To test the hypotheses, a multiple regression model was employed to explore the relationship between six predictor variables (SFGD, OFGD, TFGD, TPGD, ALGD, and FOEGD) and a single dependent variable (loss of aquaculture products by environmental factors). The results indicated that increased investments in staff funds (SFGD) and operating funds (OFGD) were not statistically significant predictors of the dependent variable, failing to support Hypothesis 1. On the other hand, enhanced training and education of fisherfolks and extension staff (TFGD and TPGD) were found to be statistically significant predictors of the dependent variable, supporting Hypothesis 2. The coefficients for TFGD and TPGD were positive, indicating that an increase in these variables is associated with a decrease in the loss of aquaculture products by environmental factors.

Moreover, the variance-inflated factor analysis showed evidence of multicollinearity between the predictor variables, indicating that the application of ordinary least squares (OLS) regression may produce biased results. To address this issue, partial least squares (PLS) regression was used instead, which identified the underlying latent variables driving the relationships between the predictor variables and the dependent variable, resulting in more accurate and reliable results.

The findings suggest that enhancing training and education of fisherfolks and extension staff can be an effective approach to reducing losses caused by environmental factors in the aquaculture industry of the Guangdong region. However, increased financial support for the industry may not necessarily lead to better environmental management practices and reduced losses, suggesting a need for more targeted investments and interventions. The study highlights the importance of addressing multicollinearity in regression analysis and the potential benefits of using PLS regression to obtain more accurate and interpretable results.

Overall, the study provides valuable insights into the relationships between investments, training, and environmental factors in the aquaculture industry in the Guangdong region, which can inform the development of more effective policies and interventions to promote sustainable development in the industry.

Figure 4 shows the plot of fitted and actual values—Loss of aquaculture products—by environmental factors. Based on the goodness-of-fit estimations, the model appears to have a reasonable fit. The mean absolute percentage error (MAPE) was found to be 0.475 and the root mean squared deviation (RMSD) was found to be 41.05%. These values suggest that the model's predictions are generally close to the actual values, with a small amount of error. Furthermore, the model was able to explain approximately 96.51% of the variation in the dependent variables, indicating a strong relationship between the predictor variables and the dependent variable. Overall, these findings suggest that the model is a reliable tool for predicting the loss of aquaculture products in the Guangdong region based on environmental factors and can provide valuable insights for improving environmental management practices in the aquaculture industry.

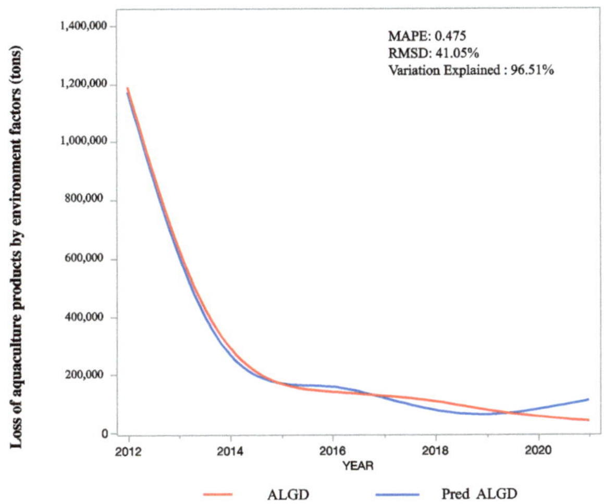

Figure 4. The plot of fitted and actual values—Loss of aquaculture products by environment factors.

4. Discussion

In this study, we delved deeper into the complex interplay between investment, training, and environmental factors in the Guangdong region's aquaculture sector, which has significant implications for policymakers and industry stakeholders. Our findings emphasize that focusing on training and education for fisherfolks and extension staff is crucial for significantly reducing losses caused by environmental factors. It is essential to understand that simply allocating more funds might not be enough to enhance environmental management practices or decrease losses in the aquaculture sector.

We also brought attention to the importance of addressing multicollinearity in regression models. The NIPALS regression technique we employed effectively mitigated the effects of multicollinearity, resulting in more precise and interpretable outcomes. This method allowed us to gain a deeper understanding of the relationships between predictor variables and the dependent variable, which, in turn, enabled us to identify the most influential factors governing these relationships. However, it is crucial to recognize that NIPALS regression may not always be the optimal method for addressing multicollinearity. Researchers should carefully evaluate their data and research objectives before choosing the most suitable technique.

Our model demonstrates a high goodness of fit (96.51%), indicating that the chosen variables significantly explain the variation in aquaculture product losses due to environmental factors. Nevertheless, it is vital to examine other external factors that might contribute to these losses, such as weather conditions, disease outbreaks, or other variables not considered in our model. This underscores the necessity for further investigation into additional factors affecting aquaculture product losses caused by environmental factors, extending beyond the Guangdong region and encompassing other regions. Recognizing the key factors linked to environmentally driven losses in aquaculture products enables decision-makers to allocate resources more effectively and create targeted strategies for enhancing environmental management practices, consequently reducing losses in the aquaculture sector. Potential interventions could involve investing in the education of fisherfolk and extension staff, exploring innovative environmental management approaches, employing technology and data-driven decision-making, and promoting public–private partnerships. Our research emphasizes the significance of accounting for regional variability and the challenges of developing universal hypotheses and models applicable to all situations. Future studies should consider the distinct characteristics of various regions in China and potentially adjust their methodology and analysis accordingly. This approach ensures that the developed interventions and policies are relevant and effective in addressing each region's unique aquaculture industry challenges.

We recognize that our study's findings may not be directly applicable to other countries and regions, given the considerable variation in aquaculture practices and environmental factors. Future research should consider comparisons with other countries or regions, using a new set of realistic data to explore the similarities and differences in aquaculture product losses due to environmental factors. Comparing data from various countries or regions would provide valuable insights into the global context of aquaculture product losses.

Based on our results, the Guangdong region faces several specific challenges in addressing the loss of aquaculture products due to environmental factors. These challenges highlight the need for targeted interventions and policy measures to promote sustainable development in the region's aquaculture industry.

Training and education for fisherfolks and extension staff: Our findings emphasize the importance of investing in training and education for fisherfolks and extension staff as a key factor in reducing losses caused by environmental factors. Guangdong should develop and implement targeted training programs focused on enhancing the knowledge and skills of fisherfolks and extension staff in environmental management practices, sustainable aquaculture techniques, and early warning systems for disease outbreaks and extreme weather events.

Utilizing technology and data-driven decision-making: The results suggest that traditional investment in staff and operating funds may not be sufficient to address the environmental challenges faced by the aquaculture industry in Guangdong. The region needs to explore innovative approaches to environmental management, such as the adoption of advanced technology like remote sensing, precision aquaculture, and water quality monitoring systems [44,45].

Public–private partnerships: Fostering public–private partnerships can play a vital role in addressing the environmental challenges faced by Guangdong's aquaculture industry. Collaborations between the government, private sector, and research institutions can facilitate the sharing of resources, knowledge, and best practices, leading to more effective and sustainable environmental management practices in the region [46].

Adaptation to regional variability: Guangdong's diverse geography and environmental conditions require tailored solutions to address the specific challenges faced by different areas within the region. Policymakers should take into account local environmental factors, such as water quality, weather patterns, and disease prevalence, when designing and implementing interventions to reduce losses in the aquaculture sector.

Comprehensive monitoring and evaluation: In order to assess the effectiveness of interventions aimed at reducing losses of aquaculture products by environmental factors, Guangdong should establish a comprehensive monitoring and evaluation system. This system should track key performance indicators related to environmental management practices, training outcomes, and aquaculture production, allowing for evidence-based decision-making and continuous improvement.

In conclusion, addressing the specific challenges faced by Guangdong's aquaculture industry requires a multifaceted approach that combines targeted investments in training and education, the adoption of innovative technology, fostering public–private partnerships, and the development of tailored solutions for regional variability. By taking these factors into account, policymakers and industry stakeholders can work together to reduce losses caused by environmental factors and promote sustainable development in the region's aquaculture sector.

5. Conclusions

Our research enhances the comprehension of the interplay between investments, training, and environmental factors within Guangdong's aquaculture sector. The results highlight the significance of focused investments in fisherfolk and extension staff training and education, and the necessity to address multicollinearity in regression models. Furthermore, this study emphasizes the importance of considering regional differences and acknowledging the challenges in creating universally applicable hypotheses and models.

There is a need for continued research to examine other factors that could influence aquaculture product losses due to environmental factors, as well as to assess the applicability of our findings across different regions and contexts.

Author Contributions: Conceptualization, P.W.; methodology, P.W.; software, P.W.; validation, P.W., I.M. and R.F.; formal analysis, P.W.; investigation, P.W.; resources, P.W.; data curation, P.W.; writing—original draft preparation, P.W.; writing—review and editing, I.M. and R.F.; visualization, P.W.; supervision, I.M.; project administration, I.M.; funding acquisition, I.M. All authors have read and agreed to the published version of the manuscript.

Funding: This work was supported by FCT, I.P., the Portuguese national funding agency for science, research and technology, under the Project UIDB/04521/2020.

Institutional Review Board Statement: Not applicable.

Informed Consent Statement: Not applicable.

Data Availability Statement: Publicly available datasets were analyzed in this study. This data can be found here: [http://fishery.ckcest.cn/featureYearbook.html].

Conflicts of Interest: The authors declare no conflict of interest.

References

1. FAO. The State of World Fisheries and Aquaculture 2020. In *Sustainability in Action*; FAO: Rome, Italy, 2020. [CrossRef]
2. Yu, J.; Wu, J. The Sustainability of Agricultural Development in China: The Agriculture–Environment Nexus. *Sustainability* **2018**, *10*, 1776. [CrossRef]
3. Wang, P.; Mendes, I. Assessment of Changes in Environmental Factors Affecting Aquaculture Production and Fisherfolk Incomes in China between 2010 and 2020. *Fishes* **2022**, *7*, 192. [CrossRef]
4. Temple, A.J.; Skerritt, D.J.; Howarth, P.E.; Pearce, J.; Mangi, S.C. Illegal, unregulated and unreported fishing impacts: A systematic review of evidence and proposed future agenda. *Mar. Policy* **2022**, *139*, 105033. [CrossRef]
5. Lin, X.; Zheng, L.; Li, W. Measurement of the contributions of science and technology to the marine fisheries industry in the coastal regions of China. *Mar. Policy* **2019**, *108*, 103647. [CrossRef]
6. Li, F.; Xing, W.; Su, M.; Xu, J. The evolution of China's marine economic policy and the labor productivity growth momentum of marine economy and its three economic industries. *Mar. Policy* **2021**, *134*, 104777. [CrossRef]
7. Bradley, D.; Merrifield, M.; Miller, K.M.; Lomonico, S.; Wilson, J.R.; Gleason, M.G. Opportunities to improve fisheries management through innovative technology and advanced data systems. *Fish Fish.* **2019**, *20*, 564–583. [CrossRef]
8. Laso, J.; Ruiz-Salmón, I.; Margallo, M.; Villanueva-Rey, P.; Poceiro, L.; Quinteiro, P.; Poceiro, L.; Quinteiro, P.; Aldaco, R. Achieving Sustainability of the Seafood Sector in the European Atlantic Area by Addressing Eco-Social Challenges: The NEPTUNUS Project. *Sustainability* **2022**, *14*, 3054. [CrossRef]
9. Xiao, R.; Wei, Y.; An, D.; Li, D.; Ta, X.; Wu, Y.; Ren, Q. A review on the research status and development trend of equipment in water treatment processes of recirculating aquaculture systems. *Rev. Aquacult.* **2019**, *11*, 863–895. [CrossRef]
10. Omer, A.M. Energy, environment and sustainable development. *Renew. Sustain. Energy Rev.* **2008**, *12*, 2265–2300. [CrossRef]
11. Knowler, D.; Chopin, T.; Martínez-Espiñeira, R.; Neori, A.; Nobre, A.; Noce, A.; Reid, G. The economics of Integrated Multi-Trophic Aquaculture: Where are we now and where do we need to go? *Rev. Aquacult.* **2020**, *12*, 1579–1594. [CrossRef]
12. Ahmad, A.; Abdullah, S.R.S.; Hasan, H.A.; Othman, A.R.; Ismail, N.I. Aquaculture industry: Supply and demand, best practices, effluent and its current issues and treatment technology. *J. Environ. Manag.* **2021**, *287*, 112271. [CrossRef] [PubMed]

13. Zheng, H.; Li, J.; Zhao, X. How does financial policy support the development of China's fishery? Characteristics, experience and prospects. *Mar. Policy* **2021**, *132*, 104678. [CrossRef]
14. Dhraief, M.Z.; Bedhiaf-Romdhania, S.; Dhehibib, B.; Oueslati-Zlaouia, M.; Jebali, O.; Ben, Y.S. Factors Affecting the Adoption of Innovative Technologies by Livestock Farmers in Arid Area of Tunisia. *FARA Res. Result* **2018**, *3*, 22.
15. Aswal, D.K. Quality Infrastructure of India and Its Importance for Inclusive National Growth. *Mapan* **2020**, *35*, 139–150. [CrossRef]
16. Zhang, S.; Chen, J. An empirical study on the relationship between agricultural scientific research investment and agricultural productivity growth. *J. Yunnan Univ. Financ. Econ.* **2013**, *29*, 83–90.
17. Deng, X.; Wang, S.Z. Research on the impact of investment in agricultural science and technology innovation on agricultural economic growth. *Dongyue Trib.* **2020**, *41*, 109–120.
18. Song, Y.; Luan, L.J. Analysis of the relationship between input and effect of agricultural science and technology in China. *Forum Sci. Technol. China* **2005**, *4*, 102–105.
19. Nordhaus, W.D. *The Challenge of Global Warming: Economic Models and Environmental Policy*; Yale University: New Haven, CT, USA, 2007; Volume 4.
20. Holdren, J.P. Science and technology for sustainable well-being. *Science* **2008**, *319*, 424–434. [CrossRef]
21. Paltasingh, K.R.; Goyari, P. Impact of farmer education on farm productivity under varying technologies: Case of paddy growers in India. *Agric. Food Econ.* **2018**, *6*, 7. [CrossRef]
22. Weir, S. *The Effects of Education on Farmer Productivity in Rural Ethiopia*; Centre for the Study of African Economies; Working Paper No. WPS/99-7; Oxford University: Oxford, UK, 1999.
23. Abdulai, A.; Huffman, W. The adoption and impact of soil and water conservation technology: An endogenous switching regression application. *Land Econ.* **2014**, *90*, 26–43. [CrossRef]
24. Biao, X.; Kaijin, Y. Shrimp farming in China: Operating characteristics, environmental impact and perspectives. *Ocean Coast. Manag.* **2007**, *50*, 538–550. [CrossRef]
25. Su, C.W.; Song, Y.; Umar, M. Financial aspects of marine economic growth: From the perspective of coastal provinces and regions in China. *Ocean Coast. Manag.* **2021**, *204*, 105550. [CrossRef]
26. He, Y.; Kuang, Y.; Zhao, Y.; Ruan, Z. Spatial Correlation between Ecosystem Services and Human Disturbances: A Case Study of the Guangdong–Hong Kong–Macao Greater Bay Area, China. *Remote Sens.* **2021**, *13*, 1174. [CrossRef]
27. Fuyan, K.E.; Zhiyuan, X.U.; Yang, L. Study on driving effects of agricultural science and technology investments on agricultural economic growth:based on the data of 60 counties in Zhejiang Province, China. *Acta Agric. Zhejiangensis* **2022**, *34*, 1537–1544. [CrossRef]
28. Yang, J.B. Econometric Analysis on the relationship between agricultural science and technology investment and grain output in China from 1978 to 2003. *Sci. Technol. Manag. Res.* **2007**, *27*, 69–71.
29. Zhang, H.; Li, W. Analysis of dynamic connection between spending on agricultural science & technology and development of China's agricultural economy. *Sci. Technol. Manag. Res.* **2013**, *33*, 149–151.
30. Long, H.; Tu, S.; Ge, D.; Li, T.; Liu, Y. The allocation and management of critical resources in rural China under restructuring: Problems and prospects. *J. Rural Stud.* **2016**, *47*, 392–412. [CrossRef]
31. Geladi, P. Notes on the history and nature of partial least squares (PLS) modelling. *J. Chemom.* **1988**, *2*, 231–246. [CrossRef]
32. Voet, H.D. Comparing the predictive accuracy of models using a simple randomization test. *Chemom. Intell. Lab. Syst.* **1994**, *25*, 313–323. [CrossRef]
33. Chu, M.T.; Funderlic, R.E.; Golub, G.H. A rank-one reduction formula and its applications to matrix factorizations. *SIAM Rev.* **1995**, *37*, 512–530. [CrossRef]
34. Asumadu-Sarkodie, S.; Owusu, P.A. A Multivariate Analysis of Carbon Dioxide Emissions, Electricity Consumption, Economic Growth, Financial Development, Industrialization and Urbanization in Senegal. *Energ Source Part B.* **2017**, *12*, 77–84. [CrossRef]
35. Administration of Fisheries and Fishery Administration (M.O.A.A.R.A.); National General Fisheries Technology Extension Station; Chinese Fisheries Society. *China Fishery Statistical Yearbook 2021 (Chinese Article)*; China Agricultural Press: Beijing, China, 2021.
36. Eriksson, L.; Johansson, E.; Kettaneh-Wold, N.; Wold, S. Multi-and megavariate data analysis: Principles and applications. *Umetrics* **2001**, *43*, 261–262.
37. JMP. Using Partial Least Squares: When Ordinary Least Squares Regression Just Won't Work. 2021. Available online: https://www.jmp.com/en_my/events/mastering/topics/using-partial-least-squares-when-ordinary-least-squares-regression-just-wont-work.html (accessed on 12 January 2023).
38. Abdel-Rahman, E.M.; Mutanga, O.; Odindi, J.; Adam, E.; Odindo, A.; Ismail, R. A comparison of partial least squares (PLS) and sparse PLS regressions for predicting yield of Swiss chard grown under different irrigation water sources using hyperspectral data. *Comput. Electron. Agric.* **2014**, *106*, 11–19. [CrossRef]
39. Crane-Droesch, A. Machine learning methods for crop yield prediction and climate change impact assessment in agriculture. *Environ. Res. Lett.* **2018**, *13*, 114003. [CrossRef]
40. Rosipal, R. *Nonlinear Partial Least Squares An Overview*; Chemoinformatics and Advanced Machine Learning Perspectives: Complex Computational Methods and Collaborative Techniques; Medical University of Vienna: Vienna, Austria, 2011; pp. 169–189. [CrossRef]

41. Yang, C.; Zhang, Y.; Huang, M.; Liu, H. Adaptive dynamic prediction of effluent quality in wastewater treatment processes using partial least squares embedded with relevance vector machine. *J. Clean. Prod.* **2021**, *314*, 128076. [CrossRef]
42. Asumadu-Sarkodie, S.; Owusu, P.A.; Hung, Y.T. The Impact Assessment of Energy, Agriculture, and Socioeconomic Indicators on Carbon Dioxide Emissions in Ghana. In *Handbook of Environment and Waste Management: Acid Rain and Greenhouse Gas Pollution Control*; World Scientific: Singapore, 2020; pp. 137–201. [CrossRef]
43. Asumadu-Sarkodie, S.; Owusu, P.A. The impact of energy, agriculture, macroeconomic and human-induced indicators on environmental pollution: Evidence from Ghana. *Environ. Sci. Pollut. Res.* **2017**, *24*, 6622–6633. [CrossRef] [PubMed]
44. Hou, Y.; Zhao, G.; Chen, X.; Yu, X. Improving Satellite Retrieval of Coastal Aquaculture Pond by Adding Water Quality Parameters. *Remote Sens.* **2022**, *14*, 3306. [CrossRef]
45. Huan, J.; Li, H.; Wu, F.; Cao, W. Design of water quality monitoring system for aquaculture ponds based on NB-IoT. *Aquac. Eng.* **2020**, *90*, 102088. [CrossRef]
46. Thorpe, J. Procedural Justice in Value Chains Through Public–private Partnerships. *World Dev.* **2017**, *103*, 162–175. [CrossRef]

fishes

MDPI

Article

Sampling Error and Its Implication for Capture Fisheries Management in Ghana

Temiloluwa Jesutofunmi Akinyemi [1,2,*], Denis Worlanyo Aheto [1,2] and Wisdom Akpalu [3]

1 Centre for Coastal Management, Africa Centre of Excellence in Coastal Resilience (ACECoR), University of Cape Coast, Cape Coast PMB TF0494, Ghana
2 Department of Fisheries and Aquatic Sciences, School of Biological Sciences, College of Agriculture and Natural Sciences, University of Cape of Coast, Cape Coast PMB TF0494, Ghana
3 School of Research and Graduate Studies, Ghana Institute of Management and Public Administration (GIMPA), Accra 039 5028, Ghana
* Correspondence: temiloluwa.akinyemi@stu.ucc.edu.gh; Tel.: +233-(0)-2-4857-4617

Abstract: Capture fisheries in developing countries provide significant animal protein and directly supports the livelihoods of several communities. However, the misperception of biophysical dynamics owing to a lack of adequate scientific data has contributed to the suboptimal management in marine capture fisheries. This is because yield and catch potentials are sensitive to the quality of catch and effort data. Yet, studies on fisheries data collection practices in developing countries are hard to find. This study investigates the data collection methods utilized by fisheries technical officers within the four fishing regions of Ghana. We found that the officers employed data collection and sampling procedures which were not consistent with the technical guidelines curated by FAO. For example, 50 instead of 166 landing sites were sampled, while 290 instead of 372 canoes were sampled. We argue that such sampling errors could result in the over-capitalization of capture fish stocks and significant losses in resource rents.

Keywords: fisheries data quality; fisheries management; Ghana

Citation: Akinyemi, T.J.; Aheto, D.W.; Akpalu, W. Sampling Error and Its Implication for Capture Fisheries Management in Ghana. *Fishes* **2022**, *7*, 333. https://doi.org/10.3390/fishes7060333

Academic Editor: Ben Drakeford

Received: 17 October 2022
Accepted: 4 November 2022
Published: 12 November 2022

1. Introduction

Millions of people around the world depend, directly or indirectly, on capture fisheries for their food security, income, and livelihoods [1]. This dependence is particularly strong in coastal communities in developing countries where the sector employs 97% of the 50 million people who make up the world's fishing workforce [2,3]. Ghana, as one of the developing countries, is home to a wide variety of biodiversity, including small pelagic species such as anchovies, sardinella, and chub mackerel and larger pelagic fish such as yellowfin, skipjack, and big-eye tuna. There are also demersal fish such as grouper and snapper, and other seafood such as shrimp and squids [4]. For sustenance and the eradication of poverty, a majority of coastal dwellers are solely dependent on the exploitation of these fisheries, with over 60% of the population relying on fish as their primary source of protein and about 10% of the population (2.6 million people out of a total population of 26 million) believed to be directly or indirectly dependent on fish resources [5].

The Ghanaian marine fishing industry is divided into three primary sectors: small-scale artisanal fishers, semi-industrial fisheries, and large industrial fisheries, with over 300 fish landing sites spread throughout its coasts [6]. The vessels used in Ghana's marine capture fishery include dugout canoes, canoes with outboard motors, trawlers, and large steel-hulled foreign-built vessels. The dugout canoes and canoes fitted with outboard motors are primarily utilized by artisanal fishers while trawlers and steel-hulled vessels are used mainly in the semi-industrial and industrial marine fisheries [7]. There is currently a total of 11,583 licensed marine artisanal canoes operating along the coast, 150 semi-industrial vessels, and 84 licensed industrial trawlers in Ghana's marine waters [8].

Despite the importance of the fisheries sectors, according to various experts, fish stocks have declining rapidly due to the overcapacity of fleets, excessive fishing quotas, illegal fishing practices, and the generally poor management of fisheries, which poses existential threats to coastal communities [9]. This has necessitated the formulation and refinement of existing management policies with the aim of limiting fishing efforts to optimize the economic, social, and ecological sustainability of capture fisheries [10]. The effectiveness of effort-limiting policies, however, depends on the availability and quality of the relevant fisheries data used for decision-making [11].

National governments and international organizations have been working hard at collecting fisheries data to inform sustainable and long-lasting management plans and strategies [12]. However, this remains a daunting task due to the complex interactions among species and marine ecosystems, and the wide distribution and migration of pelagic stocks across national jurisdictions. These complexities of biophysical dynamics make fisheries management difficult [13]. Nevertheless, management decisions must be made as livelihoods and incomes depend on wise decisions made by the managers, and they can only make wise decisions if they have sufficient knowledge of the ecosystem and fishery to understand the causes of the current fisheries situation and predict how the resource and fishery will change in response to management actions [14].

Accurate and consistent knowledge about how a fishery is doing, as well as what, where, and how much of a species is being captured requires more precise data collection and faster and more advanced reporting, processing, and analysis, as well as more efficient mechanisms to disseminate the results to enable close to real-time analysis [15]. The fisheries data collected is usually the manager's major source of information, which is essential in developing appropriate management tools to support the sustainable use of the stock [16,17]. However, the data quality is low in many developing countries owing to inadequate resources, including skills and funding.

Although the FAO Code of Conduct (Paragraph 6.4) has stated that the conservation and management of fisheries must be based on the best scientific knowledge available at any point in time. Unfortunately, many fisheries agencies lack sufficient data, making attempts at managing fisheries difficult. For instance, the reconstruction of catches carried out by [18,19] revealed that the catch and effort data compiled by FAO were deficient. As noted by [20], the unavailability and suspicion of errors in catch data due to lack of skills and resources in member countries have resulted in the complementation or replacement of countries' data with data from other sources. These omissions or errors in data collection could lead to erroneous fisheries management policies, which in turn could result in suboptimal extraction, losses in resource rents, and eventual collapse of capture fisheries. It is therefore expedient to assess how catch and effort data are collected to better inform management policies.

An analysis of the national fisheries data collection protocols in Ghana suggests that the Fisheries Scientific Survey Division (FSSD) is mandated to conduct scientific research and deploy surveys on marine environments and fisheries to inform the formulation and management of policies aimed at the sustainable management of Ghana's marine fisheries resources. The FSSD is under the Fisheries Commission (FC), which was established in 1962 with technical assistance from the Food and Agriculture Organization (FAO). Due to limited human and financial resources, the FSSD has not been able to provide adequate monitoring of the data collection activities of the technical officers. Thus, any errors that occur on the field are ignored. To the best of our knowledge, no study has been undertaken to investigate whether the recommended sampling procedures are followed by the field enumerators.

2. Materials and Methods

2.1. Study Area

This study was carried out in the following twenty-nine fish landing sites out of thirty representing the four coastal administrative regions in Ghana: Abutiakope, Lighthouse (Volta), Gbegbeyise, Botianor, Agjivompanye, Odin-nyonma, Osu alata, Teshie, Ga mashie,

Awudun (Greater Accra), Saltpond, Kromantse, Apam main, Elmina main, Elmina, Ayipey, Abrofo mpoano, Mumford main, Enfano (Central Region), Dixcove, Sekondi, Fante line (Axim), Akyinim, Ewe line, Fante line (Half Assini), Sharma Apo, Sekondi-Takoradi, Akwadae, and Adjua (Western) (See Figure 1 for the geographical location of the landing sites). These twenty-nine sites chosen for primary data collection were selected based on the total number of enumerators in Ghana and where they are assigned along the coast.

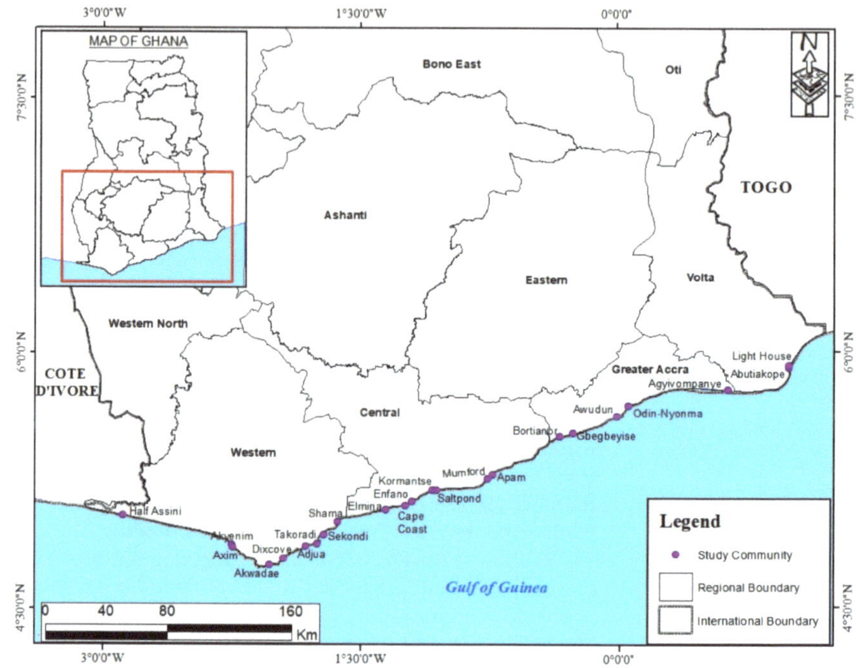

Figure 1. Map showing the twenty-nine landing beaches in Ghana.

2.2. Research Design

This study used a quantitative survey design to examine the types of data collected and the methods associated with the data collection in Ghana. The data were gathered between May and June 2022 at all 30 landing sites. A structured questionnaire was used for data collection. Field assistants were trained on the administration of the questionnaire, ethical standards, and COVID-19 safety protocols. The respondents included 29 Fisheries Commission field enumerators and 1 Field Scientific Survey Division data manager. The surveys were conducted in English and local languages, including Fante, Ga, Nzema, and Twi. Each interview lasted between 40 to 60 min.

2.3. Research Instrument

The questionnaire used was made up of three sections. The first part of the questionnaire (Section A) consisted of an introductory statement and questions about the relevant sociodemographic characteristics of respondents. Some of the variables included age, years of experience in data collection, the number of landing sites, gender, fishing experience, and level of education (basic education, secondary, and tertiary). The next two sections highlighted the types of fisheries data gathered using the FAO data collection guidance as a benchmark [21]. The data was classified as biological, ecological, economic, or social. A total of 24 questions were developed through an extensive review of the literature [14,16,22,23]. For the evaluation of the data collection procedure, the questions comprised five categories. These categories were based on the source of the data on fish production, the type of effort data gathered, the type of capture data gathered, and the frequency of data collection.

2.4. Data Analysis

Responses from interviews were coded using the IBM Statistical Package for Social Scientists (SPSS) computer software version 20.0. (2012) and analyzed for trends in response to research questions using Software for Statistical Analysis (STATA SE 15.0) (STATA Corp, College Station, TX, USA) and Microsoft excel. To understand the distributions of all relevant variables, descriptive statistics (frequencies and percentages) were generated. The summaries of the results are presented in tables (Tables 1–3). To check for sampling error, this study compared the capture fisheries data collection procedures in Ghana to the recommended best practices (i.e., the FAO guidelines) along the entire coast of Ghana using the FAO toolkit for small-scale fisheries routine data collection [22] and the FAO data collection guidelines [21]. The sample size formula developed by [24] was used to estimate the actual sample size for comparison with the number sampled.

Table 1. Catch data collection by Ghana's Fisheries Commission enumerators.

Variables	Collect (%)	Do Not Collect (%)
Biological data		
Total fish landings by major species	66	34
Total fish landings by canoes	69	31
The total effort by canoes	86	14
Length and/or age composition of fish landings	21	79
Discards of fish species per canoe	0	100
Length and/or age composition of discards	0	100
Areas fished by each canoe	17	83
Ecological data		
Total catches of bycatch species	17	83
Length and/or age composition of bycatch	3	97
Economic data		
The average income per fishing unit	52	48
The cost of premix fuel	7	93
Price of fish landed per canoe	93	7
Social data		
Crew size within each canoe	93	7

Table 2. Type and method of catch data collected by enumerators.

Variables	Freq	Percent
Type of catch data collected		
Multi-species (all species)	15	51.72
Single-species (only one species)	10	34.48
Single-species and multi-species	4	13.79
Data collection method		
By canoes	10	34.48
By gear	5	17.24
By species	10	34.48
By species and gear	4	13.79

To enhance visualization and appreciation of the study context, graphs are presented. The landing sites and canoes sampled across the whole district were compared with the landing sites and canoes that were required to be sampled. Summaries of the results are presented in Figures 2–5. The chi-square test was then used to verify whether there was a significant difference between the actual and the expected sampled landing beaches.

Table 3. Type of effort data collected by enumerators in each district (✓ = data collected; ✗ = Data not collected).

District	Number of Canoes	Size of Fishing Gear	Type of Fishing Gear	Number of Trips	Trip Duration	Size of Canoe
Keta	✓	✗	✓	✗	✓	✗
Ada East	✓	✓	✓	✗	✓	✗
Kpone Ketamanso	✓	✗	✓	✗	✓	✗
AMA	✓	✓	✓	✓	✓	✗
TMA	✓	✗	✓	✗	✓	✗
Ga South	✓	✗	✓	✓	✓	✗
Efutu Municipal	✓	✗	✓	✓	✓	✗
Gomoa West	✓	✗	✓	✓	✓	✗
Ahanta West	✓	✗	✓	✓	✓	✗
Abura-Asebu Kwamankes	✗	✗	✗	✓	✓	✗
Cape Coast	✗	✗	✓	✓	✓	✗
Nzema East	✓	✗	✓	✗	✓	✗
Jomoro	✓	✗	✓	✓	✓	✗
Komenda-Edina-Equafo	✓	✗	✓	✗	✓	✗
Ledzokuku	✓	✓	✗	✓	✓	✓
Mfantseman	✓	✗	✓	✓	✓	✗
Sekondi-Takoradi	✓	✗	✓	✓	✓	✗
Shama	✓	✗	✓	✗	✓	✗

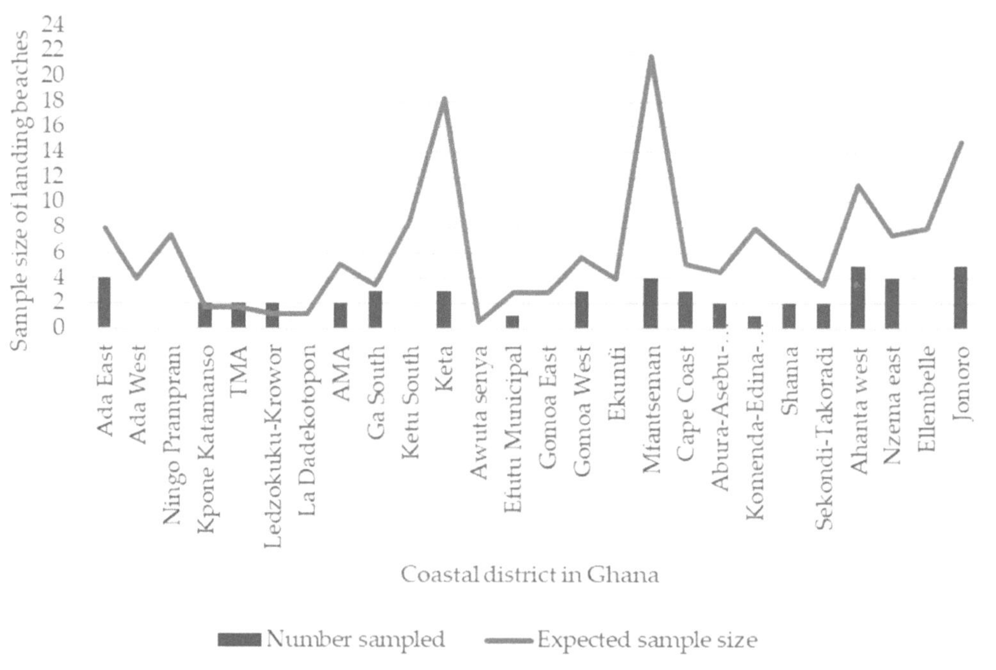

Figure 2. Comparison between the observed sample size of the Fisheries Scientific Survey Division (FSSD) and the expected sample size of landing beaches according to the districts in Ghana.

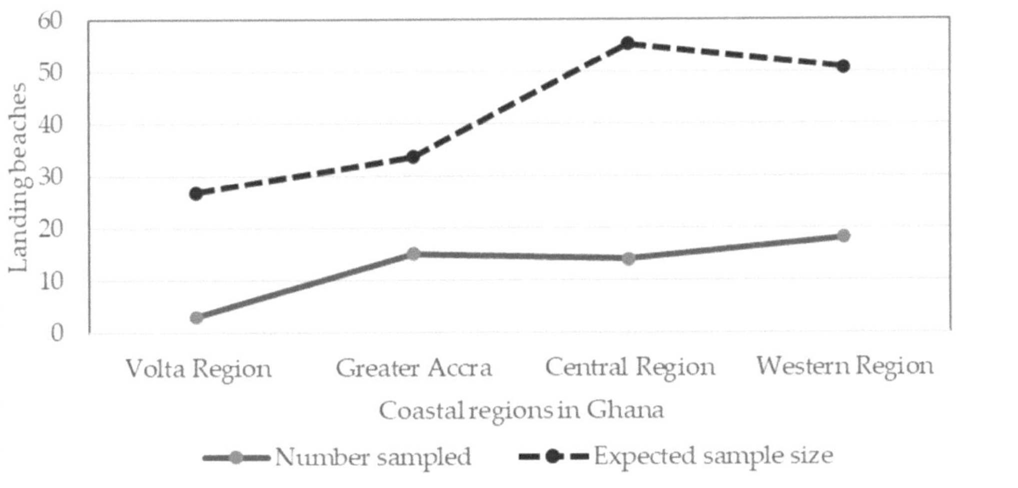

Figure 3. Comparison between the observed sample size of the Fisheries Scientific Survey Division (FSSD) and the expected sample size of landing beaches according to the coastal regions in Ghana.

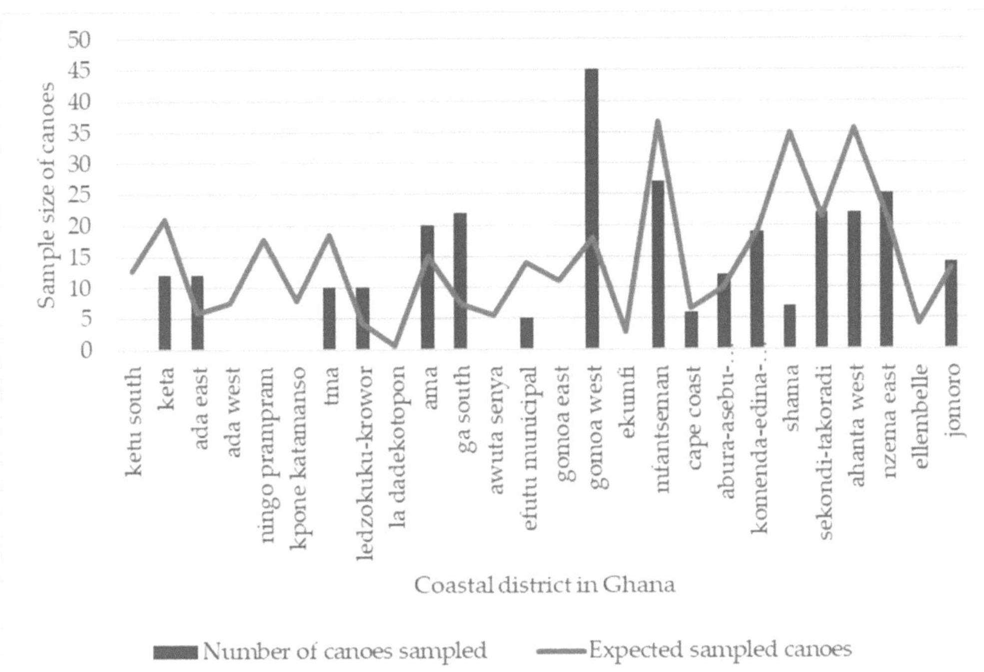

Figure 4. Comparison between the observed sample size of the Fisheries Scientific Survey Division (FSSD) and the expected sample size of canoes according to the districts in Ghana.

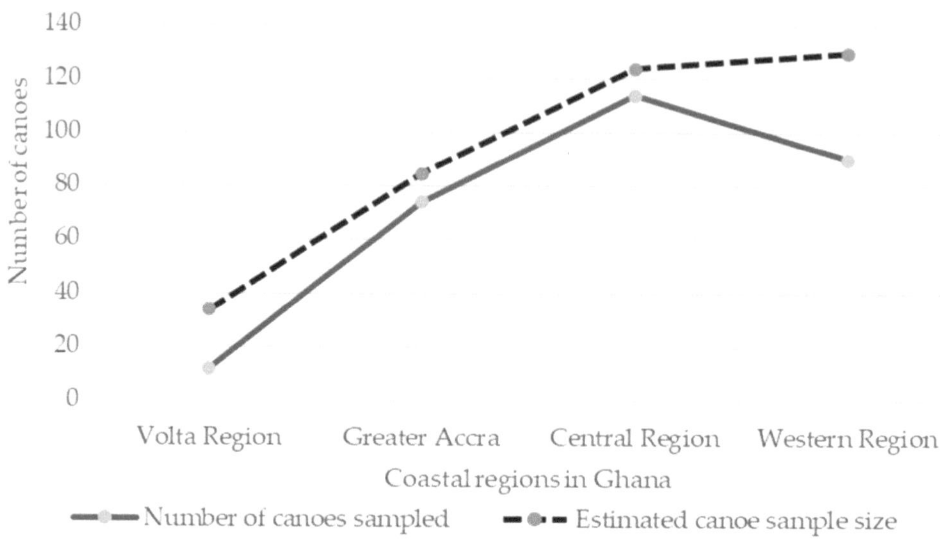

Figure 5. Comparison between the observed sample size and the expected sample size of canoes according to the coastal regions in Ghana.

3. Results

This section presents the data collection procedures of the FSSD and the results of the study. FSSD employs 30 enumerators to collect artisanal fisheries data from 50 landing beaches out of approximately 292 landing beaches [8]. These 50 landing beaches were obtained using the three-stage sampling survey by dividing the whole coastal area into four regions (i.e., Major strata) and the four regions into districts (Minor strata). Sampled canoes and landing sites were selected within the minor strata (districts) based on the canoe frame survey for the sole purpose of increasing the accuracy of the derived estimates using the proportional stratified sampling method.

The equation for the sampling is $n_k = \frac{n}{N} \times N_k$, with a maximum of 12 canoes sampled daily, where N and n are the total population and sample sizes, respectively, k is the number of strata, N_k is the number of units in stratum k, and n_k is the number of sampled units in stratum k. To calculate the sample size of the total population, sample size formula $n = \frac{N}{1 + N(e^2)}$ is used [24], where e is the level of precision. The FSSD employ 95% as the confidence interval and +/−5 as the degree of accuracy.

The sampling procedure is adopted from the FAO toolkit for small-scale fisheries routine data collection [22] and, as stated in the toolkit, enumerators at the landing beaches sample data for 14 days/gear/month with each enumerator having two gears which in some cases spill over to two landing sites each, depending on the size of the landing site and the abundance of fishing gears. For the recording of data, two forms are provided by the Fisheries Commission, Forms 1a and 1b, with each performing a different function. Form 1a is used to record daily information on fishing activity at the landing site, and Form 1b is used to record information and data collected. The data collected by FSSD are placed into three categories: the fisheries statistical data, i.e., the catch and effort data; the biological data; and the environmental data.

The fisheries data expected to be recorded at each landing beach are catch and species composition (single-species), fishing effort, price of fish, number of operating fishing crafts, types and sizes of fishing crafts, types of gears and their target species, areas of operation of fishing crafts, number of fishermen on fishing crafts, and information on landing sites.

The biological data to be collected are fish length, fish weight, gonad weight, and sex. The environmental data are salinity, temperature, and dissolved oxygen.

The data is collected by the 30 enumerators at all fifty landing sites but due to the shortage of resources and intellectual capacity, the biological data is collected at four landing sites across the four coastal regions, and the environmental data is collected at six landing sites. These sites were selected based on the abundance of fish species and the flow of water, respectively. After the data are recorded, the zonal officials (supervisors) in charge of the enumerators receive the records from each landing site and transmit them to FSSD, where they are compiled. For this study, since 1 of the enumerators had hearing loss, only 29 were interviewed. These enumerators had a male-to-female ratio of 25:4, an average age of 37 years, and an average of 10 years of data collection experience.

3.1. Sampling of Landing Beaches in the Coastal Districts of Ghana

Ghana has 292 landing beaches. This means that the calculated sample size is 166 [24]. However, only about a third of these beaches are sampled by enumerators. Each of the 26 fishing districts should have at least 1 landing beach sampled based on ratio and proportion, but as can be seen in Figure 2, the enumerators cover 18 out of the 26 fishing districts, resulting in an under-sampling of 8 districts. The data gatherers also stated that they sample a total of 50 landing beaches from the 18 districts they work in, which is 80 beaches less than what should be sampled from those 18 districts (assuming the sampling of the 18 districts is desirable). However, they over-sample in Ledzokuku-Krowor by 1 landing beach.

To determine whether there is a significant difference between the number of landing beaches sampled and the number of landing beaches expected to be sampled, a chi-square test was undertaken, and we found a significant difference (93.87276, p-value of 0.001). The low coverage of landing beaches is attributed to a lack of human and financial resources.

3.2. Sampling of Landing Beaches in the Coastal Regions of Ghana

On a regional level, we discovered a considerable discrepancy between the actual and expected landing beaches sampled, as shown graphically in Figure 3. This was found using the same methodology (sampling, ratio, and proportion). We discovered that the Central Region has a more pronounced under-sampling of 41 landing beaches as compared to Greater Accra which is under-sampled by 19 landing beaches.

3.3. Sampling of Canoes in the District of Ghana

Ghana had 11,583 canoes in total as of 2016, according to MoFAD. Out of this total, 372 canoes were to be sampled. Based on a proper sampling procedure, at least 1 canoe should be sampled from each coastal district. We also discovered from our research that the 290 canoes from 18 districts that the enumerators collectively sample are either under- or over-sampled. Figure 4 indicates that canoes are over-sampled in approximately half of the district, with Gomoa West and Ga South oversampled by 27 and 15 canoes, respectively. However, the canoes were under-sampled by 28 and 14 canoes in Sharma and Ahanta West districts, respectively. The over-sampling of canoes was found to be attributable to misalignment of incentives: i.e., compensation for the district from which data is not being collected or the district with a smaller number of canoes.

3.4. Sampling of Canoes in the Coastal Regions of Ghana

As presented in Figure 5, there are variations in the number of canoes sampled and the expected sample in each region. Clearly, there is under-sampling, with the Western region having the highest proportion of under-sampled canoes (40 canoes) as opposed to the other regions, especially the Central region which is under-sampled by 10 canoes.

3.5. Percentage Contributions

Each coastal region's contribution to the under-sampling of canoes and landing beaches is shown in Figure 6. Using the differences between the actual sampled with the estimated sample across the coastal regions, we found that the Western region contributed the most to the under-sampling of canoes (49%) and slightly less than the Central region to the under-sampling of landing beaches (28%), with the Central region contributing the most to the under-sampling of landing beaches (35%) and the least to the under-sampling of canoes (12%).

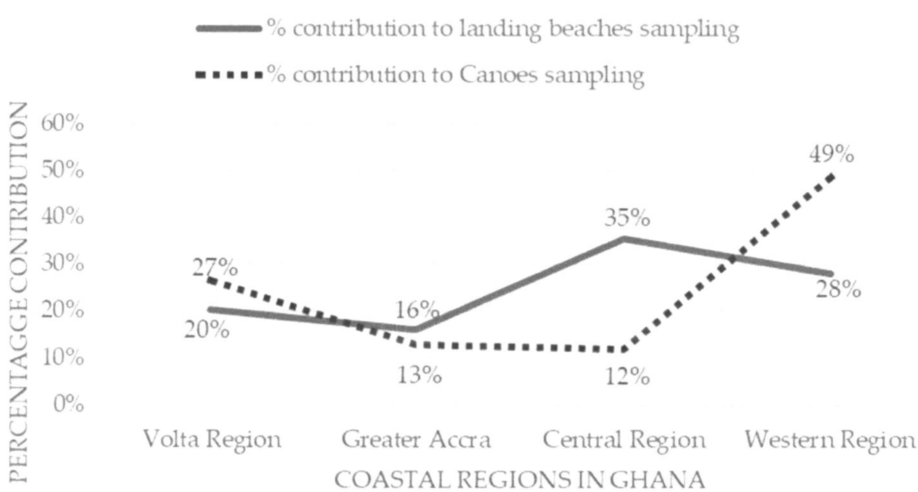

Figure 6. Percentage contribution of coastal regions to under-sampling of Canoes and Landing beaches in Ghana.

3.6. Catch Data

The four catch data categorizations (i.e., biological, ecological, economic, and social) were analyzed [14]. Each category has various components as presented in the first column of Table 1. The results revealed that none of the components under each thematic area show 100% data collection among the enumerators.

Out of the seven components under the biological category, the enumerators do not collect data on discards of fish species per canoe and the length and/or age composition of discards. However, 86% of the enumerators collect information on fishing effort. Only 17% of them indicated that they collect information on bycatch species, while 93% collect data on the price of fish and crew size of each canoe. These inconsistencies discount the reliability of the data aggregated by the FSSD for effective fisheries management.

Some of the enumerators collected data on single species and others on multiple species. The multi-species and single-species data indicate an ecosystem-based approach and a precautionary approach, respectively [25]. Due to the establishment of an ecosystem-based approach in national and international law, the authors of [26,27] suggested an ecosystem-based approach as the appropriate starting point for management; however, as indicated in Table 2, we can establish that there has been no consensus on which approach to use. About 52% of the enumerators collect multi-species (i.e., collect data by canoes and by gear) while 14% collect both single-species and multi-species (i.e., collect by either species or by both). The Fisheries Commission, on the other hand, indicated that the enumerators were instructed to collect only single-species data.

3.7. Effort Data

Regarding data on fishing effort, differences in the frequency across the enumerators were found. Comparing results from the 50 landing sites (Table 3), 86% of the enumerators collect data on the number of canoes and the type of gear, while only 16% indicated that they collect data on the size of a canoe.

4. Conclusions

From the survey on data collection practices by technical officers at the landing beaches, evidence of under-sampling and over-sampling has been found. This implies that the FAO toolkit for best practice is not being followed in practice in Ghana. This may be due to a lack of financial resources and the requisite skills to follow the desired protocol for fishery data collection. The sampling procedure deviates significantly from the ideal, which has implications for the quality of data generated.

A sample size that is too small might result in a Type I error [28], which is the likelihood of incorrectly rejecting a certain discovery when it should be accepted. Additionally, the author argued that an excessively high sample size is not appropriate due to the potential for type II error, which involves accepting a certain finding when it should be rejected. Thus, the relevant data needed for the formulation of management policies could be erroneous, thereby affecting the accuracy of the estimated catch and effort data.

The collection of catch and effort data sets and the method by which they are collected were different at some landing beaches. This discrepancy contrasts with FSSD's objective of collecting reliable data guided by scientific procedures. As noted by the authors [29], components of each thematic area should be the same at every landing site (beach) to ensure accurate data for fisheries management.

Errors in the sampling of landing beaches and canoes, as well as discrepancies in data sets gathered, could lead to the exaggeration of catch potentials, resulting in erroneous estimates of the maximum sustainable yield level (MSY) and the effort corresponding to maximum sustainable yield (F_{MSY}). These wrong estimates could lead to over-exploitation or over-capitalization of fisheries and their eventual collapse, as suggested by many studies.

To improve the quality of data collection, proper monitoring of the field enumerators should be incorporated as part of the Ministry's activities and the use of the FAO Open Data Kit (ODK) mobile phone application should be reviewed, upgraded, and its usage continued to ensure accurate collection of data. National service personnel from fisheries academic departments should also be employed to ensure better coverage of landing sites in the country. This suggestion comes with limited cost implications. In addition, there should be a balance between an understanding of the sampling techniques, the need for data, and the kind of data to be collected by the field enumerators and office staff.

Author Contributions: T.J.A.: Conceptualization, Methodology, Formal Analysis, Investigation/Field Data Collection, Writing—Original Draft, Project Administration. D.W.A.: Conceptualization, Validation, Funding, Writing—Review and Editing, Supervision. W.A.: Conceptualization, Validation, Formal Analysis, Writing—Review and Editing, Supervision. All authors have read and agreed to the published version of the manuscript.

Funding: This work was supported by the World Bank through the Africa Centre of Excellence in Coastal Resilience (ACECoR) Project at the Centre for Coastal Management, University of Cape Coast [Credit No.: 6389-GH].

Institutional Review Board Statement: As a result of the study's lack of danger, the study's sample size, and the Ministry of Fisheries and Aquaculture's consent that its staffs participate in it, ethical review and approval were waived for this study.

Informed Consent Statement: A letter was written to the Ministry of Fisheries and Aquaculture Development and to the technical officers to inform and obtain consent and consent has been obtained from them to publish this paper.

Data Availability Statement: All data herein are publicly available.

Acknowledgments: The authors would like to thank the World Bank for funding this research. We are also indebted to the Fisheries Survey Scientific Division staff and their field enumerators who dedicated their time to contribute to the research interviews. Finally, we wish to express our sincere gratitude to Mark Senanu Kudzordzi for supporting various aspects of the research.

Conflicts of Interest: The authors declare no conflict of interest.

References

1. FAO. *The State of World Fisheries and Aquaculture 2018—Meeting the Sustainable Development Goals*; FAO: Rome, Italy, 2018.
2. Béné, C.; Macfadyen, G.; Allison, E.H. *Increasing the Contribution of Small-Scale Fisheries to Poverty Alleviation and Food Security*; FAO Fisheries Technical Paper, No. 481; FAO: Rome, Italy, 2007; p. 125.
3. FAO and World Fish Center. *Small-Scale Capture Fisheries: A Global Overview with Emphasis on Developing Countries*; World Bank: Washington, DC, USA, 2008; p. 64.
4. Perry, R.I.; Sumaila, R. Marine ecosystem variability and human community responses: The example of Ghana, West Africa, Mar. *Policy* **2007**, *31*, 125–134. [CrossRef]
5. UNFAO. Fishery and Aquaculture Country Profile: The Republic of Ghana Ghana. 2019. Available online: http://www.fao.org/fishery/facp/GHA/en (accessed on 12 May 2019).
6. Ameyaw, G.A.; Tsamenyi, M.; McIlgorm, A.; Aheto, D.W. Challenges in the management of small-scale marine fisheries conflicts in Ghana. *Ocean. Coast. Manag.* **2021**, *211*, 105791. [CrossRef]
7. Afoakwah, R.; Mensah Bonsu Dan, O.; Elizabeth, E. *A Guide on Illegal Fishing Activities in Ghana*; USAID/Ghana Sustainable Fisheries Management Project; Coastal Resources Center, Graduate School of Oceanography, University of Rhode Island: Narragansett, RI, USA, 2018; 64p, Prepared by the University of Cape Coast, Ghana.
8. Amador, K.; Dovloo, E. *Report on the 2016 Ghana Marine Canoe Frame Survey*; Ministry of Fisheries and Aquaculture Development: Accra, Ghana, 2016.
9. Worm, B.; Hilborn, R.; Baum, J.K.; Branch, T.A.; Collie, J.S.; Costello, C.; Fogarty, M.J.; Fulton, E.A.; Hutchings, J.A.; Jennings, S.; et al. Rebuilding global fisheries. *Science* **2009**, *325*, 578–585. [CrossRef] [PubMed]
10. AUC-NEPAD. *The Policy Framework and Reform Strategy for Fisheries and Aquaculture in Africa*; AUC-NEPAD: Midrand, South Africa, 2014.
11. Dowling, N.A.; Dichmont, C.M.; Haddon, M.; Smith, D.C.; Smith, A.D.M.; Sainsbury, K. Guidelines for developing formal harvest strategies for data-poor species and fisheries. *Fish. Res.* **2015**, *171*, 130–140. [CrossRef]
12. Mangubhai, S.; Lawless, S. Exploring gender inclusion in small-scale fisheries management and development in Melanesia. *Mar. Policy* **2021**, *123*, 104287. [CrossRef]
13. Bradley, D.; Merrifield, M.; Miller, K.M.; Lomonico, S.; Wilson, J.R.; Gleason, M.G. Opportunities to improve fisheries management through innovative technology and advanced data systems. *Fish Fish.* **2019**, *20*, 564–583. [CrossRef]
14. Cochrane, K.L. *The Use of Scientific Information in the Design of Management Strategies*; FAO Fisheries Technical Paper, 95-130; FAO: Rome, Italy, 2002.
15. Wilson, J.R.; Lomonico, S.; Bradley, D.; Sievanen, L.; Dempsey, T.; Bell, M.; Gleason, M. Adaptive co-management to achieve climate-ready fisheries. *Conserv. Lett.* **2018**, *11*, e12452. [CrossRef]
16. De Graaf, G.; Nunoo, F.; Danson, P.O.; Wiafe, G.; Lamptey, E.; Bannerman, P. *International Training Course in Fisheries Statistics and Data Collection*; FAO Fisheries and Aquaculture Circular: Rome, Italy, 2015; (C1091).
17. Doyle, A.; Edwards, D.; Gregory, A.; Brewin, P.; Brickle, P. *T2T Montserrat. A Guide to Fisheries Data Collection and Biological Sampling*; JNCC: Peterborough, UK, 2017.
18. Nunoo, F.K.E.; Asiedu, B.; Amador, K.; Belhabib, D.; Pauly, D. *Reconstruction of Marine Fisheries Catches for Ghana, 1950–2010*; Fisheries Centre, University of British Columbia: Vancouver, BC, Canada, 2014.
19. Pauly, D.; Zeller, D. The best catch data that can possibly be? Rejoinder to Ye et al. "FAO's statistic data and sustainability of fisheries and aquaculture". *Mar. Policy* **2017**, *81*, 406–410. [CrossRef]
20. Ye, Y.; Barange, M.; Beveridge, M.; Garibaldi, L.; Gutierrez, N.; Anganuzzi, A.; Taconet, M. FAO's statistic data and sustainability of fisheries and aquaculture: Comments on Pauly and Zeller (2017). *Mar. Policy* **2017**, *81*, 401–405. [CrossRef]
21. FAO. Fishery Resources Division and Fishery Policy and Planning Division. In *Fisheries Management*; FAO Technical Guidelines for Responsible Fisheries. No. 4.; FAO: Rome, Italy, 1997; 82p.
22. De Graaf, G.J.; Stamatopulous, C.; Jarrett, T. *OPEN ARTFISH and the FAO ODK Mobile Phone Application—A Toolkit for Small Scale Fisheries Routine Data Collection*; FAO Fisheries and Aquaculture Circular: Rome, Italy, 2016; 120p.
23. Environmental Defense Fund (EDF). Fishery Solutions Catch Shares Database. 2013. Available online: http://fisherysolutionscenter.edf.org/database (accessed on 12 May 2019).
24. Yemane, M. *Elemantary Sampling Theory*; Printice-Hall Inc.: Englewood Cliffs, NJ, USA, 1967.
25. Howell, D.; Schueller, A.M.; Bentley, J.W.; Buchheister, A.; Chagaris, D.; Cieri, M.; Townsend, H. Combining ecosystem and single-species modeling to provide ecosystem-based fisheries management advice within current management systems. *Front. Mar. Sci.* **2021**, *7*, 607831. [CrossRef]

26. Collie, J.S.; Botsford, L.W.; Hastings, A.; Kaplan, I.C.; Largier, J.L.; Livingston, P.A.; Plagányi, É.; Rose, K.A.; Wells, B.K.; Werner, F.E. Ecosystem models for fisheries management: Finding the sweet spot. *Fish Fish.* **2016**, *17*, 101–125. [CrossRef]
27. Tallis, H.; Levin, P.S.; Ruckelshaus, M.; Lester, S.E.; McLeod, K.L.; Fluharty, D.L.; Halpern, B.S. The many faces of ecosystem-based management: Making the process work today in real places. *Mar. Policy* **2010**, *34*, 340–348. [CrossRef]
28. Sekaran, U. *Research Methods for Business: A Skill Building Approach*, 4th ed.; John Wiley and Sons: Hoboken, NJ, USA, 2013.
29. Fitzgerald, D.L. Landing Site Selection for UAV Forced Landings Using Machine Vision. Ph.D. Thesis, Queensland University of Technology, Brisbane, QLD, Australia, 2007.

fishes

Review

Recent Advances in Tilapia Production for Sustainable Developments in Indian Aquaculture and Its Economic Benefits

Manikandan Arumugam [1], Sudharshini Jayaraman [1], Arun Sridhar [1], Vignesh Venkatasamy [1], Paul B. Brown [2], Zulhisyam Abdul Kari [3,4], Guillermo Tellez-Isaias [5] and Thirumurugan Ramasamy [1,*]

[1] Laboratory of Aquabiotics/Nanoscience, Department of Animal Science, School of Life Sciences, Bharathidasan University, Tiruchirappalli 620024, India
[2] Department of Forestry and Natural Resources, Purdue University, West Lafayette, IN 47907, USA
[3] Department of Agricultural Sciences, Faculty of Agro-Based Industry, Universiti Malaysia Kelantan, Jeli Campus, Jeli 17600, Malaysia
[4] Advanced Livestock and Aquaculture Research Group, Faculty of Agro-Based Industry, Universiti Malaysia Kelantan, Jeli Campus, Jeli 17600, Malaysia
[5] Center of Excellence for Poultry Science, University of Arkansas, Fayetteville, NC 72701, USA
* Correspondence: ramthiru72@bdu.ac.in

Abstract: Tilapia is a widely cultured species native to Africa; these fish are prolific breeders and constitute an economically important fish species supplying higher-quality protein. To meet the global food demand and achieve the UN's Sustainable Developmental Goals (SDG), the aquaculture industry has conceived of productive solutions with the potential for adaptability, palatability, and profitability. Tilapia may play a vital role with respect to the possibility for sustainability in the nutrition and aquaculture sectors. India contributes to the promotion of aquacultural practices through a structural framework focused on agricultural, environmental, geographical, and socio-economic factors that provide opportunities for tilapia farming. Globally, the Indian aquaculture sector is currently the second largest in terms of aquacultural production but is moving toward different species that meet SDG and facilitate international marketing opportunities. The farming of aquacultural species with innovative technology constitutes an efficient use of resources. Productive research on feeding, disease management, construction, and layout helps overcome the challenges faced in aquaculture. These focused and sustained factors of the aquaculture industry offer a latent contribution to global food security. This review reports on the state of the art, the challenges regarding tilapia aquaculture in India, and the Indian government's schemes, missions, subsidies, projects and funding related to tilapia production.

Keywords: blue economy; disease management; fish nutrition; species selection; tilapia aquaculture

Key Contribution: The present review deals with the important farming strategies of tilapia aquaculture in India. Also, the policies framed by the Indian government through various programs and subsidies to expand the blue economy relating tilapia farming and their direct benefits to the aquaculture farmers were highlighted.

Citation: Arumugam, M.; Jayaraman, S.; Sridhar, A.; Venkatasamy, V.; Brown, P.B.; Abdul Kari, Z.; Tellez-Isaias, G.; Ramasamy, T. Recent Advances in Tilapia Production for Sustainable Developments in Indian Aquaculture and Its Economic Benefits. *Fishes* **2023**, *8*, 176. https://doi.org/10.3390/fishes8040176

Academic Editor: Ben Drakeford

Received: 28 February 2023
Revised: 21 March 2023
Accepted: 23 March 2023
Published: 24 March 2023

1. Introduction

Aquaculture plays a pivotal role in meeting the United Nations' SDG of alleviating poverty (SDG 1) and global hunger, ensuring food security and the provision of adequate nutrition (SDG 2), and promoting sustainable socio-economic growth (SDG 8) [1]. The farming of aquatic organisms in inland and coastal areas improves the local supply of food and the economy. Asia is the leading producer of seafood, producing at a rate of more than 6% per annum [2]. This is due to the increase in the per capita consumption of fish. To meet the SDG and provide food to those in need and economic opportunities in rural areas, culturally appropriate species of fish and production approaches must be identified. These needs are increasingly being fulfilled by tilapia. Wing-Keong et al. [3] stated that tilapia is one of the most important species of fish in aquaculture, which is capable of filling the gap of the increasing worldwide demand for protein sources. Tilapia farming is widespread, occurring in more than 135 countries and territories [4]. Production is increasing because of tilapia's large size, fast growth, prolific breeding characteristics, palatability, and relatively low cost for production [5]. Although tilapia is a freshwater species, it can tolerate osmotic and alkalinity stresses up to a particular range [6] as well as low dissolved oxygen concentrations and osmotic and alkalinity stress [7]. These fish can mature within 2–3 months of hatching and produce 75–1000 offspring every 22–40 days. Nile tilapia have been cultivated widely in many parts of the world; they are considered one of the first fish species to have been cultured and their cultivation constitutes the largest of the tilapia industries. Globally, Nile tilapia started being cultivated more than 3000 years ago [8]. The Mozambique tilapia industry is the second largest tilapia industry based on its production and exportation rates. The World Bank [9] projected that global tilapia production will reach 7.3 million tons by 2030, an increase from the 4.3 million tons reported in 2010. India's share of global fish production amounted to 5.68% from 2016–2017, corresponding to about 10.79 million tons. Tilapia are preferred over carp because of their firm, white flesh and lack of intermuscular bones. Based on their reproductive behavior, the commercial species of tilapia have been classified into three major categories: (1) maternal mouth brooders (*Oreochromis* species); (2) paternal and biparental mouth brooders (*Sarotherodon* species); and (3) substrate incubators (*Tilapia* species) [10]. The most common commercially farmed species are blue tilapia (*Oreochromis aureus*), Mozambique tilapia (*Oreochromis mossambicus*), Nile tilapia (*Oreochromis niloticus*), longfin tilapia (*Oreochromis macrochir*), redbreast tilapia (*Tilapia rendalii*), redbelly tilapia (*Coptodon zilli*), Sabaki tilapia (*Oreochromis spilurus*), three-spotted tilapia (*Oreochromis andersonii*), and Jaguar guapote (*Parachromis managuensis*). Numerous hybrids have been developed and evaluated, and monosex populations can be developed for various species. The production of various hybrids is also increasing [11]. India's contribution to the yearly annual rate of aquacultural food production amounts to 7.56%, which is greater than the global average from 2000 to 2018 [12]. Thus, this review attempts to study the state of the art and challenges of tilapia culture in India and elaborate the development of the technology that drives this critical food production system in a sustainable manner. The governmental program, Neel Kranti, also known as the blue revolution mission, is a centrally sponsored initiative with the objective of doubling the production and tripling the exportation of fish by 2022. This program began in 2014 and was designed to encourage the use of sustainable and integrated approaches for the development of aquaculture [13]. The main focus of this mission is the utilization and promotion of technological advancement in aquaculture for national food-related and nutritional security. The ultimate goal of the program is to encourage the use of sustainable and integrated approaches for the development of the fisheries sector in India. [14]. This initiative has four major components: strengthening infrastructure and security at ports and harbors, boosting skill development and training for fishermen, encouraging aquaculture, and ensuring fishermen have greater access to financial facilities. The program's infrastructure enhancement component intends to offer better facilities at ports and harbors, mobile health services, and a fishing insurance plan. This will allow fishermen to carry out their activities in a safe and secure manner and enhance their overall living conditions. The associated training program seeks to educate fishermen with respect to the optimal practices in water safety, fishing equipment maintenance, and current fishing techniques. To support aquaculture, the initiative will allow fishermen to receive improved technical support, thereby allowing them to launch their own fish-farming companies. This program will assist fishermen in purchasing higher-quality seeds and gaining access to more markets, thereby improving their revenue and providing additional job prospects in the industry. In summary, the Indian government's Neel Kranti program is a much-needed effort that can dramatically increase the country's aquacultural output and assist fishermen in securing superior economic and living conditions [15–17]. Though India is the largest producer of Indian carp (*Catla catla, Labeo rohita,* and *Cirrhinus mrigala*), the global demand and consumption of tilapia have paved the way for the enrichment of the productivity of Indian aquaculture [18]. In this regard, the government of India has set forth detailed agro- and socio-economic guidelines for the cultivation of this non-native species with the goal of protecting native inland species and their production. The guidelines encourage the amassment of collective knowledge and the undertaking of interdisciplinary efforts, including mathematical modelling systems, Internet-of-things (IoT)-based in-silico approaches, geospatial technology, fisheries and engineering technology, and management strategies, to provide an innovative and productive outcome regarding the production of tilapia from the aquaculture industry. The government of India also provides subsidies and development funds to facilitate tilapia farming based on the poverty line for farmers. In this review, specific facets such as tilapia aquaculture, the contribution of the Indian tilapia industry to global aquaculture, major production guidelines, various culturing methods, species-specific selection criteria, feed and disease management strategies, and the development of projects/schemes for tilapia production in India will be discussed.

2. Tilapia Aquaculture in India

Nile tilapia is the primary cultivable species in India. This cichlid was initially introduced in the state of Kerala, while Mozambique tilapia were imported in 1952 and stocked into reservoirs and ponds in Kerala state [19]. Due to their rapid rates of reproduction, the fish overpopulated the area and slowly migrated into the reservoirs of Tamil Nadu, Karnataka, and Rajasthan, resulting in the extinction of certain inland fish species, such as *Tor tor* and *Tor putitora.* In 2005, the Yamuna River harbored a certain quantity of Nile tilapia; due to this species' characteristic reproductive behavior, the abundance of tilapia increased in comparison with the total fish species in the river by 3.5% in 2 years (reported by the National Fisheries Developmental Board) [20]. Johnson et al. [21] reported that a drastic increase in the catch percentage of tilapia ranged from 6.7% to 85.9% from 2008 to 2018, which is expected to reach >90% according to their decadal species composition study. The experimental study also noted the species diversity of Nile tilapia from the total catch in the Halali reservoir [22]. The introduction of tilapia via a polyculture strategy also reduced the average weight of other major carp. Panikkar [23] recommended the formulation of a national policy, which led to a ban on tilapia propagation. The strict guidelines on tilapia farming in India have resulted in a renewed interest in the cultivation of several species, including *Oreochromis mossambicus, Oreochromis niloticus, Oreochromis urolepis,* and *Captodon zillii,* which are now available throughout the country [24]. Globalization, the food demand within India, and economic development opportunities precipitated the current situation, which, consequently, facilitated tilapia farming under the guidelines discussed below. The relevant regulatory entities in this regard include the Department of Fisheries, the Central Institutes for Marine and Inland Fisheries Research, the Rajiv Gandhi Centre for Aquaculture (RGCA), the National Fisheries Development Board, and other government agencies. Thus, tilapia is now farmed with sustainable farming technology by following the respective government-issued guidelines.

2.1. India's Contributions to Tilapia Production

The State of World Fisheries and Aquaculture has acknowledged the stupendous growth of the Indian fisheries sector, as it ranks, globally, fourth in terms of capture fisheries and second in terms of inland capture fisheries, contributing as much as 14% of the share of the total global inland capture [25]. The Indian government has launched a number of initiatives and projects to boost aquacultural output in the country. The Blue Revolution Plan, the National Fisheries Development Board (NFDB), and the Fish Farmers Development Agency (FFDA) are among the major projects. Reflecting and driving the global shift from capture to culture, the report underscores the fact that 57% of India's total fish production stems from aquaculture. The inland and marine sectors provide a wide range of water resources for culture and capture fisheries. In 1950–1951, India's total fish production was 0.75 million metric tons (MMT); then, it drastically increased to 9.5 MMT in 2012–2013. Moreover, the current production level has reached 16.25 MMT due to the projects and schemes funded by the Indian government [25]. The aim of the Blue Revolution Program is to boost fish output by building fish farms, hatcheries, and processing facilities. The National Fisheries Development Board promotes sustainable aquaculture methods and assists relevant businesses financially. The Fish Farmers Development Agency seeks to boost the productivity of fish farmers by offering training and assistance. These measures have resulted in tremendous development in the aquaculture sector, increasing employment and strengthening the country's export revenues [13]. The GOI aims to double the income of fishers, fish farmers, and fish workers over five years, with a 9% annual growth rate, to attain the fish production target of 22 million tons by 2025. This scheme, with reservoir fisheries as one of the focus areas, aims to create additional employment opportunities, both directly and indirectly, for six million people employed in the fisheries industry and its allied activities [26]. The Food and Agriculture Organization (FAO) has predicted that India's fish production level will grow by 26% between 2018 and 2030, which is 6.8% and 11.5% faster than the projected growth rate for Asia and the world, respectively [27].

2.2. Guidelines for Tilapia Culture in India

In aquaculture, efforts to increase the productivity of tilapia resulted in high population density, which, in turn, caused outbreaks of Tilapia Lake Virus (TLV). Although tilapia farming has resulted in adverse environmental impacts on native fish species, tilapia have also become a prominent species whose consumption allows rural communities to meet their food and nutritional requirements. Thus, the National Committee approved the introduction of exotic aquatic species such as Nile tilapia in 2006. However, farmer-friendly guidelines for tilapia were not implemented until December 2011. These guidelines were established based on the concept of the monitoring (M), control (C), and surveillance (S) of the hatchery, nursery, and farming practices of tilapia culture in India [28]. The detailed guidelines for farming tilapia in India can be found on the Department of Fisheries website maintained by the Ministry of Fisheries, Animal Husbandry, and Dairying, Government of India (www.dahd.nic.in, accessed on 20 January 2023). A Steering Committee was established at the department of fisheries at the national level to monitor tilapia seed and grow-out production. The guidelines of the committee initially dealt with cage farming, which, subsequently, requires registration and information on location, the area of culture, the type of culture and its intensity, the size of the seed to be stocked, the stocking density, and the biosecurity parameters in both cage-based and intensive culture. For subsidies and governmental funds, the guidelines should be followed strictly, with particular emphasis on stocking density and biosecurity.

3. Farming Strategies of Tilapia Culture: The State of The Art

The use of appropriate and proven farming strategies for tilapia aquaculture facilitates better yields and utilization of resources [29]. Several technological advancements are widely used in aquaculture to overcome various challenging factors, such as climate change, land availability, socio-economic concerns, and environmental barriers. Various studies have been reported concerning the strategies and efficient practices for the successful production of tilapia. These practices include Biofloc technology (BFT); backyard brackish water aquaculture; recirculatory aquaculture systems (RAS); cage culture systems for the farming of potential high-yield varieties of tilapia such as the Genetically Improved Farmed Tilapia (GIFT) strain, hybrids, and monosex populations; and Integrated Multi Tropic Aquaculture (IMTA). Polyculture (multiple species in the same production system) and integrated fish farming (fish farms integrated with terrestrial agricultural crops) provide additional income to farmers. One study reported that the integration of aquaponics with BFT applied to GIFT tilapia and bell peppers resulted in improved production without affecting growth or stress parameters [30]. This technological advancement helps overcome the challenges in the agro-aquaculture sector [31]. The integration of BFT and

RAS resulted in better resource utilization and production by providing supplementary feed for Nile tilapia [32]. Oparinde [33] developed a mathematical model to address the adaptation strategies associated with changes in the climatic conditions for aquaculture. Geographical-Information-System (GIS)- and remote-sensing-technology-based data are associated with applications for effective farming, land or site suitability assessment, or resource availability. The GIS-based (AHP-Analytical Hierarchical Process) approach facilitates geospatial mapping for the planning or construction of fish farms and the use of brackish water resources [34]. Hence, the application of these technological advancements in aquaculture paves the way for sustainable farming practices. The following strategies concern the improved farming practices applied in tilapia production.

3.1. *Recirculatory Aquaculture System (RAS)*

An RAS uses biofiltration to eliminate trash and raise oxygen levels, thus allowing for an extremely efficient and eco-friendly approach. Initially, cleansed water is treated with chemicals to remove chlorine and other hazardous compounds [35]. After this water has been treated, it then passes into the fish tanks or raceways where the fish are cultured. These tanks are often constructed with space to swim while also allowing for effective water flow through the system. The water becomes tainted with ammonia, nitrites, and nitrates as the fish create waste; this waste can be passed through a biological filter, which is a series of tanks containing beneficial microorganisms, to decrease such impurities. These bacteria convert ammonia and nitrites into nitrates, which may be utilized as fertilizer for plants. The major aerobic bacteria involved in this system belong to the genera Nitrosomonas, Nitrosococcus, Nitrosospira, or Nitrosolobus. These bacteria tend to convert nitrite to nitrate (Figure 1). The water is constantly pumped through the biological filter and back into the fish tanks, thereby ensuring that the fish have a healthy aquarium habitat. The mechanism of the recirculation system reduces water usage significantly, rendering it a more sustainable approach than standard aquacultural methods. To maintain ideal water quality, some RASs contain additional water treatment procedures, such as protein skimming, carbon dioxide level monitoring, or UV sterilization, in addition to biofiltration, thereby increasing the potency of water quality maintenance. In summary, an RAS is a highly efficient and environmentally friendly method for raising aquatic plants and animals [36–38].

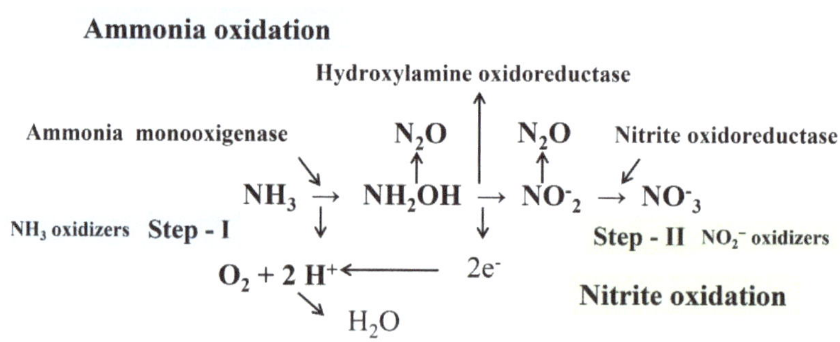

Figure 1. Reaction mechanism of ammonia–nitrite oxidation used in Recirculatory Aquaculture System (RAS).

RASs are closed systems that conserve water by recycling and are capable of affording super-intensive production levels (Figure 2). One of the plausible solutions to the water crisis and problems regarding land utilization in urban areas is RAS technology. Ye et al. [39] developed a statistically based imaging technique for tilapia farming in an RAS. Shnel et al. [40] designed the zero-discharge RAS production system for tilapia. In this method, nitrogen removal was performed by a fluidized bed reactor. A rotating biological contactor device for tilapia was used to manage water quality and remove ammonia in an RAS production system [41]. An RAS provides optimum environmental conditions year-round and may be one of the best solutions for the climatic crisis currently threatening aquaculture [42]. The production of holy basil (*Ocimum tenuiflorum*) and Nile tilapia resulted in a better growth rate of tilapia and an improved holy basil yield [43].

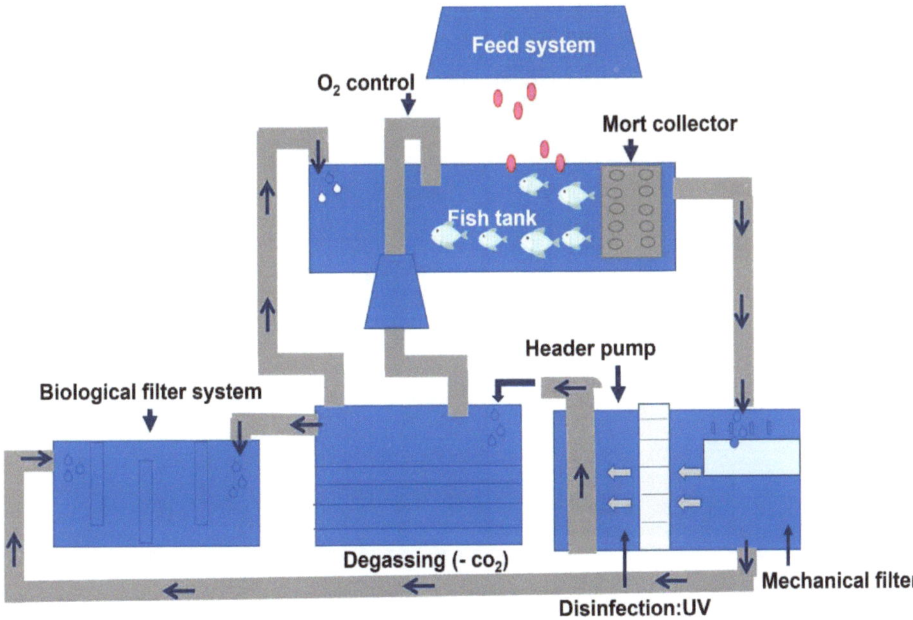

Figure 2. Typical representation of tilapia−oriented Recirculatory Aquaculture System (RAS).

3.2. Biofloc Technology (BFT)

BFT is also known as the activated suspension technique (AST); it involves the use of microbial communities to break down waste particles and transform them into a protein-rich biomass that can be easily consumed by the fish [44]. The process is reliant on the production of high levels of organic matter that results in high concentrations of suspended solids. However, these solids provide a surface for bacterial colonization, and these bacteria then serve as a food source for the fish [45]. By utilizing this biofloc technology, farmers can create a self-sustaining system that increases the efficiency of production and reduces their dependency on external inputs, thereby reducing their overall operational costs. The technology has been shown to effectively improve yields, reduce costs, and ensure the sustainable production of tilapia [46,47] (Figure 3).

This process serves as a source of food for fish [48]. The addition of carbon (C) and nitrogen (N) sources and the constant aeration and agitation of the water column result in the superior production of natural feed for the cultured aquatic species. The optimum ratio of C to N in BFT is 10:1 [49]. The use of BFT helps reduce the environmental impacts of aquaculture. The formulated diets and their ingredients can constitute an effective and sustainable farming technique for producing commercially valuable species in aquaculture [50]. Effluents from BFT can also be used in an aquaponic-based system ("flocponics"); this feasible approach enhances the growth of tilapia more than that of the plants [51]. The use of BFT has been shown to improve the quality of larvae and brood fish [52].

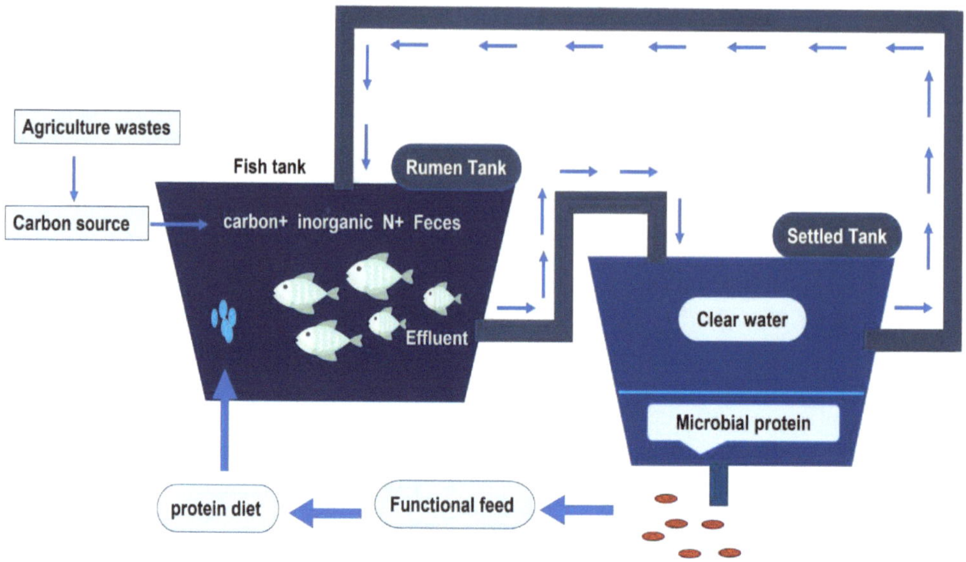

Figure 3. Typical illustration of tilapia farming using Biofloc Technology.

Based on a partial production cost analysis and a subsequent investigation, Luo et al. [53] reported that the production of GIFT tilapia was more lucrative when using BFT compared to a traditional RAS. A zero-water exchange system using BFT resulted in optimum growth and hematological and immune parameters in GIFT strain fish [54]. The indoor tank cultivation of Nile tilapia using BFT resulted in 100% survivability and an increased production rate [55]. Certain studies have suggested that the biofloc system reduces the entry of pathogens due to the recycling of nutrients and water [56,57] and that the flocs produced by this technology can enhance the amount of protein available for the tilapia to consume, leading to a reduction in the usage of feed [58]. This type of approach reduces the costs of production and generates greater profits [59].

3.3. *Cage Farming*

Cage culture in open water is another production system that is particularly well suited for the introduction of aquaculture in rural areas or for adoption by farmers with little aquaculture experience [60]. The major advantage of cage culture is that it can be implemented in existing water bodies such as rivers, lakes, ponds, seawater, etc. In addition, it provides an excellent environmental sustainability index, allowing for affordances such as lower usage of resources and reduced pollutant accumulation [61]. Formulated feed is commonly fed to fish housed in cages. In cage culture, fish require significant feed supplements, including formulated feed, to promote growth, health, and productivity. The GIFT strain is productive in cage culture systems. The use of sterile, monosex male tilapia (*Oreochromis niloticus*) is permitted in cage culture in India [62]. The farming of tilapia in ponds and cage culture is prominent and gaining popularity in India, wherein the focus is on Nile tilapia [57]. Seed, larval, and brood quality and stocking density play essential roles in the success of tilapia cage culture. Stocking density is vital for production, disease, and stress management in a fish culture environment. This intensive culture method has certain guidelines in the Indian regime for the culture of monosex tilapia, GIFT, hybridized, and hormonal sex-reversed tilapia, which have been designed to impede the prolific breeding tendency of the tilapia. Chakraborty et al. [63] evaluated the stocking density and growth of Nile tilapia in the Gangetic plains, India. They recommended a stocking density of kg/m^3 for caged-cultured mono-sex Nile tilapia in the Indian context. Another important factor in tilapia cage culture is feed management. Feeding and nutrient management in cage culture involves artificial and natural feed. Providing natural feed (phytoplankton and zooplankton) improves the nutritional quality of the fish and reduces the necessity for the supplementation of artificial feed in cage culture. Periphyton is a natural food source that is gaining popularity in cage culture as it reduces the protein requirements necessitated by commercial feed and functions as a complementary feed for the fish [64]. According to Delphino et al. [65], streptococcus-resistant tilapia cultured in cages were found to present $\leq 10\%$ mortality, which significantly increases the production rate by preventing a streptococcus infection. The major drawbacks of cage culture are environmental impacts such as the release of nitrogen, nutrients, and pollutants in waterbodies by uneaten feed [66].

3.4. Polyculture Tilapia Farming

The culturing of more than one species of aquatic organism in the same system is called polyculture. This approach facilitates the better utilization of the available natural feed in ponds by using species displaying different food habits (foraging), thereby facilitating higher fish production per unit area [67,68]. Polyculture systems can also be referred to as co-culture, multi-trophic, or integrated aquaculture farming systems. However, the systemic approaches differ in each system [69]. The primary and secondary species in a polyculture system enable cost-effective production [70]. Tilapia has a shorter growing period (a maximum of 6 months to reach 500 g in body weight) when compared to other teleost species; thus, the cultivation of tilapia with other species requires specific techniques and strategies {68]. Detailed guidelines and recommendations regarding the species cultured with tilapia polyculture facilitate better income without affecting species production. Tilapia have been successfully co-cultured with crustaceans (prawn/shrimp) and other teleost fishes such as silver carp (*Hypophthalmichthys molitrix*) and common carp (*Cyprinus carpio*) [71]. When tilapia are co-cultured with shrimp/prawns, the tilapia are able to act as an effective filter feeder by consuming zooplankton, while the leftover phytoplankton are consumed by the shrimps/prawns, thereby reducing the formation of algal blooms and enhancing economic value [72]. Hisano et al. [73] reported that the co-culturing of Nile tilapia and giant prawns (*Macrobrachium rosenbergii*) in a BFT-based RAS polyculture system resulted in better feed and protein utilization for the tilapia. However, in a polyculture system consisting of a combination of tilapia and carp, the tilapia achieved greater growth than the carp due to the reduced feed conversion ratio [74]. Similar results were obtained by Papoutsoglou et al. [75], where the ratio of 40:60 carp/tilapia production resulted in better growth with a lower FCR (Feed Conversion Ratio) and carcass lipid concentration. In fertilized ponds, the mortality rate of tilapia was higher than that of carp [76]. Previous studies have suggested that management approaches incorporating parameters such as stocking densities, species, the age of the species, and feed and niche requirements are essential in the polyculture farming of tilapia [68,77–79].

3.5. The Integrated Farming of Tilapia

Integrated fish farming involves the combination of farming fish with livestock or other terrestrial agricultural animals. In this approach, the systems are linked to each other; thus, land and water resources are efficiently used, and financial and labor costs are reduced. Integrated fish farming commonly incorporates waste or by-products from the terrestrial side for utilization on the aquatic side. The overall outcome of the integrated farming system is a high yield with low input and a limited amount of supplementary feed required for the fish [80]. Zoonotic pathogen sources and organic manure can contaminate soil and water in an IFS (environment) with dangerous chemicals and pathogens that pose a threat to human health [81]. Concerns regarding environmental risks and the bioconcentration of harmful substances should be mitigated to achieve sustainable IFSs [82]. Adverse effects on an IFS should be reduced by adopting and adapting environmentally friendly approaches that are eventually safe and hygienic and prevent further environmental degradation [83].

3.6. Integrated Multi Tropic Aquaculture (IMTA)

An expansion of the Integrated Farming System has been developed and termed integrated multi tropic aquaculture. IMTA is commonly practiced as a semi-intensive culture method that is widely used for the cultivation of animals feeding on diverse trophic grades (Figure 4). Waste nutrients are collected as sediments in this system and are utilized by other organisms. This strategy involves the use of filter feeders to remove excess feed to avoid environmental water pollution [66]. In IMTA, species from different niches consume the available resources; hence, the nutrient inputs become more efficient [72]. IMTA practices are of several kinds and have also been called Integrated Peri-Urban Aquaculture Systems (IPUASs), Integrated Agriculture Aquaculture Systems (IAASs), and Integrated Fisheries and Agriculture Systems (IFASs) [84,85]. David et al. [86] reported the results of the cultivation of Amazon River prawn (*Macrobrachium amazonicum)* and Nile tilapia using the IMTA technique. In this study, the Nile tilapia acted as a feeding organism, whereas the Amazon River prawn acted as a recycler. It has been reported that IMTA could be used as an environmental stability agent in the Sundarbans, serving as a balance between food production while also supporting the ecological security of the mangrove ecosystem [87]. When applied to floating cage systems, IMTA approaches enhance the growth and production of tilapia [88]. According to Rodrigues et al. [89], integrated farming incorporating tilapia and the Amazon River prawn results in higher growth rates when natural live feed is utilized. In an integrated farming strategy, the size of the species plays a vital role. When prawns and tilapia are cultured via integrated farming, the size of the prawn will increase due to the increased uptake of phytoplankton [90].

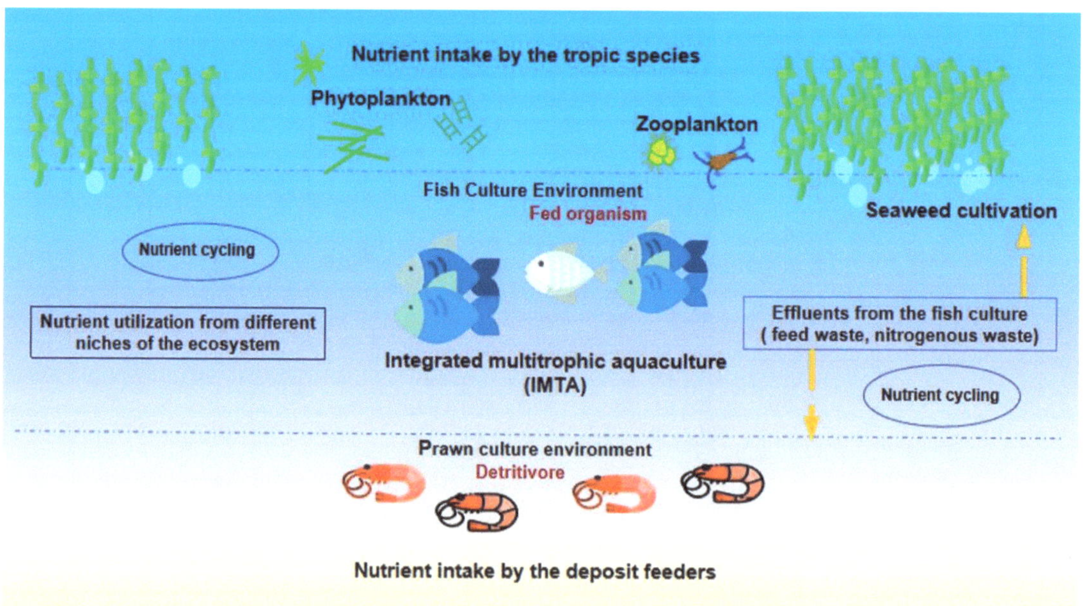

Figure 4. Graphical illustration of IMTA depicting efficient utilization of nutrient recycling system.

4. Strategies for Species Selection in Tilapia Farming

4.1. Farming of Monosex Tilapia

When undertaking the farming practices of tilapia culture, farmers face unrestrained reproduction. To overcome this limitation, monosex tilapia cultivation has been implemented [91]. The monosex production of tilapia is a rapidly growing and popular technique in the field of tilapia farming (Figure 5). This technique is widely used due to the uniform size of these fish, which are also gaining popularity among consumers [63]. The novel production and masculinization of Nile tilapia involves crossing the YY male genotype with XX females (wild). This technique is known as genetically male tilapia or YY male tilapia technology. This method is also used as a male factorial sex-determining mechanism [92]. Other methods involved in monosex production include hormonal sex reversal, interspecific hybridization, and the production of supermales and genetically improved varieties [93]. Male tilapias grow more quickly than females and use less metabolic energy to obtain a uniformly sized output; hence, these practices lead to the production of males at a higher rate for monosex populations [94]. Androgenesis, triploidy, and transgenesis methods are also available [93]. These methods have the potential to transform tilapia production by allowing farmers to produce males or females based on their preferences, removing the need to sort and eliminate fish [95]. They also have the potential to provide considerable economic advantages to farmers while contributing to the expansion of the blue economy [96]. However, the adoption of these methods raises queries about food safety, environmental effects, and ethical problems. As a result, adequate laws and standards must be implemented to reduce possible hazards related to the usage of these approaches [97]. The proper usage of these strategies may aid the expansion of the blue economy and bring economic advantages to farmers while also ensuring the industry's safety and sustainability [96]. The monosex production of tilapia is an ongoing line of research. The study conducted by Sayed and Moneeb [98] indicated that the nonsteroidal aromatase inhibitor Fadrozole could be used to produce male populations of fish. The synthetic male hormone 17α-methyltestosterone is used to reverse the sex of tilapia and produce monosex populations. Considering the negative health-related effects of using synthetic hormones for sex reversal, it has been recommended that they be substituted with pyotosterols [99]. Ghosal et al. [100] suggested that the ethanolic extract of *Basella alba* leaf and the methanolic extract of *Asparagus racemosus* can be used as safe and eco-friendly alternatives for synthetic sex reversal hormones for monosex Nile tilapia.

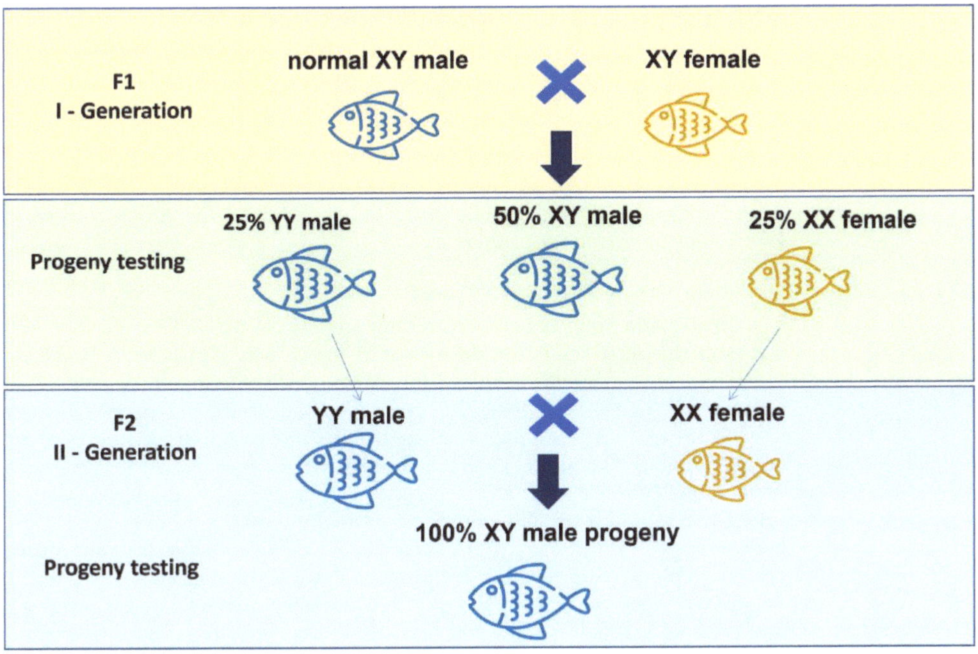

Figure 5. Male tilapia (monosex) production through YY male tilapia technology.

4.2. Farming of GIFT Tilapia

GIFT tilapia were successfully developed by the International Center for Living Aquatic Resources Management (ICLARM), which is now known as the World Fish Centre (WFC), and its allies [101]. In India, the GIFT strain is an improvement compared to other available strains. The main aim of introducing the GIFT strain is to achieve high yields, rapid growth, and high rates of survival at a low cost. The value offered by monosex GIFT strains of tilapia resulted in wide-ranging adoption in Asian countries [102]. It has been reported that the improved variety of Nile tilapia tolerates both saline and freshwater without affecting the FCR, growth, or gill conditions of the fish [103]. One study suggests that an around 27–36% faster growth rate can be achieved in GIFT tilapia compared to non-GIFT tilapia by using mono and polyculture strategies [104]. The detection and identification of potential genes for the improvement of the cultured organism offers significant potential for further improvement. Thus, genetically based selective breeding with the aid of genome sequencing and mapping will pave the way toward the improved production of GIFT tilapia. Xia et al. [105] constructed selection footprints and a genome-wide map of genetic variation in tilapia. These tools could be used to help construct new and productive varieties of genetically improved tilapia by using markers such as DNA markers, thereby constituting a significant contribution to the production of fish species.

4.3. Farming of Hybrid Tilapia

Two of the more common hybrid tilapia are the red tilapia or golden tilapia. The *Oreochromis mossambicus* × *Oreochromis niloticus* hybrid is gaining popularity because of ease with which its cultivation can be managed [106]. Beardmore et al. [93] indicated that hybridization can result in monosex populations. Based on the performance of analyzed the fish, they found the gene and regulatory pathways related to osmoregulatory tolerance in hybrid tilapia [107]. The advancement of the production of hybrid tilapia requires one to understand the genetic linkages of the parental strains (the ancestry) of these fish [108]. Gene-sequenced omics and computational investigations aid the development of productive hybrid tilapia strains. [109]. Avallone et al. [110] developed an simple and inexpensive method called Local Ancestry Inference (LAI) for a tilapia-breeding program using Digest-RAD-sequence-derid Single Nucleotide Polymorphism Markers (SNPM). The goal of their experiment was to trace ancestral genes via a fast and accurate method for the production of potential high-yield and disease-resistant varieties of hybrid tilapia. This method helps remove the unwanted traits in fish [111]. The selective breeding of tilapia to produce hybrid varieties leads to the optimal presentation of economically and environmentally favorable traits [112].

5. Management of Feed and Nutrients

To maintain optimum growth and immune functions, feed should contain energy and nutrients that meet the requirements for tilapia culture [113]. Nutrients play a vital role in the regulation of metabolism and the maintenance of homeostasis in fish [114]. Various parameters, such as body weight increase (BWI), FCR, the protein efficiency ratio (PER), specific growth rate (SGR), and weight gain (WG), are used to measure growth as a function of feed offered. Fishmeal is a major source of nutrients in fish feed. However, due to the depletion of fishmeal stock and fluctuations in its selling price, investigations are already underway to find a suitable alternative. Mostly plant-based alternatives are preferred because of their nutritional profile and abundance. Nevertheless, the antinutritional factors present in plant sources hinder the process of completely replacing fishmeal in fish feed. The dietary needs of tilapia vary based on the developmental stage, water temperature, and fish size [115]. It is critical to balance the diet with the proper macronutrients and micronutrients while avoiding overfeeding, which can cause water quality concerns such as increased fish waste and uneaten food [116]. Producers must also consider feed costs and devise feeding systems that improve economic efficiency while preserving fish growth and quality [117]. Fish feed production is extremely difficult since it frequently necessitates the exploitation of arable land to grow crops that are then transformed into fish feed [118]. This is a sizeable issue since arable land may be better employed for human food production rather than for fish feed manufacturing. The growing need for fish feed is consuming a large quantity of arable land that could otherwise be exploited to produce food for human consumption [119]. The land used to manufacture fish feed could be utilized to grow crops that could feed humans in many parts of the world, particularly in areas where food insecurity is already a serious concern [120]. As a result, it is critical to investigate sustainable alternatives for the production of fish feed, such as the utilization of insect-based protein sources [121]. This would reduce the strain on arable land and water resources while also providing a long-term source of protein for tilapia production. Researchers have conducted various studies to find plant-based alternatives to fishmeal. Nevertheless, due to antinutritional factors, plant-based alternatives have only been used to partially replace fish meal [122].

Table 1. Various feed supplements and their performance with respect to fish health.

S. No	Feed Supplement	Performance	Fish Species	References
1.	*Tridax procumbens*	Improves growth, production of antioxidants, immunity, and resistance to monogenean parasitic infection	(*Oreochromis niloticus*) Nile tilapia	[123]
2.	Caraway seed	Improves growth performance	(*Oreochromis niloticus*) Nile tilapia	[124]
3.	*Silybum marianum*	Promotes growth and enhances serum biochemical indices, antioxidant status, and gene expression	*Oreochromis niloticus* Nile tilapia	[125]
4.	*Trigonella foenum-graecum*	Improves oxidative status and immune gene expression and histopathology	(*Oreochromis niloticus*) Nile tilapia	[126]
5.	*Salvadora persica*	Improves hematoimmunological parameters and enhances antioxidant responses against *A. hydrophila* infection	(*Oreochromis niloticus*) Nile tilapia	[127]
6.	*Yucca schidigera*	Improves growth performance, hepato-renal function, and antioxidative status and effects histopathological alterations against hypoxia	(*Oreochromis niloticus*) Nile tilapia	[128]
7.	Menthol essential oil	Improves growth performance, digestive enzyme activity, immune-related genes, resistance against acute ammonia exposure	(*Oreochromis niloticus*) Nile tilapia	[129]
8.	Dietary coenzyme Q10 and Vitamin C	Enhances growth, digestive enzyme activity, immune-related genes, and resistance against acute ammonia exposure	(*Oreochromis niloticus*) Nile tilapia	[130]
9.	Soybean meal diet combined with bokashi leachate	Improves feed intake and growth performance	(*Oreochromis mossambicus* × *Oreochromis niloticus*) Red tilapia	[131]
10.	Enzymatic feather meal	Enhances growth, nutrient retention, and digestibility	(*Oreochromis niloticus* × *Oreochromis aureus*)	[132]
11.	Organic acid salt blend and protease complex combination	Improves growth and nutrient digestibility	*Oreochromis niloticus* × *Oreochromis aureus*	[133]
12.	Methylated soy protein isolates	Acts as good immune-modulating substance and improved gut health	(*Oreochromis niloticus*) Nile tilapia	[134]

Table 1. *Cont.*

S. No	Feed Supplement	Performance	Fish Species	References
13.	Whey Protein Concentrate (WPC)	Improves gut health, total weight gain, survival rate, and immune status of fish against *Aeromonas hydrophila*	(*Oreochromis niloticus*) Nile tilapia	[135]
14.	*Bacillus subtills* and *Lactobacillus plantarun*	Increases amylase (enzymatic) activity, modulates intestinal microbiota profile	(*Oreochromis niloticus*) Nile tilapia	[136]
15.	*Bacillus pumilus* and exogenous protease	Enhances growth, immunity, serum parameters, gene expression and gut bacteria	(*Oreochromis niloticus*) Nile tilapia	[137]
16.	*Enterococcus faecium*	Improves growth, hematological and biochemical parameters, and non-specific immune response	(*Oreochromis niloticus*) Nile tilapia	[138]
17.	*Aspergillus oryzae*	Improves oxidative status and immune response against hypoxia	(*Oreochromis niloticus*) Nile tilapia	[139]
18.	*Clostridium butyricum*	Improves growth, feed utilization, and gut health	*Oreochromis niloticus* × *Oreochromis aureus*	[140]
19.	Chitosan and chitosan nanoparticles	Improves health and phagocytic activity	(*Oreochromis niloticus*) Nile tilapia	[141]
20.	Zinc oxide nanoparticles	Improves health	(*Oreochromis niloticus*) Nile tilapia	[142]
21.	Dietary sodium butyrate nanoparticles	Enhances growth	(*Oreochromis niloticus*) Nile tilapia	[143]
22.	Synergized selenium and zinc oxide nanoparticles	Improves growth, hemato-biochemical profile, and immune status and reduces oxidative stress	(*Oreochromis niloticus*) Nile tilapia	[144]
23.	Cinnamon nanoparticles	Enhances antioxidant and digestive enzyme activity, growth, and health	(*Oreochromis niloticus*) Nile tilapia	[145]

Natural organisms, supplementary feed, and feed additives are widely used in commercial fish farming [146]. Depending on the culturing practices employed and the foraging behavior of the specific group, tilapia will grow rapidly when fed with fishmeal-based diets, plant-based diets, biofortified feed additives, or other natural types of feed. The use of formulated diets helps curtail unwanted chemical inputs, and the use of synthetic antibiotics naturally fosters the growth and immune status of the fish [147]. Fish meal is an excellent protein ingredient in diets but is very expensive [148]. Fish meal provides protein and essential amino acids but can also contain thiaminase, an anti-nutritional factor that can degrade thiamine [132]. The demand for fishmeal exceeds the supply and alternative protein sources are needed. Tilapia present positive results when fed with alternative protein ingredients (Table 1). Thus, feasible, balanced, low-cost, anti-nutritional-agent-free feed should be formulated for sustainable aquacultural production. Studies concerning feed formulation and nutrition technology are increasingly relying on proteomics, transcriptomics, genomics, and metabolomics to interpret the efficiency of growth- and immune-enhancing feed formulations in aquatic feed and nutrition [149].

6. Strategies for Diseases Management of Tilapia

Disease outbreaks can cause severe losses in aquaculture. Proper diagnostic advancements should be implemented to avert economic loss [150]. Several diseases are caused by poor water quality management, the high stocking of fish, and improper feeding strategies [151]. The continuous usage of antibiotics/medications leads to an increased incidence of drug-resistant bacteria; another consequent drawback is an accumulation of antibiotics in fish [152,153]. Tilapia are highly susceptible to pathogens such as bacteria, fungi, viruses, and ecto- or endoparasites or their secondary toxic metabolites. Tilapia are also highly susceptible to Motile Aeromonas Septicemia (MAS), columnaris, edwardsiellosis, francisellosis, streptococcosis, and vibriosis [5]. TLV (an ortho myxo-like virus) is a potential threat to farming and production [154], and it is ascribable to certain bacterial pathogens such as *Aeromonas, Flavobacterium,* and *Streptococcus.* Certain co-existence studies analyzed TLV and bacterial pathogens to assess the resultant epidemic disease [155]. Ectoparasites that affect tilapia farming include monogeneans (*Cichlidogyrus, Cyrodactylus* etc.,) and protozoans (*Trichodina, Vorticella*), which can result in severe monetary losses in the tilapia industry [156]. These disease-causing agents effect high mortality rates and are a menace to future production [157]. A disease outbreak in tilapia production causes adverse effects on aquaculture (Figure 6). *Streptococcus agalactiae* and *Streptococcus iniae* are the major causative agents for the endemic disease streptococcosis. This disease causes severe mortality, specifically during the summer months when the increase in water temperature favors the growth of *S. iniae* [158]. Ismail et al. [159] reported that vaccine-based diets reduce the severity of streptococcosis infection by as much as 13% and increase survival rates by up to 75%. Biocontainment measures include the quarantining of the diseased fishes, water treatment using ultraviolet light, and chemical treatment (disinfectants) to reduce the risk of diseases in the culture environment before the administration of medication. However, antibiotics, chemical agents, or chemotherapeutics are only used after the identification of sick fish [160,161]. Vaccination and improved hygiene protocols are critical to avoiding antibiotic abuse in tilapia production. Antibiotic resistance occurs when bacteria develop resistance to the actions of antibiotics, rendering them more difficult to treat. Antibiotic overuse in the fish farm industry can result in the entrance of antibiotics into the food chain, thereby potentially compromising human health. Vaccination protects farmed tilapia against infections that might harm them, thus lowering the need for antibiotics. Improved hygiene protocols can also help avert disease outbreaks by lowering the likelihood of pathogen transmission. The implementation of these strategies is assured to promote sustainable and safe tilapia farming both in terms of the environment and human health.

The sustainability of aquaculture requires the control of diseases. The government, NGOs, and various research institutes in India are focusing on this challenge and providing disease-resistant strains of fish [119]. The emerging techniques, such as the sequencing of whole genomes, provide new insights into the disease resistance of high-yield varieties of tilapia. *Oreochromis spilurus* cultured in seawater contains an antimicrobial peptide [162].

6.1. Vaccines

Fish are cold-blooded animals but respond to vaccines like warm-blooded animals [163]. Vaccinating fish may reduce the use of antibiotics in aquaculture. Duff [164] was the first to examine oral immunization against furunculosis in Atlantic salmon (*Salmo salar*). The advantage of using vaccines over antibiotics is that a vaccine stimulates the immune response and induces immunological memory, thus preventing future outbreaks by exposure to pathogens [165].

Table 2. Bacterial vaccines administered for tilapia.

S. No.	Pathogens	Type	Mode of Administration	Efficacy	Performance	References
1.	*Streptococcus iniae* improves the simulation of GALT (Gut-Associated Lymphoid Tissue) and specific antibodies	Attenuated	Intraperitoneal	79–100%	Leads to higher antibody production conferred by cell-mediated immunity	[166]
			Bath	86%	Leads to higher antibody production	
		Formalin-Inactivated	Intraperitoneal	79–100%	Provides good immunogenicity	
		DNA Vaccine Modified PCI-neo plasmid or PBS (Streptococcal α-enolase gene in pCI-neo plasmid)	Intramuscular	72.5%	Leads to increased levels of proinflammatory cytokines and *S. iniae*-specific neutralizing antibodies	[167]
2.	*Streptococcus agalactiae*	DNA Vaccine (Recombinant bacteria with surface immunogenic protein)	Oral	75%	Immunogenic	[168]
		Attenuated with erythromycin.	Intraperitoneal	82–100%	Leads to higher antibody production	[169]
3.	*Aeromonas hydrophila*	Heat-Inactivated Formalin Inactivated	Intramuscular	90%, 86.6%	Immunogenic and facilitates highest antibody production	[170]
4.	*Flavobacterium columnare*	Attenuated (Rifampicin-resistant low-virulence strains) subunit vaccine	Bath	80%	Provides good immunogenicity and cross-protection to multiple genomovar co-infections	[171]
5.	*Vibrio anguillarum*	DNA Vaccine (Recombinant flagellin A protein)	Intraperitoneal	Higher survival rate	Facilitates greater agglutination and bactericidal activity	[172]
6.	*Edwardsiella tarda*	Whole-cell formalin-inactivated + recombinant GAPDH proteins that were emulsified with Montanide adjuvant	Intraperitoneal	71.4%	Promotes greater antibody response	[172]

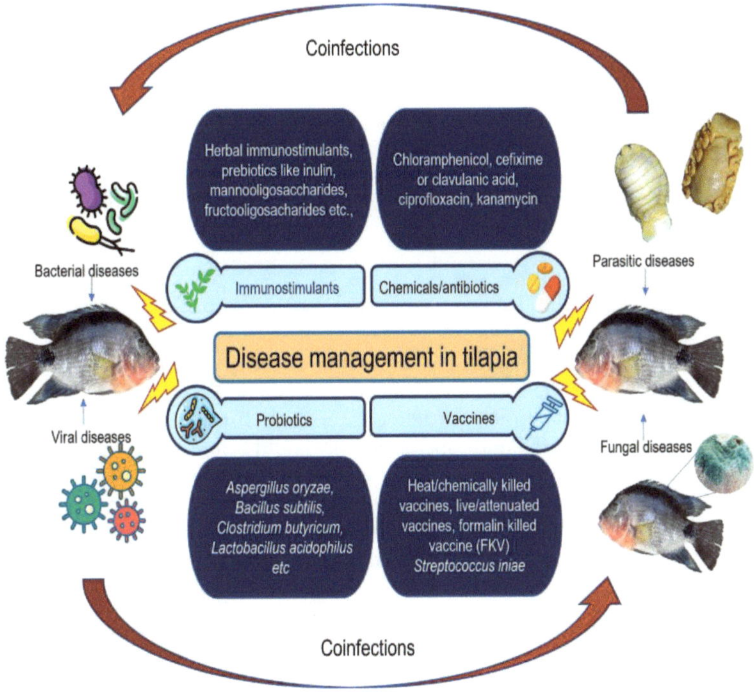

Figure 6. Steps involved in disease management in tilapia culture.

Attenuated and inactivated DNA- and RNA-type vaccines have been widely used to treat various bacterial, viral, and parasitic diseases in fish and have been experimentally tested in tilapia species. These vaccines can be monovalent, bivalent, or polyvalent [173–179]. Zhang et al. [180] suggested that understanding the mechanism of fish vaccination leads to higher defensive efficiency towards pathogens. The oral injection of an engineered formalin killed vaccine (FKV) for *Streptococcus iniae* administered to a red tilapia hybrid led to positive responses [181]. El tantawy and Ayoub. [182] reported that the inclusion of turmeric in fish feed combined with whole dead *A. hydrophila* cells led to a 100% survival rate in a group of *A. hydrophila*-infected tilapia. Table 2. shows that polyvalent vaccines consisting of formalin-inactivated *Streptococcus agalactiae, Streptococcus iniae, Enterococcus fecalis, Francisella orientalis*, and *Lactococcus garvieae* combined with the commercial adjuvant Montanide significantly increase the survival rates and immunogenicity of Nile tilapia [183,184]. However, commercial vaccines are not available in India [185].

6.2. Antibiotics

Commercial antibiotics are widely used to treat various fish diseases. Raj et al. [186] stated that *Aeromonas veronii* samples from diseased Nile tilapia exhibiting bilateral exophthalmia were sensitive to the following antibiotics: chloramphenicol, cefixime or clavulanic acid, ciprofloxacin, and kanamycin. However, the misuse or overuse of antibiotics impacts overall fish health and causes multidrug resistance in the pathogen [187]. There are also certain health concerns concerning the usage of antibiotics in aquaculture. For example, the gut microbiome of tilapia should not be altered as it promotes the growth and health status of the fish [188]. In this regard, effective technology has been developed to reduce the unwanted impacts of antibiotics by using absorbent material that delivers the antibiotics efficiently [189] (Table 3).

Table 3. Usage of antibiotics in tilapia culture.

S. No	Antibiotic	Target Disease/Causative Organisms	References
1	Oxytetracycline	Francisellosis, motile Aeromonas septicemia, and Streptococcosis	[190]
2	Florfenicol	*Aeromonas salmonicida, Aeromonas hydrophila, Flavobacterium psychrophilum, Yersinia ruckeri,* and *Vibrio anguillarum*	[191]
3	Azithromycin	*Aeromonas spp., Pseudomonas fluorescens, Vibrio anguillarum, Flavobacterium columnare, Edwardsiella tarda, Streptococcus spp.,* and *Enterococcus spp.*	[192]
4	Sulfamethoxazole	Alphaproteobacteria, cyanobacteria, Fusobacteria, and unclassified–P-proteobacteria	[193]
5	Erythromycin	Streptococcosis	[194]

Módenes et al. [195] designed a mathematical modelling system for tilapia and tetracycline using an absorbent material (biochar) capable of absorbing this antibiotic and serving as a potential delivery method. It was shown that the use of a combination of natural compounds and antibiotics could be a method for reducing antibiotic resistance and other adverse effects. This study confirmed that rutin obtained from Citrus sinensis, a flavonoid compound rich in Vitamin P, combined with gentamicin exhibited better antibacterial and anti-biofilm effects against *Pseudomonas aeruginosa* [196]. The study also reported that rutin and the antibiotic florfenicol possess potential antibacterial and anti-biofilm properties both in vitro and in vivo against *Aeromonas hydrophila* [197].

6.3. Immunostimulants

Herbal plants are promising agents as they stimulate fish immunity at low doses without any side effects [198,199]. Their potential immunostimulants have significant natural characteristics, such as possessing low molecular weight, being water-soluble and amphoteric, and containing nitrogen molecules [200]. Immunostimulants in the form of chemicals, drugs, and natural compounds from plants and other sources can activate the host defense mechanisms against various disease-causing pathogens (Table 4). Bricknell and Dalmo [201] reported that immunostimulants boost the immune system of fish during larval development. Meena et al. [202] reported that beta-glucan can be used as a potential immunostimulant in aquaculture as it enhances the immune performance of fish. Beta-glucan and other immunosaccharides such as inulin, mannooligosaccharide, and fructooligosaccharide are widely used immunostimulants and are considered prebiotics. Immunostimulants, or immunopotentiators, improve the adaptive and innate immune system of the host [203]. Immunostimulants also serve as eco-friendly feed additions that can enhance a fish's growth and immune performance.

Table 4. Usage of herbal immunostimulants in tilapia culture.

S. No	Immunostimulant	Organism	Performance	References
1	Turmeric *(Curcuma longa)*	Nile tilapia *(Oreochromis niloticus)*	Enhances growth, immunity, and antioxidant status	[204]
2	Pumpkin seed meal *(Cucurbita mixta)*	Mossambique tilapia *(Oreochromis mossambicus)*	Enhances growth, immune, and disease resistance activity	[205]
3	Velvet bean *(Mucuna pruriens)*	Mossambique tilapia *(Oreochromis mossambicus)*	Enhances innate immunity and growth performance	[206]
4	Ashwagandha *(Withania somnifera)*	Nile tilapia *(Oreochromis niloticus)*	Provides an immuno-therapeutic effect	[207]

Table 4. *Cont.*

S. No	Immunostimulant	Organism	Performance	References
5	Mangrove (*Excoecaria agallocha*)	Red hybrid tilapia (*Oreochromis niloticus*)	Enhances non-specific immune responses and disease resistance	[208]
6	Guava (*Psidium guajava*)	Nile tilapia (*Oreochromis niloticus*)	Enhances growth, nutrient utilization, and immune system	[209]
7	African wormwood (*Artemisia afra*)	Mossambique tilapia (*Oreochromis mossambicus*)	Enhances growth and disease resistance	[210]
8	Chamomile (*Matricaria chamomilla*)	Nile tilapia (*Oreochromis niloticus*)	Enhances growth and immune parameters	[211]
9	Spanish dagger (*Yucca schidigera*)	Nile tilapia (*Oreochromis niloticus*)	Enhances growth, hematology, nonspecific immune responses, and disease resistance	[212]
10	Oregano (*Origanum vulgare*)	Red belly tilapia (*Coptodon zillii*)	Enhances innate immunity	[213]
11	Peppermint (*Mentha piperita*)	Nile tilapia (*Oreochromis niloticus*)	Enhances hemato-immunological parameters	[214]

Bustamam et al. [215] reported that a 2.5% inclusion of *Isochrysis galbana* (IG) supplemented as a dietary immunostimulant enhances the immune system of red hybrid tilapia. It also increases the abundance of certain secondary metabolites such as glutamate, isoleucine, and tyrosine. Notably, immunostimulants tend to alter the metabolomics of the fish, which alters their metabolism [216].

6.4. Probiotics

Live microorganisms that can improve host health are collectively referred to as probiotics. The common probiotics used in aquaculture include the *Aeromonas, Bacillus, Clostridium, Cornybacterium, Enterococcus, Enterobacter, Lactobacillus, Lactococcus, Pseudomonas, Shewanella, Saccharomyces,* and *Vibrio* species [122]. These potential probiotics tend to enhance the growth and immune system of fish [217]. Essa et al. [218] reported that tilapia growth performance and the activity of digestive enzymes such as amylase, protease, and lipase were improved by providing *Bacillus subtilis* and *Lactobacillus plantarum* or a mixture of yeast (*Saccharomyces cerevisiae*) as an alternative feed. Moreover, these probiotics were associated with the gut microbiota and enhanced the enzymes that hydrolyze macronutrients for the better digestion and absorption of nutrients [115]. Ghosh et al. [219] investigated the probiotic and antipathogenic nature of *Bacillus sp.* Banerjee and Ray. [220] experimented with the antagonistic effects of *Bacillus megatarium* in the intestine of tilapia. Certain species of *Bacillus* can degrade cellulose. *Bacillus circulans* isolated from the gut of tilapia increased the fermentation of cellulose [221]. Lara-Flores et al. [222] stated that probiotics incorporated in a diet consisting of 40% or 27% crude protein improved feed conversion ratios and weight gain compared to a control diet. Probiotics not only promote growth but also improve the immune system, disease resistance, and survival rate of tilapia. Aly et al. [223] fed sample fish a mixture of *Bacillus subtilis* and *Lactobacillus acidophilus* as a probiotic, which resulted in a significantly higher survival rate in Nile tilapia. Samat et al. [224] attempted the administration of a probiotic via live feed. *Moina micrura* was used as the live feed and *Bacillus pocheonensis* as the probiotic. This combination resulted in the improved health and survival of the fish. Ringo et al. [225] reported that *Bacillus amyloliquefaciens* supplemented as a probiotic in feed for tilapia modifies the gut microbiome and enriches the production of secondary metabolites. The major criteria for the supplementation of probiotics to fish vary based on the species and depend on the concentration, mode of administration, etc. [122].

7. Projects Developed for The Production of Tilapia in India

Governing bodies such as the National Fisheries Development Board (NFDB) and the Rajiv Gandhi Center for Aquaculture (RGCA) have given sustained and focused priority to the fisheries sector through policies and financial support designed to support small-scale farmers, women, and various centers in order to achieve sustainable fish production in India (Table 5). The RGCA, in association with the World Fish Centre (WFC), developed genetically improved varieties of tilapia for the betterment of fish farmers and small householders, thereby helping to promote tilapia farming and improve local economies in the country. The WFC also focuses on sustainable and logical breeding programs for the tilapia industry in India [28,226].

Table 5. Projects and schemes for tilapia culture in India.

S. No.	Governing Body/Funding Agencies	Project	Target Fish Species
1.	NFDB	Brackish water cage culture for sustainable aquaculture in coastal regions of India	Milk Fish (*Chanos chanos*), Asian seabass (*Lates calcarifer*), grey mullet (*Mugil cephalus*), pearlspot (*Etroplus suratensis*), Nile tilapia (*Oreochromis niloticus*), silver pompano (*Trachinotus blochii*)
		Demonstration of azolla production for tilapia feed supplement in Madhavaram, TNJFU Campus, Tamil Nadu	GIFT Tilapia
		Backyard Recirculatory Aquaculture System	Monosex tilapia, *Pangasius valenciennes*
2.	RGCA working in association with (WFC) to enhance the genetic strains of tilapia.	Establishment of a satellite nucleus of the GIFT strain at RGCA to support tilapia production in India: Phase I (2011–2016) Establishment of a satellite nucleus of the GIFT strain at RGCA, India: Phase II (2019–2023)	GIFT Tilapia

The Indian government's policies and goals for the fisheries sector have been strengthened by FAO activities. The Bay of Bengal Program (BOBP), a regional fisheries program created by FAO, is centered in Chennai, India [13]. Through collaboration with global aquaculture and fisheries allies, India is contributing to the share of global public goods, including by sharing its expertise in agriculture (aquaculture) and rural development with other developing countries. In 2022, the Indian government launched Pradhan Mantri Matsya Sampada Yojana (PMMSY) to form a blue revolution by enhancing the sustainable development of fisheries and aquaculture (Figure 7). This program creates various employment opportunities. In addition, this program is collaborating with various private organizations such as Fountainhead Agro Farms Private Limited to enrich the production of tilapia using Israeli technology.

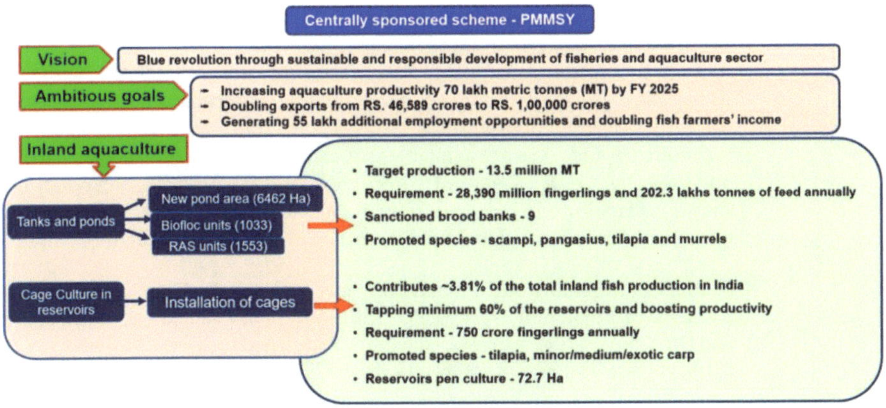

Source: Inland fisheries in India, Department of Fisheries, Government of India (https://dof.gov.in/inland-fisheries)

Figure 7. The schematic representation of aims and expected outcomes of centrally sponsored scheme (PMMSY) with the potential to increase aquacultural productivity, exportation, and employment opportunities.

8. Blue Economy—Future Perspectives

The blue economy is critical to tilapia production and is predicted to grow dramatically in the upcoming years. The tilapia sector is under pressure to enhance its productivity while reducing environmental concerns as the demand for food increases [227]. With the global population estimated to exceed 9 billion by 2050, the tilapia sector will be critical in fulfilling the increasing need for protein. The future of tilapia production is bright because of several elements that support the rise of the blue economy [228]. This review has highlighted the critical factors defining the tilapia industry's future. To begin with, technological advances in tilapia farming have revolutionized the sector. Conventional agricultural practices are no longer appropriate for today's commercial market needs. The Blue Economy has created new prospects for international commerce, which has led to greater growth in the tilapia sector. Foreign investment is being driven by rising global consumer demand, and trade agreements are simplifying market access for many nations. Furthermore, the increasing demand for live fish, such as tilapia, gives providers additional potential to develop the market beyond the commonly sold frozen fish. In summary, the future of tilapia production from the perspective of the blue economy seems promising. Technological advancements, advances in fish feed production, shifting consumer habits, and possibilities regarding international commerce are all contributing to this expansion. The industry's sustained growth should help to drive economic development and food security by fulfilling the rising consumer demand for healthy, sustainable foods.

9. Conclusions

In India, aquaculture is a promising economic activity and a rising sector with wide resources and potential. The vibrancy of the aquaculture sector could be visualized as a drastic advancement in the field of aquaculture, which India has achieved in past decades. Tilapia significantly contributes to the total share of aquaculture exports in India, which boosts the country's economy. With recent breakthroughs in aquacultural technology and improvements in the diets of tilapia, there has been constant advancement in tilapia output, leading to the sustainable development of Indian aquaculture. Tilapia cultivation may be an economically feasible choice for aquaculture production in various locations of India, as long as suitable investment and management practices are employed. The Rajiv Gandhi Centre for Aquaculture (MPEDA, Ministry of Commerce and Industry, Government of India) has established a tilapia project and breeding program focused on the use of potential GIFT strains to improve production conditions in India in collaboration with the WFC,

Malaysia. The NFDB of India was formed in 2006 and is an autonomous body under the Ministry of Fisheries, Animal Husbandry, and Dairying of the Government of India that seeks to promote and encourage tilapia farming. Still, farmers are facing difficulties related to disease management while culturing tilapia, necessitating the provision of vaccines for longer-term protection and low-cost vaccines that increase mucosal immunity. Various technologies and tools are available that can support the future of aquacultural production and the betterment of the country's economy and food supply. The policy making regarding tilapia aquaculture in India not only aspires to promote economic value but also concerns ensuring national and global food security, diminishing malnutrition, and reducing poverty.

Author Contributions: All authors contributed to this study's inception and design. M.A. and S.J. wrote the first draft of the manuscript, which was reviewed and edited by A.S., V.V., P.B.B., Z.A.K. and G.T.-I. Finally, T.R. reviewed, edited, and supervised the work. All authors have read and agreed to the published version of the manuscript.

Funding: Research funding was supported in part by funds provided by USDA-NIFA Sustainable Agriculture Systems, Grant No. 2019-69012-29905. Title of the project: Empowering US Broiler Production for Transformation and Sustainability USDA-NIFA (Sustainable Agriculture Systems): No. 2019-69012-29905.

Data Availability Statement: Data will be made available on reasonable request.

Acknowledgments: The first author is grateful to Bharathidasan University for providing University Research Fellowship (Ref. No. 026525/URF/DIR-RES/2020; dt: 04.01.2020). The authors (M.A., S.J., A.S., V.V. and T.R.) are thankful for UGC-SAP-DRS-II (F.3-9/2013(SAP-II), the Department of Science and Technology-Fund for the Improvement of Science and Technology Infrastructure (DST-FIST) Level-I (stage-II) (Ref. No. SR/FST/LSI-647/2015(C); Date.11.08.2016) and the Department of Science and Technology for the Promotion of University Research and Scientific Excellence (DST PURSE Phase - II) (Ref. No. SR/PURSE PHASE 2/16(G)/and 16(C); Date. 21.02.2017) of the Department of Animal Science, Bharathidasan University, for allowing our use of their instrumentation facility. The authors also thank "RUSA, 2.0-Biological Sciences, Bharathidasan University". The authors greatly acknowledge Paul B. Brown for critically reviewing the manuscript.

Conflicts of Interest: The authors declare they have no conflict of interest.

References

1. World Health Organization. *The State of Food Security and Nutrition in the World 2018: Building Climate Resilience for Food Security and Nutrition*; Food & Agriculture Org.: Rome, Italy, 2018.
2. Tacon, A.G.J. Trends in global aquaculture and aquafeed production: 2000–2017. *Rev. Fish. Sci. Aquac.* **2020**, *28*, 43–56. [CrossRef]
3. Ng, W.-K.; Chong, C.-Y.; Wang, Y.; Romano, N. Effects of dietary fish and vegetable oils on the growth, tissue fatty acid composition, oxidative stability and vitamin E content of red hybrid tilapia and efficacy of using fish oil finishing diets. *Aquaculture* **2013**, *372–375*, 97–110. [CrossRef]
4. Sunarto, A.; Grimm, J.; McColl, K.A.; Ariel, E.; Nair, K.K.; Corbeil, S.; Hardaker, T.; Tizard, M.; Strive, T.; Holmes, B. Bioprospecting for biological control agents for invasive tilapia in Australia. *Biol. Control* **2022**, *174*, 105020. [CrossRef]
5. Prabu, E.; Rajagopalsamy, C.; Ahilan, B.; Jeevagan, I.; Renuhadevi, M. Tilapia—An excellent candidate species for world aquaculture: A review. *Annu. Res. Rev. Biol.* **2019**, *31*, 1–14. [CrossRef]
6. Suresh, A.V.; Lin, C. Effect of stocking density on water quality and production of red tilapia in a recirculated water system. *Aquac. Eng.* **1992**, *11*, 1–22. [CrossRef]
7. Zhao, Y.; Zhang, C.; Zhou, H.; Song, L.; Wang, J.; Zhao, J. Transcriptome changes for Nile tilapia (Oreochromis niloticus) in response to alkalinity stress. *Comp. Biochem. Physiol. Part D: Genom. Proteom.* **2020**, *33*, 100651. [CrossRef]
8. Popma, T.J.; Lovshin, L.L. *Worldwide Prospects for Commercial Production of Tilapia*; International Center for Aquaculture and Aquatic Environments: Auburn, AL, USA, 1996.
9. World Bank. 2013. Available online: https://documents1.worldbank.org/curated/en/458631468152376668/pdf/831770WP0P1126OES003000Fish0to02030.pdf (accessed on 20 January 2023).
10. Popma, T.; Masser, M. Tilapia Life History and Biology. 1999. Available online: http://hdl.handle.net/20.500.12364/145 (accessed on 20 January 2023).
11. Kamal, A.H.M.M.; Mair, G.C. Salinity tolerance in superior genotypes of tilapia, Oreochromis niloticus, Oreochromis mossambicus and their hybrids. *Aquaculture* **2005**, *247*, 189–201. [CrossRef]

12. Menaga, M.; Fitzsimmons, K. Growth of the tilapia industry in India. *World Aquac.* **2017**, *48*, 49–52. Available online: https://www.researchgate.net/profile/Kevin-Fitzsimmons-4/publication/319681169_Growth_of_the_tilapia_industry_in_India/links/59b9501d458515bb9c486c4f/Growth-of-the-tilapia-industry-in-India.pdf (accessed on 20 January 2023).
13. Shenoy, L.; Rajpathak, S. *Sustainable Blue Revolution in India: Way Forward*; CRC Press: Boca Raton, FL, USA, 2021. [CrossRef]
14. Lakra, W.S.; Gopalakrishnan, A. Blue revolution in India: Status and future perspectives. *Indian J. Fish.* **2021**, *68*, 137–150. [CrossRef]
15. Mishra, S.S.; Rakesh, D.; Dhiman, M.; Choudhary, P.; Debbarma, J.; Sahoo, S.N.; Mishra, C.K.I. Present status of fish disease management in freshwater aquaculture in India: State-of-the-art-review. *J. Aquac. Fish.* **2017**, *1*, 14. [CrossRef]
16. Ngasotter, S.; Panda, S.P.; Mohanty, U.; Akter, S.; Mukherjee, S.; Waikhom, D.; Devi, L.S. Current scenario of fisheries and aquaculture in India with special reference to Odisha: A review on its status, issues and prospects for sustainable development. *Int. J. Bio-Resour. Stress Manag.* **2020**, *11*, 370–380. [CrossRef]
17. Singh, A. Towards an integrated 'Blue Economy'framework in the Bay of Bengal. *ORF Issue Brief* **2020**, *411*. Available online: https://orfonline.org/wp-content/uploads/2020/10/ORF_IssueBrief_411_BIMSTEC-BlueEconomy.pdf (accessed on 20 January 2023).
18. Naylor, R.L.; Hardy, R.W.; Buschmann, A.H.; Bush, S.R.; Cao, L.; Klinger, D.H.; Little, D.C.; Lubchenco, J.; Shumway, S.E.; Troell, M. A 20-year retrospective review of global aquaculture. *Nature* **2021**, *591*, 551–563. [CrossRef] [PubMed]
19. Sugunan, V.V.; Katiha, P.K. Impact of stocking on yield in small reservoirs in Andhra Pradesh, India. *Fish. Manag. Ecol.* **2004**, *11*, 65–69. [CrossRef]
20. National Fisheries Development Board (2015) Guidelines for Seaweed Cultivation. Available online: http://nfdb.gov.in/PDF/Draft%20National%20Policy%20on%20Mariculture%20NFDB%20-%20CMFRI%2028%20Sept.2018.pdf (accessed on 20 January 2023).
21. Johnson, C.; Sarkar, U.K.; Koushlesh, S.K.; Das, A.K.; Das, B.K.; Naskar, B.K. Population structure of Nile tilapia and its impact on fisheries of a tropical impacted reservoir, Central India. *Environ. Sci. Pollut. Res.* **2020**, *27*, 29091–29099. [CrossRef]
22. Johnson, C.; Sarkar, U.K.; Koushlesh, S.K.; Das, A.K.; Das, B.K.; Naskar, B.K. Fish assemblage, ecosystem status and potential impact of Nile Tilapia in Halali Reservoir of Central India. *Environ. Dev. Sustain.* **2022**, *24*, 7753–7775. [CrossRef]
23. Panikkar, N.K. Possibilities of further expansion of fish and prawn cultural practices in India. *Curr. Sci.* **1952**, *21*, 29–33. Available online: http://eprints.cmfri.org.in/id/eprint/6916 (accessed on 20 January 2023).
24. Singh, A.K.; Lakra, W.S. Risk and benefit assessment of alien fish species of the aquaculture and aquarium trade into India. *Rev. Aquac.* **2011**, *3*, 3–18. [CrossRef]
25. FAO. The State of Food and Agriculture. 2022. Available online: https://www.fao.org/publications/sofa/2022/en/ (accessed on 20 January 2023). [CrossRef]
26. MoFAHD; Ministry of Fisheries; Animal Husbandry and Dairying. Basic Framework of Pradhan Mantri Matsya Sampada Yojana (PMMSY)—A Scheme to Bring about Blue Revolution through Sustainable and Responsible Development of Fisheries Sector in India. 2020. Available online: http://dof.gov.in/sites/default/filess/Annexure%20Framework%20to%20statesUT%20%282%29.pdf (accessed on 20 January 2023).
27. FAO. Tracking Progress on Food and Agriculture-Related SDG Indicators 2022. Rome—A Report on the Indicators under FAO Custodianship. 2022. Available online: https://www.fao.org/sustainable-development-goals/indicators/en/ (accessed on 20 January 2023).
28. Cullhaj, M.; Gutierrez, A. Establishment of a Satellite Nucleus of the GIFT Strain at Rajiv Gandhi Center for Aquaculture (RGCA), India: Phase II. Project Brief April 2019–March 2020. 2020. Available online: https://hdl.handle.net/20.500.12348/4134 (accessed on 20 January 2023).
29. Thomas, M.; Pasquet, A.; Aubin, J.; Nahon, S.; Lecocq, T. When more is more: Taking advantage of species diversity to move towards sustainable aquaculture. *Biol. Rev.* **2021**, *96*, 767–784. [CrossRef]
30. Saseendran, S.; Dube, K.; Chandrakant, M.; Rani, A.B. Enhanced growth response and stress mitigation of genetically improved farmed Tilapia in a biofloc integrated aquaponic system with bell pepper. *Aquaculture* **2021**, *533*, 736200. [CrossRef]
31. Pinho, S.M.; Lima, J.P.; David, L.H.; Emerenciano, M.G.C.; Goddek, S.; Verdegem, M.C.J.; Keesman, K.J.; Portella, M.C. FLOCponics: The integration of biofloc technology with plant production. *Rev. Aquac.* **2022**, *14*, 647–675. [CrossRef]
32. Hisano, H.; Barbosa, P.T.L.; Hayd, L.A.; Mattioli, C.C. Evaluation of Nile tilapia in monoculture and polyculture with giant freshwater prawn in biofloc technology system and in recirculation aquaculture system. *Int. Aquat. Res.* **2019**, *11*, 335–346. [CrossRef]
33. Oparinde, L.O. Fish farmers' welfare and climate change adaptation strategies in southwest, Nigeria: Application of multinomial endogenous switching regression model. *Aquac. Econ. Manag.* **2021**, *25*, 450–471. [CrossRef]
34. Jayanthi, M.; Thirumurthy, S.; Samynathan, M.; Kumararaja, P.; Muralidhar, M.; Vijayan, K. Multi-criteria based geospatial assessment to utilize brackishwater resources to enhance fish production. *Aquaculture* **2021**, *537*, 736528. [CrossRef]
35. Shitu, A.; Liu, G.; Muhammad, A.I.; Zhang, Y.; Tadda, M.A.; Qi, W.; Liu, D.; Ye, Z.; Zhu, S. Recent advances in application of moving bed bioreactors for wastewater treatment from recirculating aquaculture systems: A review. *Aquac. Fish.* **2021**, *7*, 244–258. [CrossRef]
36. Losordo, T.M.; Westerman, P.W. An analysis of biological, economic, and engineering factors affecting the cost of fish production in recirculating aquaculture systems. *J. World Aquac. Soc.* **1994**, *25*, 193–203. [CrossRef]

37. Verdegem, M.C.J. Nutrient discharge from aquaculture operations in function of system design and production environment. *Rev. Aquac.* **2013**, *5*, 158–171. [CrossRef]
38. Deng, M.; Dai, Z.; Senbati, Y.; Li, L.; Song, K.; He, X. Aerobic denitrification microbial community and function in zero-discharge recirculating aquaculture system using a single biofloc-based suspended growth reactor: Influence of the carbon-to-nitrogen ratio. *Front. Microbiol.* **2020**, *11*, 1760. [CrossRef]
39. Ye, Z.; Zhao, J.; Han, Z.; Zhu, S.; Li, J.; Lu, H.; Ruan, Y. Behavioral characteristics and statistics-based imaging techniques in the assessment and optimization of tilapia feeding in a recirculating aquaculture system. *Trans. ASABE* **2016**, *59*, 345–355. [CrossRef]
40. Shnel, N.; Barak, Y.; Ezer, T.; Dafni, Z.; van Rijn, J. Design and performance of a zero-discharge tilapia recirculating system. *Aquac. Eng.* **2002**, *26*, 191–203. [CrossRef]
41. Brazil, B.L. Performance and operation of a rotating biological contactor in a tilapia recirculating aquaculture system. *Aquac. Eng.* **2006**, *34*, 261–274. [CrossRef]
42. Martins, C.; Eding, E.; Verdegem, M.; Heinsbroek, L.; Schneider, O.; Blancheton, J.; D'Orbcastel, E.R.; Verreth, J. New developments in recirculating aquaculture systems in Europe: A perspective on environmental sustainability. *Aquac. Eng.* **2010**, *43*, 83–93. [CrossRef]
43. Angkha, B.; Verma, A.K.; Kumar, S.H.; Prakash, C.; Thomas, R.M. Mobilization of mica by Bacillus sp. and its effect on Nile tilapia (Oreochromis niloticus) cum holy basil (Ocimum tenuiflorum)–based aquaponic system. *Aquac. Int.* **2020**, *28*, 2045–2058. [CrossRef]
44. Nisar, U.; Peng, D.; Mu, Y.; Sun, Y. A solution for sustainable utilization of aquaculture waste: A comprehensive review of biofloc technology and aquamimicry. *Front. Nutr.* **2022**, *8*, 1231. [CrossRef]
45. Khanjani, M.H.; Sharifinia, M. Biofloc technology as a promising tool to improve aquaculture production. *Rev. Aquac.* **2020**, *12*, 1836–1850. [CrossRef]
46. Choo, H.X.; Caipang, C.M.A. Biofloc technology (BFT) and its application towards improved production in freshwater tilapia culture. *Aquacult. Aquar. Conserv. Legis.* **2015**, *8*, 362–366.
47. Green, B.W.; Rawles, S.D.; Schrader, K.K.; Gaylord, T.G.; McEntire, M.E. Effects of dietary protein content on hybrid tilapia (Oreochromis aureus × O. niloticus) performance, common microbial off-flavor compounds, and water quality dynamics in an outdoor biofloc technology production system. *Aquaculture* **2019**, *503*, 571–582. [CrossRef]
48. Hargreaves, J.A. Photosynthetic suspended-growth systems in aquaculture. *Aquac. Eng.* **2006**, *34*, 344–363. [CrossRef]
49. Avnimelech, Y. Carbon/nitrogen ratio as a control element in aquaculture systems. *Aquaculture* **1999**, *176*, 227–235. [CrossRef]
50. Ng, W.K.; Romano, N. A review of the nutrition and feeding management of farmed tilapia throughout the culture cycle. *Rev. Aquac.* **2013**, *5*, 220–254. [CrossRef]
51. Pinho, S.M.; Molinari, D.; de Mello, G.L.; Fitzsimmons, K.M.; Emerenciano, M.G.C. Effluent from a biofloc technology (BFT) tilapia culture on the aquaponics production of different lettuce varieties. *Ecol. Eng.* **2017**, *103*, 146–153. [CrossRef]
52. Ekasari, J.; Rivandi, D.R.; Firdausi, A.P.; Surawidjaja, E.H.; Zairin, M., Jr.; Bossier, P.; De Schryver, P. Biofloc technology positively affects Nile tilapia (Oreochromis niloticus) larvae performance. *Aquaculture* **2015**, *441*, 72–77. [CrossRef]
53. Luo, G.; Gao, Q.; Wang, C.; Liu, W.; Sun, D.; Li, L.; Tan, H. Growth, digestive activity, welfare, and partial cost-effectiveness of genetically improved farmed tilapia (Oreochromis niloticus) cultured in a recirculating aquaculture system and an indoor biofloc system. *Aquaculture* **2014**, *422–423*, 1–7. [CrossRef]
54. Long, L.; Yang, J.; Li, Y.; Guan, C.; Wu, F. Effect of biofloc technology on growth, digestive enzyme activity, hematology, and immune response of genetically improved farmed tilapia (Oreochromis niloticus). *Aquaculture* **2015**, *448*, 135–141. [CrossRef]
55. Azim, M.E.; Little, D.C. The biofloc technology (BFT) in indoor tanks: Water quality, biofloc composition, and growth and welfare of Nile tilapia (Oreochromis niloticus). *Aquaculture* **2008**, *283*, 29–35. [CrossRef]
56. Menaga, M.; Felix, S.; Charulatha, M.; Gopalakannan, A.; Panigrahi, A. Effect of in-situ and ex-situ biofloc on immune response of Genetically Improved Farmed Tilapia. *Fish Shellfish. Immunol.* **2019**, *92*, 698–705. [CrossRef]
57. El-Sayed, A.F.M. *Tilapia Culture*; Academic Press: Cambridge, MA, USA, 2019. [CrossRef]
58. Khanjani, M.H.; Alizadeh, M.; Mohammadi, M.; Aliabad, H.S. The Effect of Adding Molasses in Different Times on Performance of Nile Tilapia (*Oreochromis niloticus*) Raised in a Low-Salinity Biofloc System. *Ann. Anim. Sci.* **2021**, *21*, 1435–1454. [CrossRef]
59. Vicente, L.R.M.; Owatari, M.S.; Mouriño, J.L.P.; Silva, B.C.; Vieira, F.D.N. Nile tilapia nursery in a biofloc system: Evaluation of different stocking densities. *Bol. Inst. Pesca* **2020**, 46. [CrossRef]
60. Coche, A.G. Cage culture of tilapia [mostly in tropical freshwaters]. In Proceedings of the International Conference on the Biology and Culture of Tilapias, Bellagio, Italy, 2–5 September 1980; Available online: https://agris.fao.org/agris-search/search.do?recordID=XB8210677 (accessed on 20 January 2023).
61. Bridson, P.B.; Stoner, J.M.; Fransen, M.H.; Ireland, J. The aquaculture sustainability continuum—Defining an environmental performance framework. *Environ. Sustain. Indic.* **2020**, *8*, 100050. [CrossRef]
62. Jayasankar, P. Present status of freshwater aquaculture in India—A review. *Indian J. Fish.* **2018**, *65*, 157–165. [CrossRef]
63. Chakraborty, S.B.; Mazumdar, D.; Banerjee, S. Determination of ideal stocking density for cage culture of monosex Nile tilapia (Oreochromis niloticus) in India. *Proc. Zool. Soc.* **2010**, *63*, 53–59. [CrossRef]
64. Azim, M.E.; Wahab, M.A.; Van Dam, A.A.; Beveridge, M.C.M.; Milstein, A.; Verdegem, M.C.J. Optimization of fertilization rate for maximizing periphyton production on artificial substrates and the implications for periphyton-based aquaculture. *Aquac. Res.* **2001**, *32*, 749–760. [CrossRef]

65. Delphino, M.; Joshi, R.; Alvarez, A.T. Economic appraisal of using genetics to control Streptococcus agalactiae in Nile tilapia under cage and pond farming system in Malaysia. *Sci. Rep.* **2022**, *12*, 8754. [CrossRef]
66. Munguti, J.M.; Nairuti, R.; Iteba, J.O.; Obiero, K.O.; Kyule, D.; Opiyo, M.A.; Abwao, J.; Kirimi, J.G.; Outa, N.; Muthoka, M.; et al. Nile tilapia (*Oreochromis niloticus* Linnaeus, 1758) culture in Kenya: Emerging production technologies and socio-economic impacts on local livelihoods. *Aquac. Fish Fish.* **2022**, *2*, 265–276. [CrossRef]
67. Boyd, C.; Zimmermann, S. Grow-out systems—Water quality and soil management. In *Freshwater Prawn Culture: The Farming of Macrobrachium Rosenbergii*; Blackwell Science: Oxford, UK, 2000; pp. 221–238. [CrossRef]
68. Wang, M.; Lu, M. Tilapia polyculture: A global review. *Aquac. Res.* **2016**, *47*, 2363–2374. [CrossRef]
69. Bunting, S. Horizontally integrated aquaculture development: Exploring consensus on constraints and opportunities with a stakeholder Delphi. *Aquac. Int.* **2008**, *16*, 153–169. [CrossRef]
70. Tian, X.; Li, D.; Dong, S.; Yan, X.; Qi, Z.; Liu, G.; Lu, J. An experimental study on closed-polyculture of penaeid shrimp with tilapia and constricted tagelus. *Aquaculture* **2001**, *202*, 57–71. [CrossRef]
71. Milstein, A. Fish-management relationships in Israeli commercial fish farming. *Aquac. Int.* **1995**, *3*, 292–314. [CrossRef]
72. Marques, H.L.A.; New, M.B.; Boock, M.V.; Barros, H.P.; Mallasen, M.; Valenti, W.C. Integrated Freshwater Prawn Farming: State-of-the-art and future potential. *Rev. Fish. Sci. Aquac.* **2016**, *24*, 264–293. [CrossRef]
73. Hisano, H.; Barbosa, P.T.L.; Hayd, L.D.A.; Mattioli, C.C. Comparative study of growth, feed efficiency, and hematological profile of Nile tilapia fingerlings in biofloc technology and recirculating aquaculture system. *Trop. Anim. Health Prod.* **2021**, *53*, 60. [CrossRef]
74. Shrestha, M.K.; Bhandari, M.P.; Diana, J.S.; Jaiswal, R.; Mishra, R.N.; Pandit, N.P. Positive impacts of Nile tilapia and predatory sahar on carp polyculture production and profits. *Aquac. Fish.* **2018**, *3*, 204–208. [CrossRef]
75. Papoutsoglou, S.; Miliou, H.; Karakatsouli, N.; Tzitzinakis, M.; Chadio, S. Growth and physiological changes in scaled carp and blue tilapia under behavioral stress in mono- and polyculture rearing using a recirculated water system. *Aquac. Int.* **2001**, *9*, 509–518. [CrossRef]
76. Knud-Hansen, C.F.; Batterson, T.R.; McNabb, C.D.; Harahat, I.S.; Sumantadinata, K.; Eidman, H. Nitrogen input, primary productivity and fish yield in fertilized freshwater ponds in Indonesia. *Aquaculture* **1991**, *94*, 49–63. [CrossRef]
77. Kang'Ombe, J.; Brown, J.A.; Halfyard, L.C. Effect of using different types of organic animal manure on plankton abundance, and on growth and survival of Tilapia rendalli (Boulenger) in ponds. *Aquac. Res.* **2006**, *37*, 1360–1371. [CrossRef]
78. Junior, A.P.B.; Azevedo, C.M.D.S.B.; Pontes, F.S.T.; Henry-Silva, G.G. Polyculture of Nile tilapia and shrimp at different stocking densities. *Rev. Bras. de Zootec.* **2012**, *41*, 1561–1569. [CrossRef]
79. Iliyasu, A.; Mohamed, Z.A.; Terano, R. Comparative analysis of technical efficiency for different production culture systems and species of freshwater aquaculture in Peninsular Malaysia. *Aquac. Rep.* **2016**, *3*, 51–57. [CrossRef]
80. Petersen, A.; Andersen, J.S.; Kaewmak, T.; Somsiri, T.; Dalsgaard, A. Impact of integrated fish farming on antimicrobial resistance in a pond environment. *Appl. Environ. Microbiol.* **2002**, *68*, 6036–6042. [CrossRef]
81. Hollas, C.E.; Rodrigues, H.C.; Oyadomari, V.M.A.; Bolsan, A.C.; Venturin, B.; Bonassa, G.; Tápparo, D.C.; Abilhôa, H.C.Z.; da Silva, J.F.F.; Michelon, W.; et al. The potential of animal manure management pathways toward a circular economy: A bibliometric analysis. *Environ. Sci. Pollut. Res.* **2022**, *29*, 73599–73621. [CrossRef]
82. Ulloa-Murillo, L.M.; Villegas, L.M.; Rodríguez-Ortiz, A.R.; Duque-Acevedo, M.; Cortés-García, F.J. Management of the organic fraction of municipal solid waste in the context of a sustainable and circular model: Analysis of trends in Latin America and the Caribbean. *Int. J. Environ. Res. Public Health* **2022**, *19*, 6041. [CrossRef]
83. Fu, H.; Tan, P.; Wang, R.; Li, S.; Liu, H.; Yang, Y.; Wu, Z. Advances in organophosphorus pesticides pollution: Current status and challenges in ecotoxicological, sustainable agriculture, and degradation strategies. *J. Hazard. Mater.* **2022**, *424*, 127494. [CrossRef]
84. Biswas, G.; Raja, R.A.; De, D.; Sundaray, J.K.; Ghoshal, T.K.; Anand, S.; Kumar, S.; Panigrahi, A.; Thirunavukkarasu, A.R.; Ponniah, A.G. Evaluation of productions and economic returns from two brackish water polyculture systems in tide-fed ponds. *J. Appl. Ichthyol.* **2012**, *28*, 116–122. [CrossRef]
85. Chakravartty, D.; Mondal, A.; Raychowdhury, P.; Bhattacharya, S.B.; Mitra, A. Role of aquaponics in the sustenance of coastal India-Aquaponics is a solution for modern agriculture in ecologically sensitive Indian mangrove Sundarbans: A review. *Int. J. Fish. Aquat. Sci.* **2017**, *5*, 441–448. Available online: https://www.researchgate.net/profile/Abhijit-Mitra4/publication/344799587_Livelihood_through_Aquaponics/links/5f90c3ba299bf1b53e3a1ed2/Livelihood-through-Aquaponics.pdf (accessed on 20 January 2023).
86. David, F.S.; Proença, D.C.; Valenti, W.C. Phosphorus budget in integrated multitrophic aquaculture systems with Nile Tilapia, *Oreochromis niloticus*, and Amazon River Prawn, *Macrobrachium amazonicum*. *J. World Aquac. Soc.* **2017**, *48*, 402–414. [CrossRef]
87. Chaudhuri, A.B.; Choudhury, A. *Mangroves of the Sundarbans. Volume 1: India*; International Union for Conservation of Nature and Natural Resources (IUCN): Gland, Switzerland, 1994; ISBN 978-2-8317-0209-4.
88. Nurfadillah, N.; Hasri, I.; Purnama, M.R.; Damora, A.; Mellisa, S. The application of integrated multi-trophic aquaculture (IMTA) using floating net cages on Tilapia fish with native fish (Peres, Lemeduk, and Depik). *Depik* **2021**, *10*, 219–224. [CrossRef]
89. Rodrigues, C.G.; Garcia, B.F.; Verdegem, M.; Santos, M.R.; Amorim, R.V.; Valenti, W.C. Integrated culture of Nile tilapia and Amazon river prawn in stagnant ponds, using nutrient-rich water and substrates. *Aquaculture* **2019**, *503*, 111–117. [CrossRef]
90. David, F.S.; Proença, D.C.; Flickinger, D.L.; Bueno, G.W.; Valenti, W.C. Carbon budget in integrated aquaculture systems with Nile tilapia (*Oreochromis niloticus*) and Amazon river prawn (*Macrobrachium amazonicum*). *Aquac. Res.* **2021**, *52*, 5155–5167. [CrossRef]

91. Bardhan, A.; Sau, S.K.; Khatua, S.; Bera, M.; Paul, B.N. A Review on the production and culture techniques of Monosex tilapia. *Int. J. Curr. Microbiol. Appl. Sci.* **2021**, *10*, 565–577. [CrossRef]
92. Mair, G.C.; Abucay, J.S.; Beardmore, J.A.; Skibinski, D.O. Growth performance trials of genetically male tilapia (GMT) derived from YY-males in *Oreochromis niloticus* L.: On station comparisons with mixed sex and sex reversed male populations. *Aquaculture* **1995**, *137*, 313–323. [CrossRef]
93. Beardmore, J.; Mair, G.; Lewis, R. Monosex male production in finfish as exemplified by tilapia: Applications, problems, and prospects. In *Reproductive Biotechnology in Finfish Aquaculture*; Elsevier: Amsterdam, The Netherlands, 2001; pp. 283–301. [CrossRef]
94. Edea, G.O.; Hinvi, L.C.; Abou, Y.; Gbangboche, A.B. Growth and Body Development of *Oreochromis niloticus* (Linnaeus, 1758) Fattened in Floating Cages Based on Commercial Feed in Benin. *J. Biol. Life Sci.* **2020**, *11*, 36–56. [CrossRef]
95. Budd, A.M.; Banh, Q.Q.; Domingos, J.A.; Jerry, D.R. Sex control in fish: Approaches, challenges and opportunities for aquaculture. *J. Mar. Sci. Eng.* **2015**, *3*, 329–355. [CrossRef]
96. Wenhai, L.; Cusack, C.; Baker, M.; Tao, W.; Mingbao, C.; Paige, K.; Xiaofan, Z.; Levin, L.; Escobar, E.; Amon, D.; et al. Successful blue economy examples with an emphasis on international perspectives. *Front. Mar. Sci.* **2019**, *6*, 261. [CrossRef]
97. Basavaraju, Y. Monosex population in aquaculture. In *Frontiers in Aquaculture Biotechnology*; Academic Press: Cambridge, MA, USA, 2023; pp. 89–101. [CrossRef]
98. Sayed, A.E.-D.H.; Moneeb, R.H. Hematological and biochemical characters of monosex tilapia (Oreochromis niloticus, Linnaeus, 1758) cultivated using methyltestosterone. *J. Basic Appl. Zool.* **2015**, *72*, 36–42. [CrossRef]
99. Abaho, I.; Masembe, C.; Akoll, P.; Jones, C.L.W. The use of plant extracts to control tilapia reproduction: Current status and future perspectives. *J. World Aquac. Soc.* **2022**, *53*, 593–619. [CrossRef]
100. Ghosal, I.; Mukherjee, D.; Chakraborty, S.B. The effects of four plant extracts on growth, sex reversal, immunological and haemato-biochemical parameters in Nile tilapia, *Oreochmomis niloticus* (Linnaeus, 1758). *Aquac. Res.* **2021**, *52*, 559–576. [CrossRef]
101. Kumar, P.; Dey, M.M.; Barik, N.K. Farm-Economics of Genetically Improved Carp Strains in Major Asian Countries and Carp Seed Price Policy Model. 2008. Available online: https://hdl.handle.net/20.500.12348/1541 (accessed on 20 January 2023).
102. Nguyen, N.H. Genetic improvement for important farmed aquaculture species with a reference to carp, tilapia and prawns in Asia: Achievements, lessons and challenges. *Fish* **2016**, *17*, 483–506. [CrossRef]
103. Thoa, N.P.; Ninh, N.H.; Knibb, W.; Nguyen, N.H. Does selection in a challenging environment produce Nile tilapia genotypes that can thrive in a range of production systems? *Sci. Rep.* **2016**, *6*, 21486. [CrossRef] [PubMed]
104. Tran, N.; Shikuku, K.M.; Rossignoli, C.M.; Barman, B.K.; Cheong, K.C.; Ali, M.S.; Benzie, J.A. Growth, yield and profitability of genetically improved farmed tilapia (GIFT) and non-GIFT strains in Bangladesh. *Aquaculture* **2021**, *536*, 736486. [CrossRef]
105. Xia, J.H.; Bai, Z.; Meng, Z.; Zhang, Y.; Wang, L.; Liu, F.; Jing, W.; Wan, Z.Y.; Li, J.; Lin, H.; et al. Signatures of selection in tilapia revealed by whole genome resequencing. *Sci. Rep.* **2015**, *5*, srep14168. [CrossRef]
106. Hashim, R.; Chong, A.S.C.; Fatan, N.A.; Layman, N.; Ali, A. Production of hybrid red tilapia, Oreochromis mossambicus × O. niloticus, at varying stocking densities in portable canvas tanks. *J. Appl. Aquac.* **2002**, *12*, 1–12. [CrossRef]
107. Yu, X.; Setyawan, P.; Bastiaansen, J.W.; Liu, L.; Imron, I.; Groenen, M.A.; Komen, H.; Megens, H.-J. Genomic analysis of a Nile tilapia strain selected for salinity tolerance shows signatures of selection and hybridization with blue tilapia (Oreochromis aureus). *Aquaculture* **2022**, *560*, 738527. [CrossRef]
108. Yáñez, J.M.; Joshi, R.; Yoshida, G.M. Genomics to accelerate genetic improvement in tilapia. *Anim. Genet.* **2020**, *51*, 658–674. [CrossRef] [PubMed]
109. Bian, C.; Li, J.; Lin, X.; Chen, X.; Yi, Y.; You, X.; Zhang, Y.; Lv, Y.; Shi, Q. Whole genome sequencing of the blue tilapia (Oreochromis aureus) provides a valuable genetic resource for biomedical research on tilapias. *Mar. Drugs* **2019**, *17*, 386. [CrossRef] [PubMed]
110. Avallone, A.; Bartie, K.L.; Selly, S.-L.C.; Taslima, K.; Mendoza, A.C.; Bekaert, M. Local ancestry inference provides insight into Tilapia breeding programmes. *Sci. Rep.* **2020**, *10*, 18613. [CrossRef]
111. Ansah, Y.B.; Frimpong, E.A.; Hallerman, E.M. Genetically-improved tilapia strains in Africa: Potential benefits and negative impacts. *Sustainability* **2014**, *6*, 3697–3721. [CrossRef]
112. Moses, M.; Chauka, L.J.; de Koning, D.J.; Palaiokostas, C.; Mtolera, M.S.P. Growth performance of five different strains of Nile tilapia (Oreochromis niloticus) introduced to Tanzania reared in fresh and brackish waters. *Sci. Rep.* **2021**, *11*, 11147. [CrossRef]
113. da Silva, M.A.; de Alvarenga, R.; da Costa, F.F.B.; Turra, E.M.; Alves, G.F.D.O.; Manduca, L.G.; de Sales, S.C.M.; Leite, N.R.; Bezerra, V.M.; Moraes, S.G.D.S.; et al. Feeding management strategies to optimize the use of suspended feed for Nile tilapia (*Oreochromis niloticus*) cultivated in bioflocs. *Aquac. Res.* **2020**, *51*, 605–615. [CrossRef]
114. Lall, S.P.; Dumas, A. Nutritional requirements of cultured fish: Formulating nutritionally adequate feeds. In *Feed and Feeding Practices in Aquaculture*; Woodhead publishing: Cambridge, UK, 2022; pp. 65–132. [CrossRef]
115. Azaza, M.S.; Dhraief, M.N. Modeling the effects of water temperature on growth rates, gastric evacuation and the return of appetite in juvenile nile tilapia, *Oreochromis niloticus* L. *J. Agric. Sci.* **2020**, *12*, p191. [CrossRef]
116. Herath, S.S.; Satoh, S. 15.1 Overview of the problem. In *Feed and Feeding Practices in Aquaculture*; Woodhead publishing: Cambridge, UK, 2020; p. 427.

117. Hassan, H.U.; Ali, Q.M.; Khan, W.; Masood, Z.; Abdel-Aziz, M.A.; Shah, M.I.A.; Gabol, K.; Wattoo, J.; Chatta, A.M.; Kamal, M.; et al. Effect of feeding frequency as a rearing system on biological performance, survival, body chemical composition and economic efficiency of Asian seabass Lates calcarifer (Bloch, 1790) reared under controlled environmental conditions. *Saudi J. Biol. Sci.* **2021**, *28*, 7360–7366. [CrossRef]
118. Boyd, C.E.; D'Abramo, L.R.; Glencross, B.D.; Huyben, D.C.; Juarez, L.M.; Lockwood, G.S.; McNevin, A.A.; Tacon, A.G.J.; Teletchea, F.; Tomasso, J.R., Jr.; et al. Achieving sustainable aquaculture: Historical and current perspectives and future needs and challenges. *J. World Aquac. Soc.* **2020**, *51*, 578–633. [CrossRef]
119. Fry, J.P.; Love, D.C.; MacDonald, G.K.; West, P.C.; Engstrom, P.M.; Nachman, K.E.; Lawrence, R.S. Environmental health impacts of feeding crops to farmed fish. *Environ. Int.* **2016**, *91*, 201–214. [CrossRef]
120. Connor, D.; Minguez, M.I. Evolution not revolution of farming systems will best feed and green the world. *Glob. Food Secur.* **2012**, *1*, 106–113. [CrossRef]
121. Rumbos, C.; Mente, E.; Karapanagiotidis, I.; Vlontzos, G.; Athanassiou, C. Insect-based feed ingredients for aquaculture: A case study for their acceptance in Greece. *Insects* **2021**, *12*, 586. [CrossRef] [PubMed]
122. Gule, T.T.; Geremew, A. Dietary strategies for better utilization of aquafeeds in Tilapia farming. *Aquac. Nutr.* **2022**, *2022*, 9463307. [CrossRef]
123. Adeshina, I.; Abdel-Tawwab, M.; Tijjani, Z.A.; Tiamiyu, L.O.; Jahanbakhshi, A. Dietary Tridax procumbens leaves extract stimulated growth, antioxidants, immunity, and resistance of Nile tilapia, Oreochromis niloticus, to monogenean parasitic infection. *Aquaculture* **2021**, *532*, 736047. [CrossRef]
124. Ahmad, M.H.; Abdel-Tawwab, M. The use of caraway seed meal as a feed additive in fish diets: Growth performance, feed utilization, and whole-body composition of Nile tilapia, *Oreochromis niloticus* (L.) fingerlings. *Aquaculture* **2011**, *314*, 110–114. [CrossRef]
125. Hassaan, M.S.; Mohammady, E.Y.; Soaudy, M.R.; El-Garhy, H.A.; Moustafa, M.M.; Mohamed, S.A.; El-Haroun, E.R. Effect of Silybum marianum seeds as a feed additive on growth performance, serum biochemical indices, antioxidant status, and gene expression of Nile tilapia, *Oreochromis niloticus* (L.) fingerlings. *Aquaculture* **2019**, *509*, 178–187. [CrossRef]
126. Moustafa, E.M.; Dawood, M.A.; Assar, D.H.; Omar, A.A.; Elbialy, Z.I.; Farrag, F.A.; Shukry, M.; Zayed, M.M. Modulatory effects of fenugreek seeds powder on the histopathology, oxidative status, and immune related gene expression in Nile tilapia (*Oreochromis niloticus*) infected with Aeromonas hydrophila. *Aquaculture* **2020**, *515*, 734589. [CrossRef]
127. El-Latif, A.M.A.; El-Gawad, E.A.A.; Soror, E.I.; Shourbela, R.M.; Zahran, E. Dietary supplementation with miswak (Salvadora persica) improves the health status of Nile tilapia and protects against Aeromonas hydrophila infection. *Aquac. Rep.* **2021**, *19*, 100594. [CrossRef]
128. Elbialy, Z.I.; Rizk, M.; Al-Hawary, I.I.; Salah, A.S.; Mohammed, R.A.; Assar, D.H.; Almeer, R.; Dawood, M.A.O. *Yucca schidigera* extract mediated the growth performance, hepato-renal function, antioxidative status and histopathological alterations in Nile tilapia (*Oreochromis niloticus*) exposed to hypoxia stress. *Aquac. Res.* **2020**, *52*, 1965–1976. [CrossRef]
129. Magouz, F.I.; Mahmoud, S.A.; El-Morsy, R.A.; Paray, B.A.; Soliman, A.A.; Zaineldin, A.I.; Dawood, M.A. Dietary menthol essential oil enhanced the growth performance, digestive enzyme activity, immune-related genes, and resistance against acute ammonia exposure in Nile tilapia (*Oreochromis niloticus*). *Aquaculture* **2021**, *530*, 735944. [CrossRef]
130. El Basuini, M.F.; Shahin, S.A.; Teiba, I.I.; Zaki, M.A.; El-Hais, A.M.; Sewilam, H.; Almeer, R.; Abdelkhalek, N.; Dawood, M.A. The influence of dietary coenzyme Q10 and vitamin C on the growth rate, immunity, oxidative-related genes, and the resistance against Streptococcus agalactiae of Nile tilapia (*Oreochromis niloticus*). *Aquaculture* **2021**, *531*, 735862. [CrossRef]
131. Lim, L.-S.; Tan, K.S.; Fu, M.-Y.; Au, H.-L.; Ebi, I.; Lal, M.T.M.; Kawamura, G.; Shapawi, R.; Lam, S.S. Valorization of Bokashi leachate as feed additive in tilapia farming. *Environ. Res.* **2020**, *198*, 110472. [CrossRef] [PubMed]
132. Poolsawat, L.; Yang, H.; Sun, Y.-F.; Li, X.-Q.; Liang, G.-Y.; Leng, X.-J. Effect of replacing fish meal with enzymatic feather meal on growth and feed utilization of tilapia (*Oreochromis niloticus* × O. aureus). *Anim. Feed. Sci. Technol.* **2021**, *274*, 114895. [CrossRef]
133. Huan, D.; Li, X.; Chowdhury, M.A.K.; Yang, H.; Liang, G.; Leng, X. Organic acid salts, protease and their combination in fish meal-free diets improved growth, nutrient retention and digestibility of tilapia (*Oreochromis niloticus* × *O. aureus*). *Aquac. Nutr.* **2018**, *24*, 1813–1821. [CrossRef]
134. Amer, S.A.; Ahmed, S.A.; Ibrahim, R.E.; Al-Gabri, N.A.; Osman, A.; Sitohy, M. Impact of partial substitution of fish meal by methylated soy protein isolates on the nutritional, immunological, and health aspects of Nile tilapia, *Oreochromis niloticus* fingerlings. *Aquaculture* **2020**, *518*, 734871. [CrossRef]
135. Amer, S.A.; Osman, A.; Al-Gabri, N.A.; Elsayed, S.A.M.; El-Rahman, G.I.A.; Elabbasy, M.T.; Ahmed, S.A.A.; Ibrahim, R.E. The effect of dietary replacement of fish meal with whey protein concentrate on the growth performance, fish health, and immune status of nile tilapia fingerlings, *Oreochromis niloticus*. *Animals* **2019**, *9*, 1003. [CrossRef]
136. Guimarães, M.C.; Guimarães, A.I.C.D.S.; Natori, M.M.; Alarcon, M.F.F.; Dias, D.D.C.; Ishikawa, C.M.; Tapia-Paniagua, S.; Moriñigo, M.; Moyano, F.J.; Tachibana, L. Oral administration of Bacillus subtilis and Lactobacillus plantarum modulates the gut microbiota and increases the amylase activity of Nile tilapia (*Oreochromis niloticus*). *Aquac. Int.* **2021**, *29*, 91–104. [CrossRef]
137. Hassaan, M.S.; Mohammady, E.Y.; Soaudy, M.R.; Elashry, M.A.; Moustafa, M.M.; Wassel, M.A.; El-Garhy, H.A.; El-Haroun, E.R.; Elsaied, H.E. Synergistic effects of Bacillus pumilus and exogenous protease on Nile tilapia (Oreochromis niloticus) growth, gut microbes, immune response and gene expression fed plant protein diet. *Anim. Feed. Sci. Technol.* **2021**, *275*, 114892. [CrossRef]

138. Tachibana, L.; Telli, G.S.; Dias, D.D.C.; Gonçalves, G.S.; Ishikawa, C.M.; Cavalcante, R.B.; Natori, M.M.; Ben Hamed, S.; Ranzani-Paiva, M.J.T. Effect of feeding strategy of probiotic Enterococcus faecium on growth performance, hematologic, biochemical parameters and non-specific immune response of Nile tilapia. *Aquac. Rep.* **2020**, *16*, 100277. [CrossRef]
139. Dawood, M.A.; Eweedah, N.M.; Moustafa, E.M.; Farahat, E.M. Probiotic effects of Aspergillus oryzae on the oxidative status, heat shock protein, and immune related gene expression of Nile tilapia (*Oreochromis niloticus*) under hypoxia challenge. *Aquaculture* **2020**, *520*, 734669. [CrossRef]
140. Poolsawat, L.; Li, X.; He, M.; Ji, D.; Leng, X. Clostridium butyricumas probiotic for promoting growth performance, feed utilization, gut health and microbiota community of tilapia (*Oreochromis niloticus* × *O. aureus*). *Aquac. Nutr.* **2020**, *26*, 657–670. [CrossRef]
141. El-Naggar, M.; Salaah, S.; El-Shabaka, H.; El-Rahman, F.A.; Khalil, M.; Suloma, A. Efficacy of dietary chitosan and chitosan nanoparticles supplementation on health status of Nile tilapia, *Oreochromis niloticus* (L.). *Aquac. Rep.* **2021**, *19*, 100628. [CrossRef]
142. Kurian, A.; Elumalai, P. Study on the impacts of chemical and green synthesized (Leucas aspera and oxy-cyclodextrin complex) dietary zinc oxide nanoparticles in Nile tilapia (*Oreochromis niloticus*). *Environ. Sci. Pollut. Res.* **2021**, *28*, 20344–20361. [CrossRef] [PubMed]
143. Abdel-Tawwab, M.; Shukry, M.; Farrag, F.A.; El-Shafai, N.M.; Dawood, M.A.; Abdel-Latif, H.M. Dietary sodium butyrate nanoparticles enhanced growth, digestive enzyme activities, intestinal histomorphometry, and transcription of growth-related genes in Nile tilapia juveniles. *Aquaculture* **2021**, *536*, 736467. [CrossRef]
144. Ghazi, S.; Diab, A.M.; Khalafalla, M.M.; Mohamed, R.A. Synergistic Effects of Selenium and Zinc Oxide Nanoparticles on Growth Performance, Hemato-biochemical Profile, Immune and Oxidative Stress Responses, and Intestinal Morphometry of Nile Tilapia (*Oreochromis niloticus*). *Biol. Trace Element Res.* **2021**, *200*, 364–374. [CrossRef]
145. Abdel-Tawwab, M.; Samir, F.; El-Naby, A.S.A.; Monier, M.N. Antioxidative and immunostimulatory effect of dietary cinnamon nanoparticles on the performance of Nile tilapia, *Oreochromis niloticus* (L.) and its susceptibility to hypoxia stress and Aeromonas hydrophila infection. *Fish Shellfish. Immunol.* **2018**, *74*, 19–25. [CrossRef] [PubMed]
146. Sribounoy, U.; Pirarat, N.; Solval, K.M.; Sathivel, S.; Chotiko, A. Development of pelleted feed containing probiotic Lactobacillus rhamnosus GG and Jerusalem artichoke for Nile Tilapia and its biocompatibility studies. *3 Biotech* **2021**, *11*, 279. [CrossRef]
147. Sgnaulin, T.; Durigon, E.G.; Pinho, S.M.; Jerônimo, G.T.; Lopes, D.L.D.A.; Emerenciano, M.G.C. Nutrition of Genetically Improved Farmed Tilapia (GIFT) in biofloc technology system: Optimization of digestible protein and digestible energy levels during nursery phase. *Aquaculture* **2020**, *521*, 734998. [CrossRef]
148. Duodu, C.P.; Adjei-Boateng, D.; Amponsah, A.K.; Andrews, P.; Obirikorang, K.A. The effect of plant protein-based diets on apparent nutrient digestibility, growth response, egesta quantity, postprandial ammonia excretion rate and serum quality of Nile tilapia. *Aquac. Res.* **2020**, *51*, 1152–1161. [CrossRef]
149. Nyadjeu, P.; Djopnang, J.D.; Ndjuissi, N.A.T.; Ayamba, A.E.; Alim, H.; Tabi-Tomedi, M.E. Replacing fishmeal with pretreated Lima bean improves growth, feed utilization, whole-body composition and nutrient retention in Nile-tilapia fingerlings. *J. Appl. Aquac.* **2021**, *34*, 922–937. [CrossRef]
150. Kelly, A.M.; Renukdas, N.N. Disease management of aquatic animals. In *Aquaculture Health Management*; Academic Press: Cambridge, MA, USA, 2020; pp. 137–161. [CrossRef]
151. Chitmanat, C.; Lebel, P.; Whangchai, N.; Promya, J.; Lebel, L. Tilapia diseases and management in river-based cage aquaculture in northern Thailand. *J. Appl. Aquac.* **2016**, *28*, 9–16. [CrossRef]
152. Manyi-Loh, C.; Mamphweli, S.; Meyer, E.; Okoh, A. Antibiotic use in agriculture and its consequential resistance in environmental sources: Potential public health implications. *Molecules* **2018**, *23*, 795. [CrossRef] [PubMed]
153. Polianciuc, S.I.; Gurzău, A.E.; Kiss, B.; Ștefan, M.G.; Loghin, F. Antibiotics in the environment: Causes and consequences. *Med. Pharm. Rep.* **2020**, *93*, 231–240. [CrossRef] [PubMed]
154. Behera, B.; Pradhan, P.; Swaminathan, T.; Sood, N.; Paria, P.; Das, A.; Verma, D.; Kumar, R.; Yadav, M.; Dev, A.; et al. Emergence of tilapia lake virus associated with mortalities of farmed Nile Tilapia *Oreochromis niloticus* (Linnaeus 1758) in India. *Aquaculture* **2018**, *484*, 168–174. [CrossRef]
155. Lu, T.-H.; Chen, C.-Y.; Liao, C.-M. Aeromonas hydrophila as an environmental indicator to detect TiLV-infected tilapia under coinfection threat. *J. Environ. Sustain.* **2021**, *11*, 100135. [CrossRef]
156. Paredes-Trujillo, A.; Velázquez-Abunader, I.; Papiol, V.; del Rio-Rodríguez, R.E.; Vidal-Martínez, V.M. Negative effect of ectoparasite burdens on the condition factor from farmed tilapia *Oreochromis niloticus* in the Yucatan, Mexico. *Veter-Parasitol.* **2021**, *292*, 109393. [CrossRef]
157. Lafferty, K.D.; Harvell, C.D.; Conrad, J.M.; Friedman, C.S.; Kent, M.L.; Kuris, A.M.; Powell, E.N.; Rondeau, D.; Saksida, S.M. Infectious diseases affect marine fisheries and aquaculture economics. *Annu. Rev. Mar. Sci.* **2015**, *7*, 471–496. [CrossRef]
158. Kayansamruaj, P.; Pirarat, N.; Hirono, I.; Rodkhum, C. Increasing of temperature induces pathogenicity of Streptococcus agalactiae and the up-regulation of inflammatory related genes in infected Nile tilapia (*Oreochromis niloticus*). *Veter-Microbiol.* **2014**, *172*, 265–271. [CrossRef] [PubMed]
159. Ismail, M.; Syafiq, M.; Siti-Zahrah, A.; Fahmi, S.; Shahidan, H.; Hanan, Y.; Amal, M.; Saad, M.Z. The effect of feed-based vaccination on tilapia farm endemic for streptococcosis. *Fish Shellfish. Immunol.* **2017**, *60*, 21–24. [CrossRef]

160. Kent, M.L.; Feist, S.W.; Harper, C.; Hoogstraten-Miller, S.; Mac Law, J.; Sánchez-Morgado, J.M.; Tanguay, R.L.; Sanders, G.E.; Spitsbergen, J.M.; Whipps, C.M. Recommendations for control of pathogens and infectious diseases in fish research facilities. *Comp. Biochem. Physiol. Part C Toxicol. Pharmacol.* **2009**, *149*, 240–248. [CrossRef]
161. Opiyo, M.; Mziri, V.; Musa, S.; Kyule, D.; Hinzano, S.; Wainaina, M.; Ombwa, V. Fish Disease Management and Biosecurity Systems. 2021. Available online: https://www.researchgate.net/profile/Mary-Opiyo/publication/351050775_Fish_Disease_Management_and_Biosecurity_Systems/links/60815e688ea909241e17589b/Fish-Disease-Management-and-Biosecurity-Systems.pdf (accessed on 1 June 2020).
162. Aljahdali, M.; Molla, M.; Filfilan, W. Whole Genome Sequence of the Newly Prescribed Subspecies *Oreochromis spilurus saudii*: A Valuable Genetic Resource for Aquaculture in Saudi Arabia. *J. Mar. Sci. Eng.* **2021**, *9*, 506. [CrossRef]
163. Embregts, C.W.; Forlenza, M. Oral vaccination of fish: Lessons from humans and veterinary species. *Dev. Comp. Immunol.* **2016**, *64*, 118–137. [CrossRef] [PubMed]
164. Duff, D.C.B. The oral immunization of trout against Bacterium salmonicida. *J. Immunol.* **1942**, *44*, 87–94. [CrossRef]
165. Ma, J.; Bruce, T.J.; Jones, E.M.; Cain, K.D. A review of fish vaccine development strategies: Conventional methods and modern biotechnological approaches. *Microorganisms* **2019**, *7*, 569. [CrossRef] [PubMed]
166. Pridgeon, J.W.; Klesius, P.H. Development and efficacy of a novobiocin-resistant Streptococcus iniae as a novel vaccine in Nile tilapia (Oreochromis niloticus). *Vaccine* **2011**, *29*, 5986–5993. [CrossRef]
167. Kayansamruaj, P.; Dong, H.T.; Pirarat, N.; Nilubol, D.; Rodkhum, C. Efficacy of α-enolase-based DNA vaccine against pathogenic Streptococcus iniae in Nile tilapia (*Oreochromis niloticus*). *Aquaculture* **2017**, *468*, 102–106. [CrossRef]
168. HHuang, L.Y.; Wang, K.Y.; Xiao, D.; Chen, D.F.; Geng, Y.; Wang, J.; He, Y.; Wang, E.L.; Huang, J.L.; Xiao, G.Y. Safety and immunogenicity of an oral DNA vaccine encoding Sip of Streptococcus agalactiae from Nile tilapia *Oreochromis niloticus* delivered by live attenuated Salmonella typhimurium. *Fish Shellfish. Immunol.* **2014**, *38*, 34–41. [CrossRef]
169. Liu, L.; Lu, D.; Xu, J.; Luo, H.; Li, A. Development of attenuated erythromycin-resistant *Streptococcus agalactiae* vaccine for tilapia (*Oreochromis niloticus*) culture. *J. Fish Dis.* **2019**, *42*, 693–701. [CrossRef]
170. Bactol, I.D.C.; Padilla, L.V.; Hilario, A.L. Immune response of tilapia (Oreochromis niloticus) after vaccination with autoclave killed, heat-killed, and formalin-killed whole cell Aeromonas hydrophila vaccines as possible serotype-independent vaccines. *Int. J. Agric. Biol.* **2018**, *20*, 846–850. [CrossRef]
171. Kwon, H.C.; Kang, Y.J. Effects of a subunit vaccine (FlaA) and immunostimulant (CpG-ODN 1668) against Vibrio anguillarum in tilapia (*Oreochromis niloticus*). *Aquaculture* **2016**, *454*, 125–129. [CrossRef]
172. Cao, T.T.; Tsai, M.-A.; Yang, C.-D.; Wang, P.-C.; Kuo, T.-Y.; Chen, H.-C.G.; Chen, S.-C. Vaccine efficacy of glyceraldehyde-3-phosphate dehydrogenase (GAPDH) from Edwardsiella ictaluri against E. *tarda in tilapia. J. Gen. Appl. Microbiol.* **2014**, *60*, 241–250. [CrossRef]
173. Aly, S.M.; Albutti, A.S.; Rahmani, A.H.; Atti, N.M.A. The response of new-season Nile tilapia to Aeromonas hydrophila vaccine. *Int. J. Clin. Exp. Med.* **2015**, *8*, 4508. Available online: http://www.ncbi.nlm.nih.gov/pmc/articles/pmc4443210/ (accessed on 20 January 2023).
174. Chen, M.; Li, L.-P.; Wang, R.; Liang, W.-W.; Huang, Y.; Li, J.; Lei, A.-Y.; Huang, W.-Y.; Gan, X. PCR detection and PFGE genotype analyses of streptococcal clinical isolates from tilapia in China. *Veter- Microbiol.* **2012**, *159*, 526–530. [CrossRef] [PubMed]
175. Eldar, A.; Bejerano, Y.; Livoff, A.; Horovitcz, A.; Bercovier, H. Experimental streptococcal meningo-encephalitis in cultured fish. *Vet. Microbiol.* **1995**, *43*, 33–40. [CrossRef]
176. Klesius, P.H.; Shoemaker, C.; Evans, J.J. Efficacy of single and combined Streptococcus iniae isolate vaccine administered by intraperitoneal and intramuscular routes in tilapia (Oreochromis niloticus). *Aquaculture* **2000**, *188*, 237–246. [CrossRef]
177. Osman, K.M.; Pires, D.S.; Franco, O.L.; Saad, A.; Hamed, M.; Naim, H.; Ali, A.H.; Elbehiry, A. Nile tilapia (Oreochromis niloticus) as an aquatic vector for Pseudomonas species of medical importance: Antibiotic Resistance Association with Biofilm Formation, Quorum Sensing and Virulence. *Aquaculture* **2020**, *532*, 736068. [CrossRef]
178. Ruangpan, L.; Kitao, T.; Yoshida, T. Protective efficacy ofAeromonas hydrophila vaccines in Nile tilapia. *Veter- Immunol. Immunopathol.* **1986**, *12*, 345–350. [CrossRef] [PubMed]
179. Shoemaker, C.A.; LaFrentz, B.R.; Klesius, P.H. Bivalent vaccination of sex reversed hybrid tilapia against Streptococcus iniae and Vibrio vulnificus. *Aquaculture* **2012**, *354–355*, 45–49. [CrossRef]
180. Zhang, Z.; Xu, L.; Liu, W.; Yang, Y.; Du, Z.; Zhou, Z. Effects of partially replacing dietary soybean meal or cottonseed meal with completely hydrolyzed feather meal (defatted rice bran as the carrier) on production, cytokines, adhesive gut bacteria, and disease resistance in hybrid tilapia (*Oreochromis niloticus* (female) × *Oreochromis aureus* (male)). *Fish Shellfish. Immunol.* **2014**, *41*, 517–525. [CrossRef]
181. Hayat, M.; Yusoff, M.; Samad, M.; Razak, I.A.; Yasin, I.; Thompson, K.; Hasni, K. Efficacy of Feed-Based Formalin-Killed Vaccine of *Streptococcus iniae* Stimulates the Gut-Associated Lymphoid Tissues and Immune Response of Red Hybrid Tilapia. *Vaccines* **2021**, *9*, 51. [CrossRef]
182. El Tantawy, M.M.; Ayoub, H.F. Efficiency of oral Aeromonas hydrophila vaccine and Tumeric powder mixture on immune response of Nile tilapia (Oreochromis niloticus). *J. Aquat. Biol. Fish* **2016**, *6*, 47–66. [CrossRef]
183. Abu-Elala, N.M.; Samir, A.; Wasfy, M.; Elsayed, M. Efficacy of Injectable and Immersion Polyvalent Vaccine against Streptococcal Infections in Broodstock and Offspring of Nile tilapia (Oreochromis niloticus). *Fish Shellfish. Immunol.* **2019**, *88*, 293–300. [CrossRef] [PubMed]

184. Shahin, K.; Shinn, A.P.; Metselaar, M.; Ramirez-Paredes, J.G.; Monaghan, S.J.; Thompson, K.D.; Hoare, R.; Adams, A. Efficacy of an inactivated whole-cell injection vaccine for nile tilapia, *Oreochromis niloticus*(L), against multiple isolates of Francisella noatunensis subsp. orientalis from diverse geographical regions. *Fish Shellfish. Immunol.* **2019**, *89*, 217–227. [CrossRef] [PubMed]
185. Sood, N.; Pradhan, P.K.; Ravindra; Verma, D.K.; Yadav, M.K.; Mishra, R.K.; Kumar, U.; Swaminathan, T.R.; Sood, N.K. Large-scale mortality in cultured tilapia *Oreochromis niloticus* due to infection with *Shewanella putrefaciens* in India. *J. World Aquac. Soc.* **2020**, *51*, 563–570. [CrossRef]
186. Raj, N.S.; Swaminathan, T.R.; Dharmaratnam, A.; Raja, S.A.; Ramraj, D.; Lal, K. Aeromonas veronii caused bilateral exophthalmia and mass mortality in cultured Nile tilapia, *Oreochromis niloticus*(L.) in India. *Aquaculture* **2019**, *512*, 734278. [CrossRef]
187. Pepi, M.; Focardi, S. Antibiotic-resistant bacteria in aquaculture and climate change: A challenge for health in the Mediterranean area. *Int. J. Environ. Res. Public Health* **2021**, *18*, 5723. [CrossRef]
188. Hallali, E.; Kokou, F.; Chourasia, T.K.; Nitzan, T.; Con, P.; Harpaz, S.; Mizrahi, I.; Cnaani, A. Dietary salt levels affect digestibility, intestinal gene expression, and the microbiome, in Nile tilapia (*Oreochromis niloticus*). *PLoS ONE* **2018**, *13*, e0202351. [CrossRef]
189. Lima, R.; Fiol, F.D.S.D.; Balcão, V. Prospects for the use of new technologies to combat multidrug-resistant bacteria. *Front. Pharmacol.* **2019**, *10*, 692. [CrossRef]
190. Payne, C.J.; Turnbull, J.F.; MacKenzie, S.; Crumlish, M. Investigating the Effect of an Oxytetracycline Treatment on the Gut Microbiome and Antimicrobial Resistance Gene Dynamics in Nile Tilapia (*Oreochromis niloticus*). *Antibiotics* **2021**, *10*, 1213. [CrossRef]
191. Shiroma, L.S.; Soares, M.P.; Cardoso, I.L.; Ishikawa, M.M.; Jonsson, C.M.; Queiroz, S.C.N. Evaluation of health and environmental risks for juvenile tilapia (*Oreochromis niloticus*) exposed to florfenicol. *Heliyon* **2020**, *6*, e05716. [CrossRef]
192. Shiogiri, N.S.; Ikefuti, C.V.; Carraschi, S.P.; da Cruz, C.; Fernandes, M.N. Effects of azithromycin on tilapia (*Oreochromis niloticus*): Health status evaluation using biochemical, physiological and morphological biomarkers. *Aquac. Res.* **2017**, *48*, 3669–3683. [CrossRef]
193. Fang, L.; Chen, X.; Shan, X.; Qiu, L.; Fan, L.; Meng, S.; Song, C. Antibiotic accumulation, growth performance, intestinal diversification, and function of Nile tilapia (*Oreochromis niloticus*) feed by diets supplemented with different doses of sulfamethoxazole. *Environ. Sci. Pollut. Res.* **2021**, *28*, 65255–65264. [CrossRef] [PubMed]
194. Aisyhah, M.A.S.; Amal, M.N.A.; Zamri-Saad, M.; Siti-Zahrah, A.; Shaqinah, N.N. Streptococcus agalactiae isolates from cultured fishes in Malaysia manifesting low resistance pattern towards selected antibiotics. *J. Fish Dis.* **2015**, *38*, 1093–1098. [CrossRef] [PubMed]
195. Módenes, A.N.; Bazarin, G.; Borba, C.E.; Locatelli, P.P.P.; Borsato, F.P.; Pagno, V.; Pedrini, R.; Trigueros, D.E.G.; Espinoza-Quiñones, F.R.; Scheufele, F.B. Tetracycline adsorption by tilapia fish bone-based biochar: Mass transfer assessment and fixed-bed data prediction by hybrid statistical-phenomenological modeling. *J. Clean. Prod.* **2021**, *279*, 123775. [CrossRef]
196. Deepika, M.S.; Thangam, R.; Sakthidhasan, P.; Arun, S.; Sivasubramanian, S.; Thirumurugan, R. Combined effect of a natural flavonoid rutin from Citrus sinensis and conventional antibiotic gentamicin on Pseudomonas aeruginosa biofilm formation. *Food Control* **2018**, *90*, 282–294. [CrossRef]
197. Deepika, M.S.; Thangam, R.; Vijayakumar, T.S.; Sasirekha, R.; Vimala, R.; Sivasubramanian, S.; Arun, S.; Babu, M.D.; Thirumurugan, R. Antibacterial synergy between rutin and florfenicol enhances therapeutic spectrum against drug resistant Aeromonas hydrophila. *Microb. Pathog.* **2019**, *135*, 103612. [CrossRef] [PubMed]
198. Hoseinifar, S.H.; Esteban, M.; Cuesta, A.; Sun, Y.-Z. Prebiotics and fish immune response: A review of current knowledge and future perspectives. *Rev. Fish. Sci. Aquac.* **2015**, *23*, 315–328. [CrossRef]
199. Ringø, E.; Song, S. Application of dietary supplements (synbiotics and probiotics in combination with plant products and β-glucans) in aquaculture. *Aquac. Nutr.* **2016**, *22*, 4–24. [CrossRef]
200. Marangaby, M. Intraspecific Variation in Life History Traits of the Panamanian Electric Fish Brachyhypopomus Occidentalis. Ph.D. Thesis, Laurentian University of Sudbury, Sudbury, ON, Canada, 2021. Available online: https://zone.biblio.laurentian.ca/handle/10219/3899 (accessed on 20 January 2023).
201. Bricknell, I.; Dalmo, R.A. The use of immunostimulants in fish larval aquaculture. *Fish Shellfish. Immunol.* **2005**, *19*, 457–472. [CrossRef]
202. Meena, D.K.; Das, P.; Kumar, S.; Mandal, S.C.; Prusty, A.; Singh, S.K.; Akhtar, M.S.; Behera, B.K.; Kumar, K.; Pal, A.K.; et al. Beta-glucan: An ideal immunostimulant in aquaculture (a review). *Fish Physiol. Biochem.* **2013**, *39*, 431–457. [CrossRef]
203. Nya, E.J.; Austin, B. Use of garlic, *Allium sativum*, to control *Aeromonas hydrophila* infection in rainbow trout, *Oncorhynchus mykiss* (Walbaum). *J. Fish Dis.* **2009**, *32*, 963–970. [CrossRef]
204. Mahmoud, H.K.; Al-Sagheer, A.A.; Reda, F.M.; Mahgoub, S.A.; Ayyat, M.S. Dietary curcumin supplement influence on growth, immunity, antioxidant status, and resistance to Aeromonas hydrophila in Oreochromis niloticus. *Aquaculture* **2017**, *475*, 16–23. [CrossRef]
205. Musthafa, M.S.; Ali, A.R.J.; Kumar, M.S.A.; Paray, B.A.; Al-Sadoon, M.K.; Balasundaram, C.; Harikrishnan, R. Effect of *Cucurbita mixta* (L.) seed meal enrichment diet on growth, immune response and disease resistance in Oreochromis mossambicus. *Fish Shellfish. Immunol.* **2017**, *68*, 509–515. [CrossRef] [PubMed]

206. Musthafa, M.S.; Asgari, S.M.; Kurian, A.; Elumalai, P.; Ali, A.R.J.; Paray, B.A.; Al-Sadoon, M.K. Protective efficacy of *Mucuna pruriens* (L.) seed meal enriched diet on growth performance, innate immunity, and disease resistance in *Oreochromis mossambicus* against Aeromonas hydrophila. *Fish Shellfish. Immunol.* **2018**, *75*, 374–380. [CrossRef] [PubMed]
207. Zahran, E.; El-Gawad, E.A.A.; Risha, E. Dietary Withania sominefera root confers protective and immunotherapeutic effects against Aeromonas hydrophila infection in Nile tilapia (*Oreochromis niloticus*). *Fish Shellfish. Immunol.* **2018**, *80*, 641–650. [CrossRef] [PubMed]
208. Laith, A.; Mazlan, A.; Effendy, A.; Ambak, M.; Nurhafizah, W.; Alia, A.; Jabar, A.; Najiah, M. Effect of Excoecaria agallocha on non-specific immune responses and disease resistance of *Oreochromis niloticus* against Streptococcus agalactiae. *Res. Veter- Sci.* **2017**, *112*, 192–200. [CrossRef] [PubMed]
209. Omitoyin, B.O.; Ajani, E.K.; Orisasona, O.; Bassey, H.E.; Kareem, K.O.; Osho, F.E. Effect of guava *Psidium guajava* (L.) aqueous extract diet on growth performance, intestinal morphology, immune response and survival of *Oreochromis niloticus* challenged with *Aeromonas hydrophila*. *Aquac. Res.* **2019**, *50*, 1851–1861. [CrossRef]
210. Mbokane, E.; Moyo, N. A preliminary investigation into the potential effect of Artemisia afra on growth and disease resistance in sub-adults of Oreochromis mossambicus. *Aquaculture* **2018**, *482*, 197–202. [CrossRef]
211. Nordin, S.F.; Nordin, M.L.; Osman, A.Y.; Hamdan, R.H.; Shaari, R.; Arshad, M.M.; Aziz, A.R. The Effect of *Matricaria chamomilla* L. on the Growth Performance of Red Hybrid Tilapia. *Biomed. Pharmacol. J.* **2017**, *10*, 1905–1915. [CrossRef]
212. Njagi, G.W.; Lee, S.; Won, S.; Hong, J.; Hamidoghli, A.; Bai, S.C. Effects of dietary Yucca meal on growth, haematology, non-specific immune responses and disease resistance of juvenile Nile tilapia, *Oreochromis niloticus* (Linnaeus, 1758). *Aquac. Res.* **2017**, *48*, 4399–4408. [CrossRef]
213. Mabrok, M.A.E.; Wahdan, A. The immune modulatory effect of oregano (Origanum vulgare L.) essential oil on Tilapia zillii following intraperitoneal infection with Vibrio anguillarum. *Aquac. Int.* **2018**, *26*, 1147–1160. [CrossRef]
214. Silva, L.T.D.S.; Pereira, U.D.P.; de Oliveira, H.M.; Brasil, E.M.; Pereira, S.A.; Chagas, E.C.; Jesus, G.F.A.; Cardoso, L.; Mouriño, J.L.P.; Martins, M.L. Hemato-immunological and zootechnical parameters of Nile tilapia fed essential oil of Mentha piperita after challenge with Streptococcus agalactiae. *Aquaculture* **2019**, *506*, 205–211. [CrossRef]
215. Bustamam, M.S.A.; Pantami, H.A.; Azam, A.A.; Shaari, K.; Min, C.C.; Ismail, I.S. The immunostimulant effects of Isochrysis galbana supplemented diet on the spleen of red hybrid tilapia (Oreochromis spp.) evaluated by nuclear magnetic resonance metabolomics. *Aquac. Nutr.* **2022**, *2022*, 1154558. [CrossRef]
216. Elkatatny, N.M.; El Nahas, A.F.; Helal, M.A.; Fahmy, H.A.; Tanekhy, M. The impacts of seasonal variation on the immune status of Nile tilapia larvae and their response to different immunostimulants feed additives. *Fish Shellfish. Immunol.* **2020**, *96*, 270–278. [CrossRef] [PubMed]
217. Owatari, M.S.; da Silva, L.R.; Ferreira, G.B.; Rodhermel, J.C.B.; de Andrade, J.I.A.; Dartora, A.; Jatobá, A. Body yield, growth performance, and haematological evaluation of Nile tilapia fed a diet supplemented with Saccharomyces cerevisiae. *Anim. Feed. Sci. Technol.* **2022**, *293*, 115453.
218. Essa, M.A.; El-Serafy, S.S.; El-Ezabi, M.M.; Daboor, S.M.; Esmael, N.A.; Lall, S.P. Effect of different dietary probiotics on growth, feed utilization and digestive enzymes activities of Nile tilapia, *Oreochromis niloticus*. *J. Arab. Aquac. Soc.* **2010**, *5*, 143–162.
219. Ghosh, K.; Banerjee, S.; Moon, U.M.; Khan, H.A.; Dutta, D. Evaluation of gut associated extracellular enzyme-producing and pathogen inhibitory microbial community as potential probiotics in Nile Tilapia, *Oreochromis niloticus*. *Int. J. Aquac.* **2017**, *7*. [CrossRef]
220. Banerjee, G.; Ray, A.K. Bacterial symbiosis in the fish gut and its role in health and metabolism. *Symbiosis* **2017**, *72*, 1–11. [CrossRef]
221. Talukdar, S.; Ringø, E.; Ghosh, K. Extracellular tannase-producing bacteria detected in the digestive tracts of freshwater fishes (Actinopterygii: Cyprinidae and Cichlidae). *Acta Ichthyol. Et Piscat.* **2016**, *46*, 201–210. [CrossRef]
222. Lara-Flores, M.; Olivera-Castillo, L.; Olvera-Novoa, M.A. Effect of the inclusion of a bacterial mix (Streptococcus faecium and Lactobacillus acidophilus), and the yeast (*Saccharomyces cerevisiae*) on growth, feed utilization and intestinal enzymatic activity of Nile tilapia (Oreochromis niloticus). *Int. J. Fis. Aquac.* **2010**, *2*, 93–101. [CrossRef]
223. Aly, S.M.; Ahmed, Y.A.-G.; Ghareeb, A.A.-A.; Mohamed, M.F. Studies on Bacillus subtilis and Lactobacillus acidophilus, as potential probiotics, on the immune response and resistance of Tilapia nilotica (*Oreochromis niloticus*) to challenge infections. *Fish Shellfish. Immunol.* **2008**, *25*, 128–136. [CrossRef]
224. Samat, N.A.; Yusoff, F.M.; Rasdi, N.W.; Karim, M. The Efficacy of Moina micrura Enriched with Probiotic *Bacillus pocheonensis* in Enhancing Survival and Disease Resistance of Red Hybrid Tilapia (*Oreochromis* spp.) Larvae. *Antibiotics* **2021**, *10*, 989. [CrossRef]
225. Ringø, E.; Harikrishnan, R.; Soltani, M.; Ghosh, K. The Effect of Gut Microbiota and Probiotics on Metabolism in Fish and Shrimp. *Animals* **2022**, *12*, 3016. [CrossRef] [PubMed]
226. Nguyen, N.H.; Ponzoni, R.W.; Vijayakumar, S.; Raj, T.S. Establishment of a satellite nucleus of the gift strain at Rajiv Gandhi, Center for aquaculture (RGCA) to support tilapia production in India (37–41). Marine Products Export Development Authority (MPEDA). 2011. Available online: https://agris.fao.org/agris-search/search.do?recordID=QW2016000355 (accessed on 20 January 2023).

227. Henriksson, P.J.G.; Troell, M.; Banks, L.K.; Belton, B.; Beveridge, M.C.M.; Klinger, D.H.; Pelletier, N.; Phillips, M.J.; Tran, N. Interventions for improving the productivity and environmental performance of global aquaculture for future food security. *One Earth* **2021**, *4*, 1220–1232. [CrossRef]
228. Béné, C.; Barange, M.; Subasinghe, R.; Pinstrup-Andersen, P.; Merino, G.; Hemre, G.-I.; Williams, M. Feeding 9 billion by 2050—Putting fish back on the menu. *Food Secur.* **2015**, *7*, 261–274. [CrossRef]

MDPI AG
Grosspeteranlage 5
4052 Basel
Switzerland
Tel.: +41 61 683 77 34

Fishes Editorial Office
E-mail: fishes@mdpi.com
www.mdpi.com/journal/fishes

www.ingramcontent.com/pod-product-compliance
Lightning Source LLC
Chambersburg PA
CBHW041401210926
51850CB00013B/3148
* 9 7 8 3 7 2 5 8 3 1 7 5 3 *